Tectonic Studies in the Talladega and Carolina Slate Belts, Southern Appalachian Orogen

Edited by

Denny N. Bearce
William W. Black
Stephen A. Kish
James F. Tull

SPECIAL PAPER

191

THE GEOLOGICAL SOCIETY OF AMERICA
P.O. Box 9140 · 3300 Penrose Place · Boulder, Colorado 80301

Published by The Geological Society of America, Inc.
3300 Penrose Place, P.O. Box 9140, Boulder, Colorado 80301

Printed in U.S.A.

Library of Congress Cataloging in Publication Data
Main entry under title:

Tectonic studies in the Talladega and Carolina slate belts, southern Appalachian orogen.

(Special paper / Geological Society of America ; 191)
Includes bibliographies.
1. Geology, Structural—Addresses, essays, lectures. 2. Geology—Appalachian Mountains—Addresses, essays, lectures. I. Bearce, Denny N., 1934- II. Series: Special paper (Geological Society of America) ; 191.
QE601.T3915 1982 551.8′0975 82-24264
ISBN 0-8137-2191-1

Contents

Part 1

Part 2

Foreword

This Special Paper focuses on the contrasting tectonic evolution of two regions of low-grade metamorphic rocks on opposite sides of the Appalachian orogen: the Talladega slate belt on the northwest in Alabama and Georgia and the Carolina slate belt on the southeast in Virginia, North Carolina, South Carolina, and Georgia. The volume is an outgrowth of symposia convened at recent annual Southeastern Section meetings of the Geological Society of America. Each of these belts provides unique insight into the tectonic evolution of the Appalachian orogen. The collection of papers presented in this volume is intended to highlight the significance of recent research efforts in the understanding of these regions. The low grade of metamorphism in both belts, combined with the preservation of sedimentary and volcanic textures, provides an opportunity to reconstruct the original geologic environments of parts of the Appalachians which elsewhere have been masked by high-grade metamorphism and plutonic activity.

Isotopic dating methods and rare fossil occurrences have established the ages of the sequences in these two belts. Briefly, the geology of the two belts can be contrasted as follows: The Talladega slate belt ranges in age from probably as old as Early Cambrian to as young as Early Devonian, whereas the Carolina slate belt records deposition spanning an earlier time, from Late Precambrian to Early Cambrian. Rocks of the Talladega slate belt exhibit numerous affinities to cratonic North America throughout their evolution, whereas the relationship of the Carolina slate belt with cratonic North America is less obvious; it may have some marginal association with the North American continent or some other continental mass, or it may have formed in an interoceanic region. Times of peak metamorphism varied between the two regions as well. Dynamothermal activity in the Talladega slate belt may be restricted to the Devonian Acadian orogeny, whereas the Carolina slate belt experienced an earlier Taconic metamorphic maximum followed by possible Acadian and Hercynian metamorphic events. Deformation in the Talladega slate belt was pervasive during the Acadian orogeny but was also intense and widespread during the late Paleozoic Alleghanian orogeny. In the Carolina slate belt Taconic deformation is well documented but later Acadian and Alleghanian structures are also present. Volcanic rocks are distinctive and important in each belt and are interpreted to be related to subductive events. In the Talladega slate belt the volcanic sequence formed along the ancient continental margin of North America during middle Paleozoic time, whereas in the Carolina slate belt volcanic activity has been interpreted as intraoceanic in origin and occurred during latest Precambrian to Cambrian times.

Geological Society of America
Special Paper 191
1982

Part 1
Talladega slate belt

Introduction

The Talladega slate belt, which forms the frontal crystalline thrust sheet of the southernmost exposed Appalachians of Alabama and Georgia, contains a unique stratigraphic sequence of much significance in understanding Appalachian tectonic history. During the past decade this belt has been the topic of considerable research efforts. These efforts have focused upon (1) definition and correlation of stratigraphic sequences within the Talladega belt and correlation of these sequences with sequences to the northwest in the Foreland fold and thrust belt and to the northeast in the Blue Ridge belt; (2) determination of absolute and relative age relationships within the belt; (3) determination of the structural framework and developmental history; and (4) understanding the tectonic history of the belt.

On March 19, 1981, a symposium on the "Geology of the Talladega Slate Belt" was held during the 30th Annual Meeting of the Southeastern Section of the Geological Society of America in Hattiesburg, Mississippi. Eleven oral presentations were given in this symposium. These papers highlighted recent research and the resulting significantly expanded data base, and emphasized new interpretations of the regional geology of this belt of rocks. The collection of papers incorporated within this Special Paper forms a symposium volume intended to present a modern comprehensive synthesis of the geology of the Talladega slate belt.

Geological Society of America
Special Paper 191
1982

Stratigraphic framework of the Talladega slate belt, Alabama Appalachians

James F. Tull
Department of Geology
Florida State University
Tallahassee, Florida 32306

ABSTRACT

The Talladega slate belt is a thick sequence of lower greenschist facies rocks that occurs in an allochthonous position immediately above and to the southeast of sedimentary rocks of the Paleozoic miogeoclinal sequence in the Appalachian foreland fold and thrust belt in east Alabama and west Georgia. The Talladega rocks are overlain tectonically by amphibolite facies metamorphic rocks of the Ashland Supergroup and the Wedowee Group which form the high-grade belts northwest of the Brevard Zone in Alabama. The Talladega belt can be divided into (1) a lower clastic sequence—the Kahatchee Mountain Group, (2) a middle carbonate sequence—the Sylacauga Marble Group, (3) an upper clastic sequence—the Talladega Group, and (4) an overlying metavolcanic complex—the Hillabee Greenstone. Limited facing data imply that this gross stratigraphy is currently upright. The Kahatchee Mountain Group is discontinuous along the base (northwest) of the belt and is composed dominantly of pelitic rocks, but also includes psammitic rocks and a minor carbonate horizon. The stratigraphic position and lithologies in the Kahatchee Mountain Group are similar to those of the late Precambrian to Cambrian Ocoee-Chilhowee sequences in the Blue Ridge belt to the northeast in north Georgia, North Carolina, and Tennessee. The Sylacauga Marble Group stratigraphically overlies the basal clastic sequence and locally forms a thick (1.5 km) sequence having affinities with the Cambrian-Ordovician carbonate sequence in the foreland fold and thrust belt. A major low-angle unconformity occurs at the base of the Talladega Group and locally cuts through the Sylacauga Marble Group into the Kahatchee Mountain Group. Boulders in diamictites above the unconformity indicate that the unconformity penetrated as deeply as the Precambrian basement in a nearby area. Above the unconformity is a thick (2 to 5 km), dominantly clastic sequence, the Talladega Group, containing fossiliferous Early to Middle Devonian rocks near the top. These rocks are overlain by the Hillabee Greenstone, a metavolcanic complex dominated by mafic rocks but containing subordinate felsic metavolcanics. The Kahatchee Mountain and Sylacauga Marble Groups appear to represent an allochthonous fragment of the foreland shelf sequence that formed along the stable trailing margin of the proto-North American continent during early Paleozoic time. The major unconformity and the diamictite facies in the Talladega Group above the shelf sequence mark a period of erosion, collapse, and subsidence of the ancient continental shelf, associated with nearby basement uplifts. Such tectonic activity may mark the onset of continent-directed subduction in the southernmost Appalachians. Arc system volcanic rocks of the Hillabee Greenstone prograded onto the collapsed shelf region, probably during

Early to Middle Devonian time, and were followed rapidly by major dynamothermal events associated with Acadian orogenesis.

INTRODUCTION

In the southernmost exposed Appalachians of Alabama and Georgia, the Talladega slate belt, a sequence of low-grade metamorphic rocks, occurs between unmetamorphosed rocks of the foreland fold and thrust belt on the northwest and higher-grade rocks on the southeast (Fig. 1). Outcrop of the Talladega belt extends from the unconformable overlap of Mesozoic and Cenozoic sediments of the Gulf Coastal Plain in central Alabama, northeastward to at least Cartersville, Georgia, and covers an area of some 3,900 km^2. This belt includes clastic, carbonate, and volcanic rocks that have been metamorphosed to lower greenschist facies, and is named (Smith, 1888) for slaty and phyllitic rocks occurring within the Talladega Mountains, along Talladega Creek, and within eastern Talladega County, Alabama (Fig. 2). Much of this region has rather rugged topography and contains unusually good exposures for the southeastern Piedmont.

Debate concerning several fundamental aspects of the Talladega slate belt has centered upon differences in interpretation of (1) the nature of essentially all major unit boundaries both within the belt and between this belt and those to the northwest and southeast, as well as the structural and stratigraphic relationships between various units; (2) the ages of the various units within the belt; and (3) the regional continuity and correlation of units within the belt.

The purpose here is to provide a regional synthesis of available data and to focus upon the major points of debate in an effort to provide a better understanding of the tectonic history of this unique and potentially significant sequence of rocks.

BOUNDARY RELATIONSHIPS OF THE TALLADEGA SLATE BELT

Northwestern Boundary

The northwestern boundary of the Talladega slate belt is represented by a sinuous line separating low-grade metamorphic rocks of the Talladega slate belt on the southeast and unmetamorphosed or possibly very low-grade sedimentary rocks of the Appalachian foreland fold and thrust belt on the northwest. In most areas this line is marked by a significant northwest-facing topographic escarpment along the edge of the Talladega slate belt. This escarpment is formed by relatively resistant slates and quartzites on the southeast, which are commonly in contact with Cambrian-Ordovician carbonate rocks of the foreland to the northwest. Along most of this boundary there is a clear metamorphic discontinuity, but in some regions the relationships are not clear, and stratigraphic continuity between the foreland sequence and that within the Talladega slate belt has been suggested (Kesler, 1950; Carrington, 1973; Hurst, 1973).

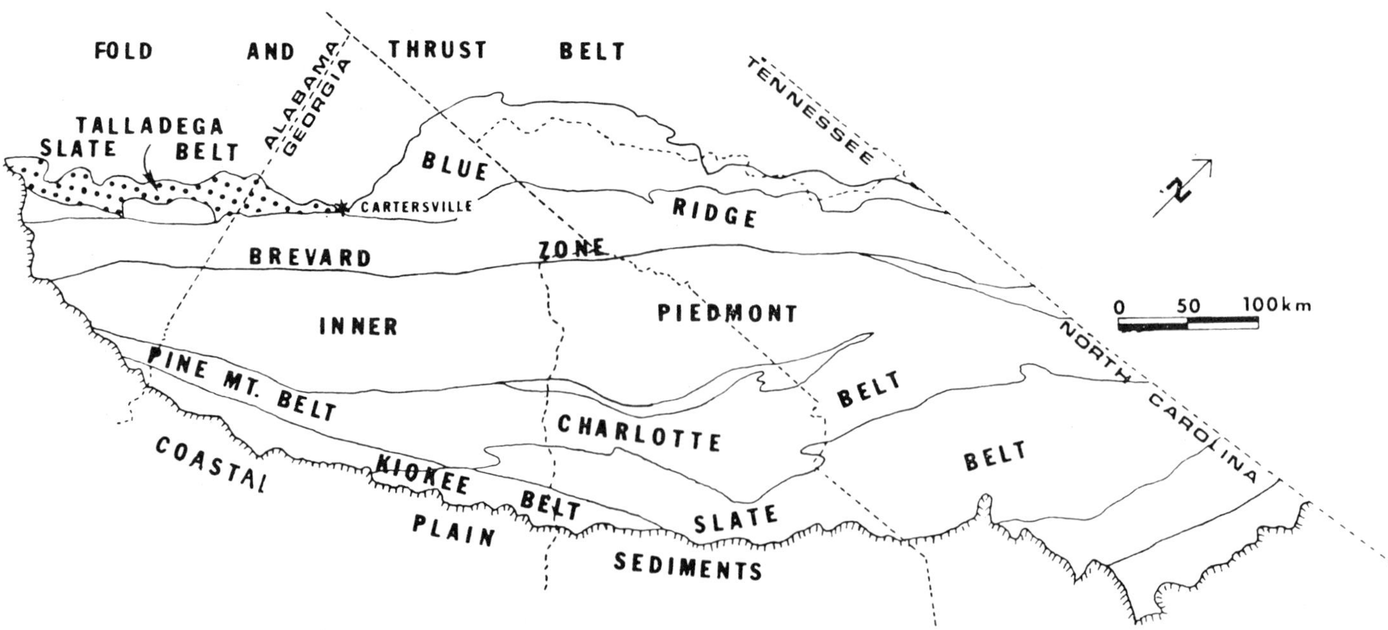

Figure 1. Major lithotectonic provinces in the Southern Appalachians showing the location of the Talladega slate belt.

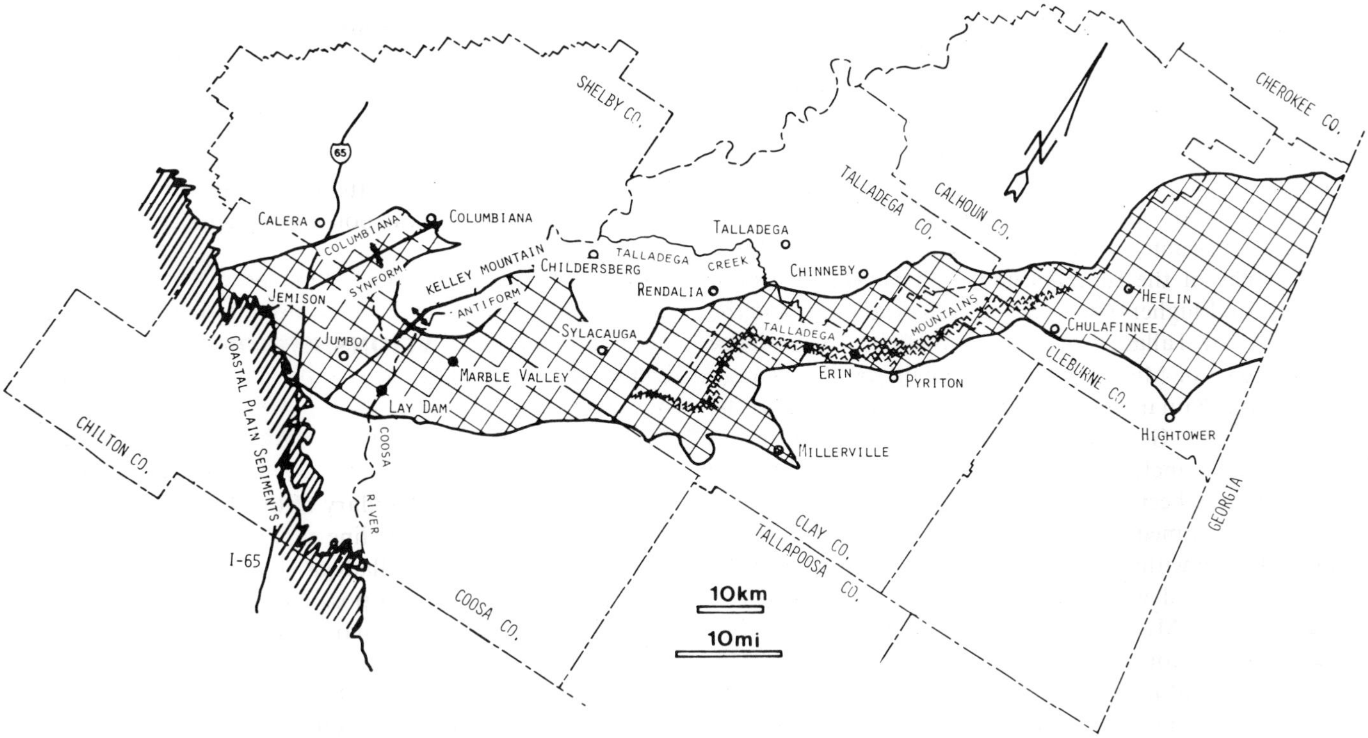

Figure 2. Geographic location map of the Talladega slate belt in Alabama.

Regional map patterns indicate a significant discordance between rock units within the fold and thrust belt and the northwest boundary of the Talladega slate belt (Butts, 1926, 1940). Foreland fold and thrust belt rocks ranging in age from Early Cambrian to Late Mississippian occur juxtaposed against the contact. Likewise, there is regional discordance between rock units in the Talladega slate belt above the contact and this boundary (Fig. 3). These relationships suggest that this boundary is a major fault system.

Some debate has existed as to the location of this fault system, and a number of names have been used to refer to all or portions of it. Hays (1891) originally referred to a portion of this fault system in Georgia southwest of Cartersville as the Cartersville fault. In Alabama the name Talladega fault

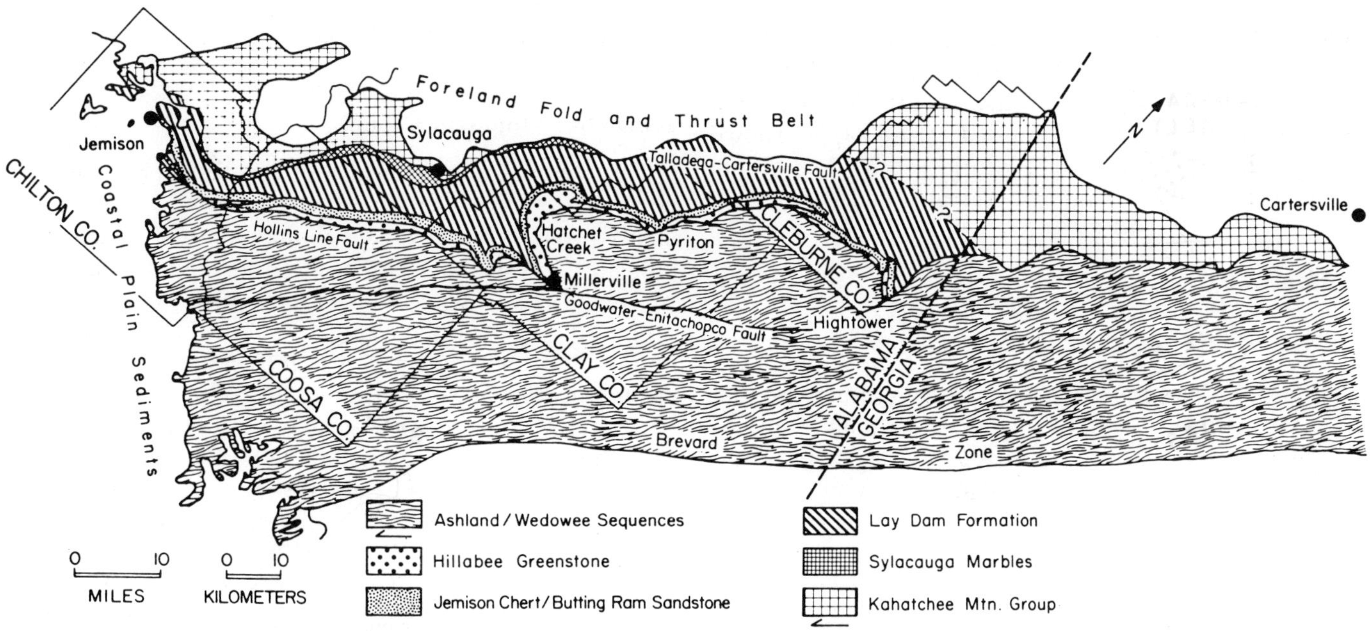

Figure 3. Generalized geologic map of the Talladega slate belt.

was originally used (McCalley, 1897) for this fault system. Local portions of this fault system have also been named Metamorphic Front fault, Kingdom Cross Roads fault, Wilsonville fault (Carrington, 1973), Columbiana fault, Indian Mountain fault (Rodgers, 1970), and Kahatchee Mountain fault (Shaw, 1970). Recently, Thomas and Neathery (1980) applied the name Talladega Front fault to the boundary between low-grade metamorphic rocks of the Talladega belt and essentially unmetamorphosed rocks of the foreland. Because of the general historical usage and the advantage of retaining terminology in current widespread use, the dual name Talladega-Cartersville fault system is proposed for the boundary fault system at the base of the Talladega slate belt nappe. The location of this fault system is outlined in Figures 2 and 3 and is distinguished by the criteria described above. The trace of this fault extends from the Coastal Plain overlap 10 km west of Jemison at the western boundary of Chilton County northeastward to the east of Calera to Columbiana, then southward and eastward to the Coosa River, then eastward and northward to Childersburg, then southeastward to north of Sylacauga, then north to east of Rendalia and Chinneby, then northeast and east to Cartersville, Georgia along a trace outlined by Bearce (1982) (Fig. 3).

The sinuous trace of this fault system suggests a generally very low regional dip angle. Half windows (reentrants) along the fault indicate a minimum horizontal component of the net slip of 23 km along this fault system in central Alabama (Fig. 3). This fault system is in a similar structural position to the Great Smoky fault system to the north, in north Georgia and Tennessee, along the northwest boundary of the Blue Ridge. One recent interpretation of the COCORP seismic reflection profile from that region suggests displacements of greater than 200 km for the Blue Ridge rocks along the Great Smoky fault system (Cook and others, 1980). A similar interpretation for the Talladega-Cartersville fault would imply that the Talladega slate belt rocks represent a very far traveled thrust nappe, rooted somewhere east of the present location of the Pine Mountain belt (Fig. 1) in central east Alabama and Georgia.

Southeastern Boundary

The southeastern boundary of the Talladega slate belt, like the northwestern boundary, is marked by a significant metamorphic discontinuity. Rocks to the southeast of this boundary are generally in the middle amphibolite facies of regional metamorphism. A northwest-facing topographic escarpment formed by schists and quartzites that are commonly in contact with low-grade mafic metavolcanic rocks of the Talladega slate belt is also present along this boundary, although it is generally not as prominent as that developed along the northwestern boundary of the belt.

It has been suggested that the southeastern boundary of the Talladega slate belt is a stratigraphic boundary (Crawford and Medlin, 1973), perhaps along a zone of steep metamorphic gradient (Neathery and Reynolds, 1973), but abundant evidence indicates that this boundary, like the lower, northwestern, boundary of the belt is a major low-angle thrust fault. For example, the regional discordance between this boundary and rock units and structures that occur on either side of this boundary, the presence of an abrupt and significant metamorphic discontinuity at the boundary, and the presence of phyllonitic rocks along the boundary support the fault interpretation (Tull, 1977), which was originally suggested by Prouty (1923). Throughout most of its trace in Alabama, it is marked by the Hollins Line fault, but northeast of Hightower, Alabama, it is marked by a second, younger fault system known as the Goodwater-Enitachopco fault system (Tull, 1978) (Figs. 2, 3).

REGIONAL LITHOSTRATIGRAPHY OF THE TALLADEGA SLATE BELT

The rocks of the Talladega slate belt were originally referred to by Smith (1888) as the Talladega group and later by Butts (1926) as the Talladega Slate, but have more recently been differentiated into a number of units. The purpose of this section is to discuss the regional Talladega slate belt lithostratigraphy, to review the stratigraphic terminology that has been applied to the belt, and to propose a regional stratigraphic framework based upon recent studies. These studies suggest that the Talladega slate belt can most effectively be subdivided into (1) a lower clastic sequence—the Kahatchee Mountain Group, (2) a central carbonate sequence—the Sylacauga Marble Group, (3) an upper clastic sequence—the Talladega Group, and (4) an overlying volcanic sequence—the Hillabee Greenstone.

Kahatchee Mountain Group

The regional outcrop width of the Talladega slate belt between the two boundary fault systems in Alabama is highly variable, ranging from about 6 km to as much as 30 km (Fig. 3). Because of this relationship and because units within the belt are discordant with the lower boundary fault system, the structurally lower parts of the belt are not continuous, but appear to be present only within the wider parts of the belt. The most intensely studied area containing this structurally lower sequence is in the southwestern part of the Talladega slate belt in southeast Shelby and northeast Chilton Counties, within the regionally significant Columbiana Synform (Fig. 2). Parts of this area have been investigated by Butts (1926, 1940), Warren (1969), Carrington (1973), Telle (1981), and recently by the present author. It is proposed here that this entire lower structural sequence, and possibly equivalent ones cropping out within the Talladega slate belt along strike farther to the northeast, be referred to as the Kahatchee Mountain Group, for the mountainous portions of this sequence to the northwest of Sylacauga (Fig. 3).

The rocks within the Kahatchee Mountain Group in this area are dominantly pelitic slate and phyllite, and metasandstone, but also include a minor carbonate unit known as the Sawyer Limestone (Butts, 1926, 1940). From a few kilometers east of Interstate 65 to several kilometers north of the city of Sylacauga (Fig. 2), where they appear to pinch out against the Talladega-Cartersville fault, these rocks are bounded above and to the south by a sequence of carbonate rocks referred to in the present study as the Sylacauga Marble Group (See below; Fig. 3). To the west of Interstate 65 (Fig. 2), the rocks of this lowermost sequence appear to be in contact with a clastic sequence, the Lay Dam Formation (Carrington, 1973), which regionally overlies the marbles and represents the lower unit of the Talladega Group.

A number of fundamentally different interpretations have been offered concerning the stratigraphic and structural relationships of the Kahatchee Mountain Group and the surrounding sequences. Prouty (1922) originally proposed that this structurally lowest sequence of rocks was Cambrian in age, and most subsequent workers (Butts, 1940; Rodgers and Shaw, 1962; Shaw, 1970) have generally agreed. However, on the basis of lithologic similarities with part of the adjacent sequence in the foreland fold and thrust belt, as well as on the basis of the preliminary interpretation of plant spores from parts of this sequence, Carrington (1973) suggested that this sequence, which he referred to as the Metamorphic Front Block, is Carboniferous in age.

A significant factor bearing on the age of this structurally lowermost sequence is the nature of its southern and eastern (upper) boundary. In Chilton County the Sylacauga Marble Group is represented by a relatively thin (approximately 65 m thick) dolomitic marble unit known as the Jumbo Dolomite (Butts, 1926) (Fig. 4). This marble is bounded above by a major unconformity referred to in the present study as the pre–Lay Dam Formation unconformity. The marble unit is conformable at its base and interlaminated with a variably

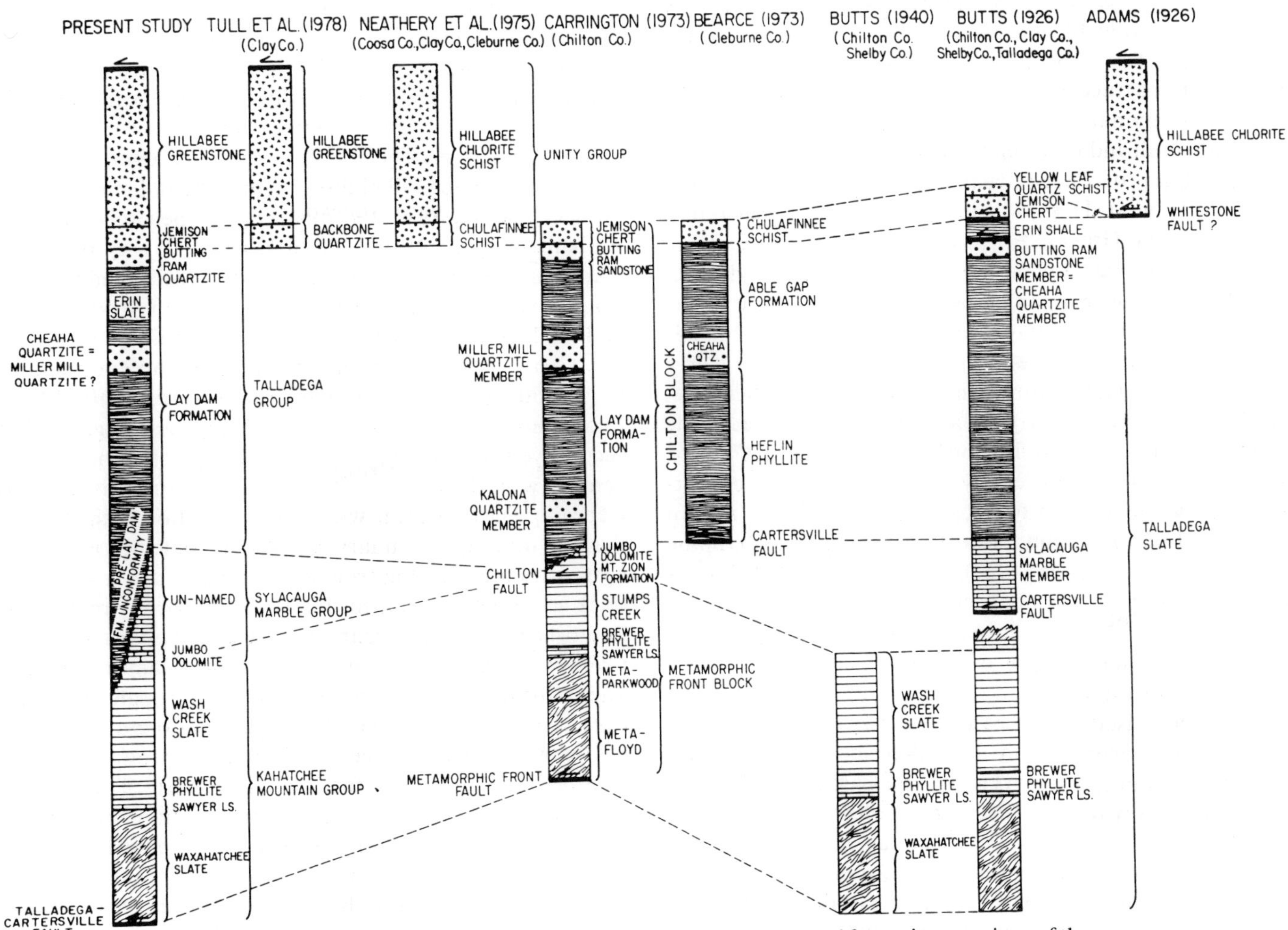

Figure 4. Correlation of structural frameworks previously proposed for various portions of the Talladega slate belt.

graphitic black and tan slate, which has locally been referred to as the Mount Zion Formation (Carrington, 1973) (Fig. 4). Carrington (1973) has proposed that a major fault, the Chilton fault, occurs below the Mount Zion Formation slates and above the structurally lowermost sequence. Recent work by the present author and by Telle (1981) in northeastern Chilton County has shown, however, that the Mount Zion Formation slates are stratigraphically continuous with the upper part of the Kahatchee Mountain Group, which Butts (1940) originally mapped as Wash Creek Slate. Unfortunately, the region mapped by Butts (1940) did not include the uppermost portion of the Wash Creek Slate near the pre–Lay Dam Formation unconformity.

Several hundred meters below its upper contact with the Jumbo Dolomite, the Mount Zion Formation of Carrington (1973) contains interlayered resistant arkosic metasandstones that form discontinuous ridges in the area. These quartzites can be traced discontinuously for 27 km westward from near Jumbo, at the type locality of the Mount Zion Formation, to the Coastal Plain overlap some 10 km west of Jemison. Near Interstate 65, Butts (1940) differentiated the metasandstones and the Mount Zion slates as a "massive sandstone" and a "black slate" member, respectively, of the Wash Creek Slate. Along trend to the north of Jemison, however, the metasandstones were named the Kalona Sandstone Member by Carrington (1973) and placed within the Lay Dam Formation. In fact, the pre–Lay Dam Formation unconformity must be above the Kalona Sandstone because the Kalona can be traced eastward below the upper boundary of the Mount Zion Formation slates and the overlying Sylacauga Marble sequence. Thus, it appears that distinctive units can be mapped across the proposed trace of the Chilton fault. Facing data are rare from this interval, but it is clear that these rocks are in stratigraphic continuity below the pre–Lay Dam Formation unconformity. Therefore, stratigraphic evidence argues against the presence of the proposed Chilton fault in northeastern Chilton County. Rock units below the pre–Lay Dam Formation unconformity appear to be conformable with one another and to exhibit gradational boundaries. North of this area, mapping by Butts (1940), Warren (1969), and Carrington (1973) indicated that no major faults occur within this region represented by the structurally lowermost sequence.

A second major region of exposure of the Kahatchee Mountain Group is in southern Talladega and northwestern Coosa Counties to the west of Sylacauga (Fig. 3). These rocks are continuous with those described above and by Butts (1926, 1940), Warren (1969), and Carrington (1973) in southern Shelby and northeastern Chilton Counties, and can be traced across the hinge of the Kelley Mountain antiform. The Kahatchee Mountain Group in this region is overlain by a much thicker and more variable sequence of the Sylacauga Marble Group than that within the Columbiana synform (see below). Rocks of the Kahatchee Mountain Group here are dominantly pelitic slates and phyllites, but also include thick units of metasandstone which form prominent ridges such as Kahatchee Mountain northwest of Sylacauga.

North of Sylacauga, the Kahatchee Mountain Group–Sylacauga Marble Group contact appears to be truncated by the Talladega-Cartersville fault, so that farther northeastward rocks of the foreland fold and thrust belt are in contact with the Sylacauga Marble Group because the Kahatchee Mountain Group has been eliminated by faulting (Fig. 3). Farther northward the Talladega-Cartersville fault apparently occurs at an even higher stratigraphic level, and the marble sequence has also been eliminated by faulting. At these localities, the Lay Dam Formation appears to be in thrust contact against rocks of the foreland fold and thrust belt (Gilbert, 1973) (Fig. 3).

The possible reappearance of portions of the Kahatchee Mountain group farther northeastward in the Talladega slate belt is a matter of much significance for regional correlations. The Sylacauga Marble Group apparently reappears above the Talladega-Cartersville fault northeast of Sylacauga near Rendalia (Prouty, 1916), and again at Chenneby (Gilbert, 1973) (Figs. 2, 3). The narrowest outcrop width (6 km) of the Talladega slate belt is in an area approximately 17 km southwest of Heflin in southwestern Cleburne County. Northeast of this area the outcrop width increases considerably to approximately 30 km in central Cleburne County (Fig. 3). A major subregional reentrant indents the southeastern boundary of the Talladega slate belt at Hightower, Alabama (Fig. 2). Northwest of Hightower, the lithostratigraphy in the structurally upper part of the Talladega slate belt curves southeastward parallel with the trace of the Hollins Line fault and is juxtaposed with rocks of the Wedowee Group northeast of Hightower along an extension of the Goodwater-Enitachopco fault system (Tull, 1979). Reconnaissance studies north of Hightower suggest that successively lower units in the Talladega slate belt are in contact with the fault system as it is traced northward into Georgia. These relationships, in conjunction with the fact that the outcrop width of the Talladega slate belt is a maximum in central Cleburne County, suggest that a relatively thick section of the Kahatchee Mountain Group may occur in northwestern Cleburne County and eastern Calhoun County. If erosion along the pre–Lay Dam Formation unconformity has penetrated entirely through the Sylacauga Marble Group, then the Lay Dam Formation would lie directly upon the Kahatchee Mountain Group, as it does southwest of Interstate 65 in Chilton County. This contact may correspond to the northwestern limit of the Heflin Phyllite (Bearce, 1973) in northwestern Cleburne County. No marbles have been reported within the Talladega slate belt northeast of Chenneby (Fig. 2). Bearce (1973) correlated the units northwest of the Heflin Phyllite in this region with his Able Gap Formation, which regionally occurs near the structural top of the Talladega slate belt (Fig. 4), but the lithologies are also quite similar to those within the Kahatchee Mountain Group southwest of Sylacauga.

Following the suggestion that the upper units of the Talladega slate belt are successively faulted out northeastward from Hightower along the trace of the Goodwater-Enitachopco fault system, it seems likely that the Kahatchee Mountain group may make up a dominant portion of the Talladega slate belt farther northeast in Georgia.

Debate has existed for some time concerning the degree of stratigraphic equivalence of rocks of the Talladega slate belt and those within the southern Blue Ridge belt north of Cartersville, Georgia (Jonas, 1932; Park, 1935; Crickmay, 1936; Stose and Stose, 1944; McConnell and Costello, 1981). In the southern Blue Ridge belt the rocks range in age from late Precambrian to Cambrian. Rocks as young as Ordovician may occur within the Murphy Marble belt (McLaughlin and Hathaway, 1973), but this correlation is uncertain. The Talladega slate belt, on the other hand, is known to be in part at least as young as Devonian (see below). If the Kahatchee Mountain Group is Cambrian or older, then it may be correlative with the Ocoee (late Precambrian)-Chilhowee (Cambrian) rocks of the Blue Ridge belt. Data relating to the age of the Kahatchee Mountain Group will be discussed below.

Sylacauga Marble Group

East of the Coosa River the carbonate sequence, represented to the southwest in Chilton County by only the Jumbo Dolomite, appears to expand into a variable series of calcite and dolomite marbles, which are locally cherty and are interlayered with phyllites. The marble exposed in the vicinity of the city of Sylacauga was referred to by Butts (1926) as the Sylacauga marble member, but he interpreted other carbonate units in areas immediately to the west and southwest as windows into the carbonate sequence of the foreland fold and thrust belt. More recently, however, it has been suggested (Shaw, 1970; Gilbert, 1973) that this entire interlayered carbonate sequence is stratigraphically continuous, and it is proposed here that this sequence, including the Jumbo Dolomite, be termed the Sylacauga Marble Group. This carbonate sequence appears to be stratigraphically above the Kahatchee Mountain Group.

Debate has existed concerning whether or not the Jumbo Dolomite is correlative with part of the marbles near Sylacauga. For example, Carrington (1973) suggested that the Jumbo Dolomite occurs above the proposed Chilton fault, whereas he suggested that the Sylacauga Marble as defined by Butts (1926) occurs below this fault. Alternatively, it has been suggested that the Jumbo Dolomite is a westward continuation of the Sylacauga Marble sequence (Shaw, 1970). Recent work by Pendexter (1981) confirmed that the Jumbo Dolomite is a unit within the Sylacauga Marble Group, most probably at the base of this sequence. In Chilton County, pre–Lay Dam Formation erosional relief existed on the Jumbo Dolomite. From a point 6 km east of Interstate 65 in northeastern Chilton County westward to the Coastal Plain overlap, the pre–Lay Dam Formation erosion surface penetrated entirely through the Jumbo Dolomite, and the Lay Dam Formation rests upon black slates and arkosic metasandstones in the upper part of the Wash Creek Slate of the Kahatchee Mountain Group.

Published studies dealing with the Sylacauga Marble Group have treated the internal stratigraphy in only a general fashion, but it is obvious that a number of different carbonate as well as clastic units exist within this sequence (Prouty, 1916; Shaw, 1970). Both dolomite and calcite marbles occur within this sequence, and the dolomite units are commonly cherty. The calcite marbles have been quarried for statuary, building stone, and aggregate since the last century, and several quarries remain active today. No fossils have been reported from the marble sequence, but most geologists who have worked in the area have correlated these rocks with the Cambrian to Ordovician carbonate sequence of the fold and thrust belt to the northwest of the Talladega slate belt (McCalley, 1897; Prouty, 1916; Shaw, 1970; Gilbert, 1973). These correlations are based upon lithologic similarities, similarities in overall lithologic sequence, and topographic expression (Shaw, 1970). The clastic rocks interlayered with the marbles are mostly sericite phyllite, but also include minor quartzite.

Northeastward from the Coosa River, the upper contact of the Sylacauga Marble Group appears to be slightly discordant with units within the group. The outcrop of the marble belt broadens to the northeast, additional units are present, and at Sylacauga the group is approximately 1,700 m thick (Shaw, 1970). Just a few kilometers east of the Coosa River the marbles are overlain unconformably by coarse, boulder-bearing diamictites of the Lay Dam Formation. At Sylacauga, diamictites are not present above the marbles, but it has been suggested that this contact is also an unconformity (Shaw, 1970; Gilbert, 1973). At Sylacauga the marbles are overlain by phyllite of the Lay Dam formation.

Talladega Group

At the structural top of the Talladega slate belt is a regionally developed metavolcanic sequence known as the Hillabee Greenstone (Tull and Stow, 1980). The dominantly clastic sequence occurring beneath the Hillabee metavolcanic rocks and above the pre–Lay Dam Formation unconformity is referred to here as the Talladega Group. Smith (1888) used the term "Talladega group" to refer to phyllites and metasandstones exposed along Talladega Creek in Clay and Talladega Counties. The rocks included within the Talladega Group as defined here correlate very closely with those originally named to the "Talladega group" by Smith (1888).

Lay Dam Formation

The Lay Dam Formation was defined by Carrington (1973) to include a sequence of slates and metasandstones lying above a major unconformity, above the Jumbo Dolomite, and below the Butting Ram Quartzite (Butts, 1926; Carrington, 1973) in Chilton County, Alabama. The Lay Dam Formation rocks were originally undifferentiated as part of the Talladega Slate (Butts, 1926). In the type area of this formation in Chilton County, a large percentage of the sequence is composed of unsorted, unbedded diamictite (Telle, 1981). The diamictite is interlayered with large quantities of metasiltstones and with zones of metagraywacke and rare beds of dark dolomitic marble. The diamictites contain cobbles and boulders of diverse provenance. These include sedimentary protoliths such as various carbonate lithologies, pelitic lithologies, quartzite, and chert, as well as felsic plutonic and high-grade metamorphic rocks. The granitoid boulders yield U/Pb zircon ages of 1.1 b.y., suggesting that they were derived from a Grenville basement source (Telle and others, 1979). The size and angularity of these fragments imply that the source was nearby.

At the Coosa River the Lay Dam Formation is approximately 4.5 km thick. To the west in northern Chilton County this sequence thins rather abruptly in the hinge of the Columbiana synform to 1 to 1.5 km in thickness and undergoes a significant facies change. West of Jemison the diamictite facies is absent and the Lay Dam Formation is represented by finely laminated metasiltstones, which appear to be turbidites. East of Jemison, similar metasiltstones are interlayered with the diamictite facies and with arkosic metasandstones and metagraywacke. The diamictite facies is less pronounced in Coosa than in Chilton County and is more restricted to the medial portions of the formation. The upper and lower parts of the formation are represented by pelitic facies; the upper part commonly is marked by intervals of graphitic phyllite.

The base of the Lay Dam Formation in Chilton County east of Interstate 65, and in western Coosa County, is clearly an unconformity, with bouldery diamictite resting directly upon the Jumbo Dolomite and in places upon the Wash Creek Slate. This contact has local erosional relief but maintains a relatively restricted stratigraphic position near the contact between the Kahatchee Mountain Group and the Sylacauga Marble Group from 6 km east of Interstate 65 to 5 km east of the Coosa River. The basal diamictites contain abundant carbonate boulders immediately above the Jumbo Dolomite. West of Interstate 65 the lower Lay Dam Formation contact lies below the former level of the Sylacauga Marble Group, and metasiltstones of the Lay Dam Formation are in contact with black slates and arkosic sandstones in the upper part of the Wash Creek Slate. Because of somewhat similar lithologies both above and below the contact, evidence for the presence of the unconformity in this region is more obscure. Diamictites at the base of the Lay Dam Formation extend as far northeast as Marble Valley (Fig. 2) but have only been observed some thickness above the basal contact farther northeast of there. Shaw (1970) suggested, however, that the "Sylacauga unconformity" occurs above the Sylacauga Marble Group northeast of Marble Valley in Talladega County. In this region the lower part of the Lay Dam Formation appears to be represented by phyllites. Farther northeastward the unconformity is intersected by the Talladega-Cartersville fault near Talladega (Fig. 2), and the Lay Dam Formation is juxtaposed with unmetamorphosed rocks of the fold and thrust belt. As mentioned above, the unconformity may reappear above this thrust plane to lie above the Sylacauga Marble Group at Chinneby in Talladega County (Fig. 2). Northeast of this location the presence of the unconformity is unknown, but it may occur at a stratigraphic level below the Sylacauga Marble Group, and metaclastic rocks of the Lay Dam Formation may occur in contact with metaclastic rocks of the Kahatchee Mountain Group (see above discussion).

The relationships described above for the pre–Lay Dam Formation unconformity document significant erosion along this contact. Present levels of exposure indicate that this unconformity penetrates as low as the Kalona Quartzite Member of the Wash Creek Slate in northern Chilton County and occurs as high in the section as the uppermost exposures of the Sylacauga Marble Group near Sylacauga. This implies that the unconformity bevels over 1,800 m of stratigraphic section across a strike distance of approximately 66 km and indicates a regional angular unconformity of as little as 1.5 degrees.

The northeastward continuation of the Lay Dam Formation from the type area in Chilton County has been a point of debate in the past, principally because a large portion of the belt in Coosa and Clay Counties has not been mapped in sufficient detail. In the type area in Chilton County, the Lay Dam Formation grades upward into the Butting Ram Quartzite. The Butting Ram Quartzite thins eastward across Coosa County from about 850 m in thickness to the west of Jemison to only 30 m in western Coosa County, 4 km east of the Coosa River. Because this is a prominent ridge-forming arkosic metasandstone in the upper part of the Talladega slate belt, most geologists (e.g., Butts, 1926; Rodgers and Shaw, 1962) have correlated it with the Cheaha Quartzite in Clay, Talladega, and Cleburne Counties. The Cheaha is a similar arkosic metasandstone that forms the Talladega Mountains (Fig. 2) (Carter, 1981), the highest mountain chain in Alabama. This correlation would suggest that the Cheaha Quartzite occurs above the Lay Dam Formation. The Cheaha Quartzite appears to be truncated in southeastern Talladega County by faulting (Cook, 1981) so that it is not possible to trace this unit into the Butting Ram Quartzite. The stratigraphic sequences immediately above the Cheaha Quartzite and the Butting Ram Quartzite appear to be dissimilar, however. The Butting Ram grades stratigraphically upward into the Jemison Chert (Butts, 1926) (Fig. 4), which is in turn overlain by the Hillabee Greenstone (Tull, 1979). The Cheaha, however, grades stratigraphically upward into phyllites through a thick interval of interlayered phyllites and metasandstone. The phyllites are commonly graphitic and have been locally termed the Erin Shale (Butts, 1926) (Fig. 4) for exposures in western Clay County. Above the thick interval of phyllites is a discontinuous metasandstone interval that is overlain by phyllitic quartzite and, in turn, the Hillabee Greenstone. This quartzite correlates well with the Jemison Chert interval in Chilton County (Tull, 1979), and it is likely that the metasandstone at its base is a facies of the Butting Ram Quartzite. Therefore, I think that it is more likely that the Butting Ram Quartzite and the Cheaha Quartzite are not correlative units, but that the Cheaha occurs stratigraphically below the Butting Ram Quartzite. Conglomeratic arkosic metasandstones are known to occur within the medial portions of the Lay Dam Formation in the type area in Chilton County. One such unit has been mapped by Carrington (1973) as the Miller Mill Quartzite Member. I believe that units such as the Miller Mill more closely correlate with the Cheaha Quartzite than does the Butting Ram Quartzite.

In Cleburne County, Bearce (1973) has included the Cheaha Quartzite and the overlying rocks up to the Chulafinee Schist within the Able Gap Formation. The Able Gap Formation would include the Erin Shale of Butts (1926) in the upper part. The Chulafinee Schist is a phyllitic quartzite unit that is similar lithologically and has a similar stratigraphic position below the Hillabee Greenstone to the Jemison Chert interval in Chilton and Coosa Counties. Metasandstones immediately below the Chulafinee Schist may correlate with the Butting Ram Quartzite. These relationships suggest that with the exception of the uppermost metasandstone units, the Able Gap Formation would correlate approximately with the upper half of the Lay Dam Formation in Chilton and Coosa Counties.

In Cleburne County the rocks below the Cheaha Quartzite are somewhat different in lithologic character from the Able Gap Formation and have been placed within the Heflin Phyllite by Bearce (1973) (Fig. 4). The Heflin Phyllite is composed dominantly of calcareous sandy siltstone but also contains arkosic metasandstone. The Lay Dam Formation diamictite facies can be traced in the medial portions of the sequence northeastward across Coosa and southern Talladega Counties into western Clay County where it underlies the Cheaha Quartzite (Cook, 1981). In this region the size of the clasts in the diamictite is much reduced from that in the type area in Chilton County, but the matrix is identical. Similar rocks underlie the Cheaha Quartzite at least as far north as northwestern Clay County. These rocks are in a similar stratigraphic position to Bearce's Heflin Phyllite, and it thus seems likely that the Heflin Phyllite is equivalent to the lower portion of the Lay Dam Formation. Subdivision of the Lay Dam Formation in the type area in Chilton County into the Able Gap Formation and Heflin Phyllite on the basis of lithologic character does not seem possible. The diamictite facies occurs intermittently throughout the Lay Dam Formation section in Chilton County. This facies occurs directly above the Jumbo Dolomite at the base of the formation and directly below the Butting Ram Quartzite at the top.

Butting Ram Quartzite

The Butting Ram Sandstone Member was named by Butts (1926) for a prominent arkosic metasandstone unit at Butting Ram Shoals on the Coosa River. It is a thick-bedded, coarse-grained metasandstone that is commonly conglomeratic in the basal portion. Its lower contact is gradational into the Lay Dam Formation. Across this contact zone, metasandstone beds are intercalated with slates and layers of diamictite. The Butting Ram Quartzite, although generally much thicker, is similar to other conglomeratic arkosic metasandstones within the Lay Dam Formation and appears to simply represent the final phase of coarse clastic deposition that occurred intermittently throughout the history of deposition of the Lay Dam Formation. The upper contact of the Butting Ram is also a gradational stratigraphic contact, representing a zone of interlayering of medium- to fine-grained metasandstone layers and metachert layers. This contact zone is overlain by the main mass of the Jemison Chert. The thickness of the Butting Ram Quartzite is highly variable, ranging from a maximum of 850 m near the Coastal Plain overlap in Chilton County to very thin or absent in portions of western Clay County. Carrington (1973) reported invertebrate fossils, including bryozoa, gastropods, corals, and brachiopods from a metasiltstone within the upper part of the Butting Ram and suggested a Devonian age, but a definitive age for these fossils could not be obtained.

Jemison Chert

The Jemison Chert is a massive, thick-bedded, fine-grained, grayish-white metachert that has long been known to contain marine invertebrate fossils of Early to Middle Devonian age (Butts, 1926; Carrington, 1973; Sutley, 1977). As it is traced eastward from the type area, the mesoscopic appearance of the Jemison Chert changes as a result of several factors. In places it is slightly graphitic and thus is gray or black instead of the typical grayish-white of the type area. It also is slightly more argillaceous to the east, so that sericite represents a few percentage of the mode. This is accompanied by a change in the fabric from the typical interlocking mosaic texture of the microquartz and chalcedony of the chert west of Jemison to a well-developed tectonite fabric that includes highly foliated mica and ribbon quartz in a mosaic of polygonal quartz grains. These changes are associated with only a slight increase in grain size. The resulting mesoscopic changes are different enough from the type area of the Jemison Chert that Butts (1926) extended the unit to the east only 11 km beyond the type locality. Fine-grained micaceous quartzites and quartz-rich sericite phyllites dominate the Jemison interval from eastern Chilton County, northeastward to Cleburne County, where the equivalent interval has been termed the Chulafinnee Schist (Bearce, 1973). No fossils have been discovered in this unit east of a locality approximately 4 km east of Jemison, probably because of the greatly increased strain and recrystallization.

In the type area of the Jemison Chert the upper contact is covered by Coastal Plain sediments. East of Interstate 65, however, the Jemison Chert interval and the correlative Chulafinnee Schist are overlain by the Hillabee Greenstone. The contact between these units is gradational through a thickness of several tens of meters. This gradation is accomplished by interlayering, generally on a scale of meters, of pelitic and quartzitic rocks of the Jemison interval with mafic phyllites of the Hillabee Greenstone. The quartz sericite phyllites of the Jemison interval in this transition zone are commonly chloritic. On both local and regional scales the Jemison and its equivalents to the northeast are concordant with the basal Hillabee contact.

Hillabee Greenstone

The Hillabee Greenstone occurs at the top of the Talladega slate belt and crops out discontinuously from inliers within the Gulf Coastal Plain in Chilton County for approximately 170 km to eastern Cleburne County near Hightower. The variable stratigraphic position of the Hollins Line fault along its trace results in a variable thickness for the Hillabee Greenstone. At some localities the Hillabee Greenstone is missing and the fault is within rocks of the Talladega Group. Throughout most of the strike length, the Hillabee is only a few hundred meters thick, but in the central portion of the belt near Millerville it reaches a thickness of 2.6 km.

More than 80 percent of the exposed area of the Hillabee Greenstone is made up of mafic rocks that can be subdivided into greenstones and mafic phyllites ("greenschists") based upon the degree of development of foliation. The greenstones are generally massive, fine-grained (0.1 to 1 mm), pale green to light olive-brown rocks that weather to a dark red or brownish yellow color. Mafic phyllites are well foliated slaty rocks that commonly display a laminar compositional layering. These two mafic lithologies are very similar chemically and mineralogically but differ texturally. These rocks are chemically equivalent to low potassium tholeiites (Tull and Stow, 1980). The metamorphic mineralogy consists of albite, actinolite, epidote, zoisite, clinozoisite, and chlorite. Dynamothermal metamorphism has resulted in almost complete recrystallization of the Hillabee rocks so that primary mineral phases are very rare. In some localities, however, relict igneous textures, such as diabasic, gabbroic, and porphyritic textures can be identified. Relict shapes of igneous pyroxene and plagioclase grains are commonly preserved in some areas. With the exception of volcanic bombs and/or lithic ejecta that have been recognized at a few localities, other primary volcanic textures and features have apparently been obliterated by the dynamothermal events.

The greenstones, which make up the bulk of the metavolcanic sequence, are interpreted to be dominantly basaltic lava flows, locally intermixed with basaltic pyroclastic rocks and epiclastic rocks. The mafic phyllites appear to include both strongly deformed greenstones and meta-tuffs. Because of the effects of the dynamothermal metamorphic overprint on the rocks and the limited extent of exposures in this region, individual lava flows have not been delineated.

Felsic metavolcanic rocks commonly are interlayered with the greenstones and mafic phyllites in areas where thickness of the Hillabee exceeds approximately 200 m. Chemically these siliceous units are in the range of quartz dacite (Long, 1981). The size of the individual dacite units is quite spectacular. They range in thickness from a few tens of meters to approximately 300 m and can be traced continuously in outcrop for distances as great as 17 km. Even more laterally extensive felsic volcanic units must have been present prior to faulting, because the thicker units all terminate against boundary faults. In the area to the northwest and southwest of Millerville, the lower quartz dacite unit is approximately 200 m above the basal Hillabee contact zone. This unit maintains a thickness between 150 and 300 m throughout most of its outcrop trace. The continuity of this lower dacite is broken locally to the southwest of Millerville by the upper boundary fault, but it appears to have once been continuous into extreme southwestern Clay County over a strike length of at least 45 km. The extent of this unit in the dip direction is not known, but the two-dimensional extent suggests an enormous outpouring of dacitic material to form this single unit.

The quartz dacites appear to be internally homogenous; no features have been observed that would allow for subdivision of these units. They are bounded at the top and the bottom by mafic phyllite that is in turn bordered by greenstone. The contacts between the siliceous lithologies and the mafic lithologies are knife sharp, planar, parallel to slaty cleavage, and show no evidence of injection. The various quartz dacite units are not connected in outcrop plan. Locally, near the upper contact of the lower quartz dacite, the dacite contains a 5-m-thick layer of mafic phyllite.

These observations suggest that the siliceous rocks are extrusive rather than intrusive. Their overall geometry is that of tabular sheets which are very thin in comparison to their strike length, rather than that of domal bodies. Because of the high viscosities of rhyodacite magmas, it is believed that the Hillabee quartz dacites must have originated as explosive ash flows (ignimbrites) of great lateral extent and thickness. Because of the high degree of metamorphic alteration, however, primary volcanic textures in support of this hypothesis, such as fiamme structure, have not been observed. The quartz dacites consist of metamorphic assemblages containing quartz, sericite, albite, epidote, actinolite, and potassium feldspar. Deformed igneous phenocrysts (xenocrysts?) of plagioclase and hornblende (up to 1 cm in length) are commonly present within this lithology. The hornblende commonly contains inclusions of zircon crystals (0.02 mm in length).

AGE RELATIONSHIPS

Stratigraphic Facing and Paleontologic Data

Fossils have been reported from various parts of the Talladega slate belt and provide a unique opportunity for determining age relationships within the above-described stratigraphic framework. Interpretation of the age of nonfossiliferous parts of the sequence is dependent upon proper interpretation of radiometric age dates and regional structure and upon determination of stratigraphic facing of the various parts of the belt. Primary facing criteria are not common within most of the Talladega belt sequence and are well developed locally only within the metasandstone units.

Several phases of thrust faulting and folding have affected the Talladega slate belt. Tight to isoclinal fold phases have resulted in local overturning of portions of the stratigraphy, but in all localities thus far examined where definitive top and bottom of sequence can be determined, the beds are currently upright. Such localities are not as abundant as would be desirable but are common along many major unit boundaries. For example, upright strata have been observed (1) at the top and the base of the Butting Ram Quartzite in Chilton County; (2) within the Miller Mill Quartzite in Chilton County; (3) at the base of the Lay Dam Formation in Chilton and Talladega Counties; (4) within the upper part of the Wash Creek Slate in Chilton County; (5) at the base, within, and at the top of the Cheaha Quartzite in Clay and Cleburne Counties; and (6) at the base of the Erin Slate in Clay County.

In addition to facing information, the nature of the major unit contacts must also be known. Stratigraphic contacts can be demonstrated at a number of key unit boundaries. These include (1) the Hillabee Greenstone–Jemison Chert (Chulafinee Schist) interval boundary; (2) the Jemison Chert–Butting Ram Quartzite boundary; (3) the Butting Ram Quartzite–Lay Dam Formation boundary; (4) the Erin Slate–Jemison Chert (locally underlain by the Butting Ram Quartzite) boundary; (5) the Cheaha Quartzite–Erin Slate boundary; (6) the Lay Dam Formation–Sylacauga Marble Group boundary; and (7) the Sylacauga Marble Group (and locally the Lay Dam Formation)–Wash Creek Slate boundary. Faults have also been observed to occur locally along a few of the above-mentioned boundaries, but in the examples cited, stratigraphic boundaries have been observed at other localities, so that the original stratigraphic nature of the boundaries can be established.

Fossils were originally reported from the central part of the belt at Erin. Smith (1903) reported the presence of plant fossils in the Erin Slate and interpreted the fossils to be Carboniferous in age, possibly Pennsylvanian. Recently, Gastaldo and Cook (1981) have reported possible plant fossil fragments from the Erin and have agreed with the late Carboniferous age assignment. Key to this interpretation are two well-preserved plant specimens from the Geological Survey of Alabama collection, which are purported to be part of the original Erin collection of Smith (1903). It is not certain that these specimens belong to the original collection, and if they do it is not certain that they were collected from the outcrop by Smith. The most perfect specimens were originally mailed to him after being collected by a local resident (Smith, 1903).

Another fossiliferous unit, the Jemison Chert, was reported in the Talladega slate belt by Butts (1926). The shallow water marine invertebrate fauna has been well documented and studied by several workers (Butts, 1926; Carrington, 1973; Sutley, 1977) and can range in age from Early to Middle Devonian. Carrington (1973) has reported a similar Dovonian(?) fauna from the upper part of the Butting Ram Quartzite just below the Jemison interval. Regional correlations suggest that these Devonian units stratigraphically overlie the Erin Slate (Tull, 1979) (Fig. 4). If the Erin fossils can be interpreted as being as old as Early to Middle Devonian, then much of the Lay Dam Formation may also be Devonian. The diamictite facies occurs sporadically throughout the Lay Dam Formation section, and it is possible that most of this sequence of rocks was deposited very rapidly.

Rocks below the pre–Lay Dam Formation unconformity would therefore appear to be pre-Devonian in age. The Sylacauga Marble Group has lithologic, sedimentary sequence, and thickness affinities to the Cambrian-Ordovician carbonate sequence in the foreland fold and thrust belt in Alabama (Tull and Thomas, 1980). An earlier Paleozoic age for this sequence would be in agreement with that suggested by previous workers (Prouty, 1916; Butts, 1926; Shaw, 1970; Gilbert, 1973) and would also be compatible with the regional stratigraphy outlined in the above interpretation. The Kahatchee Mountain Group appears to occur stratigraphically below the Sylacauga Marble Group and therefore probably ranges from late Precambrian to Cambrian in age, making it stratigrpahically equivalent to the Ocoee-Chilhowee sequence in the Blue Ridge belt to the northeast. The Kahatchee Mountain Group would, therefore, represent the only part of the Talladega slate belt sequence that is well represented in the Blue Ridge belt. This would account for much of the controversy associated with attempts to correlate the bulk of the Blue Ridge and Talladega slate belt stratigraphies (see Jonas, 1932; Park, 1935; Crickmay, 1936; Stose and Stose, 1944; McConnell and Costello, 1981).

Radiometric Dating

Radiometric age dating in the Talladega slate belt has been rather limited and consists principally of conventional K-Ar studies. However, samples from throughout the sequence are represented in these studies, and the available data can provide significant constraints on the minimum age of metamorphism and thus indicate a minimum age for the various lithostratigraphic units.

The rocks of the Talladega slate belt contain evidence of only one Paleozoic progressive regional dynamothermal metamorphic event. In this belt the maximum metamorphic grade obtained was lower greenschist facies. Biotite has been reported from portions of the Kahatchee Mountain Group by Warren (1969) and from the Hillabee Greenstone by Carrington and Wigley (1967), but these occurrences have not been confirmed by subsequent studies. Biotite may be easily confused with stilpnomelane, which occurs at a slightly lower grade. Detrital biotite has been observed within the Lay Dam Formation.

In Chilton County, sericite phyllites occurring within the Kahatchee Mountain Group, the Lay Dam Formation, and the Jemison Chert interval were radiometrically analyzed by Kish and Harper (1973). These analyses were dominantly whole-rock K-Ar determinations, but the sole K-bearing phase in the samples was metamorphic sericite (Kish, 1982, personal commun.). Whole-rock K-Ar ages derived from low metamorphic grade phyllites and slates are believed to be very near to the age of metamorphism (Butler, 1972; Kish and Harper, 1973). The K-Ar ages from sericite phyllites of the Talladega slate belt range beteen 417 and 329 m.y. (Table 1).

The Jemison Chert interval has clearly experienced the lower greenschist facies regional dynamothermal event that is evident throughout the Talladega slate belt. This implies that regional metamorphism was Early to Middle Devonian or younger. The K-Ar phyllite ages (Table 1) suggest that the peak of metamorphism could have been Early to Middle Devonian, but not younger.

These data, in addition to the stratigraphic correlations cited in the preceding discussions, bring the proposed assignment of a Pennsylvanian age to the Erin Slate plant fossils (Gastaldo and Cook, 1981) into question. No thermal ages as young as Pennsylvanian have been recorded in the Talladega slate belt, nor has evidence been recorded of two separate and distinct regional metamorphic events. The metamorphic event affecting the Erin Slate must certainly have been represented elsewhere in the belt. Mapping in the region around Erin demonstrates that the fabrics associated with this event are continuous from the Erin Slate into the surrounding sequences and are developed on a regional scale.

Additional isotopic studies have been conducted on a metaquartz dacite unit within the Hillabee Greenstone in Clay County. Igneous hornblende xenocrysts(?) yield a conventional K-Ar age of 382 ± 14 m.y. (Russell, 1978). The high argon blocking temperature of hornblende combined with the low metamorphic grade (lower greenschist facies) of these rocks suggests that this age may represent a magmatic cooling age. It is also conceivable that this age could represent an age partially reset by metamorphism. U/Pb analyses of zircons from the quartz dacite are nearly concordant at 457 m.y. (Russell, 1978) and could suggest an Ordovician age for the Hillabee Greenstone. However, a significant number of the zircons are zoned and contain rounded and discolored cores so that the apparent Ordovician age may represent a mixed age and the true age of the Hillabee Greenstone may be younger. As mentioned in the preceding section, detailed mapping studies suggest that the Hillabee lies stratigraphically above the Jemison Chert.

DISCUSSION

Modern concepts of tectonics suggest that certain types of orogenies develop after passive or trailing continental margins are converted into active or convergent continental margins. The change from a passive continental margin to an active lithospheric plate boundary is accompanied by significant changes in the tectonic framework of a region. These changes result in important changes in sedimentation, structural development, and thermal regime. A change from a trailing to a convergent continental margin is believed to result in collapse of the continental rise, continental slope, and the edge of the continental shelf.

Plate-tectonics models proposed to explain the development of the Appalachian orogenic system involve the early (late Precambrian) formation of a passive (Atlantic-type) margin along the eastern edge of the proto–North American continent and persistence of a stable carbonate shelf during the early Paleozoic (Bird and Dewey, 1970; Hatcher, 1972, 1978; Odom and Fullagar, 1973; Rankin, 1976; Thomas, 1977; Cook and others, 1980; Hatcher and Odom, 1980). This early Paleozoic continental shelf extended eastward at least to the edge of the present foreland fold and thrust belt and perhaps more than 200 km farther eastward in the southern Appalachians, depending on the interpretation of

TABLE 1. RADIOMETRIC AGE DATES, TALLADEGA SLATE BELT

Lithologic Unit	Location	Dating Method	Age	Reference
Brewer Phyllite (sericite phyllite)	NE ¼ Sec. 31, T.24 N , R.15 E Chilton Co.	K-Ar Whole Rock	380 ± 7 m.y.	Kish and Harper (1973); Kish (1982, Personal Communication)
Wash Creek Slate (sericite phyllite)	SE ¼ NW ¼ Sec. 6, T.23 N , R.15 E Chilton Co.	K-Ar Whole Rock	396 ± 8 m.y.	Kish and Harper (1973); Kish (1982, Personal Communication)
Wash Creek Slate (sericite phyllite)	NW ¼ Sec. 1, T.23 N , R.14 E Chilton Co.	K-Ar Whole Rock	417 ± 8 m.y.	Kish and Harper (1973); Kish (1982, Personal Communication)
Lay Dam Formation (meta-diamictite)	SW ¼ SW ¼ Sec. 7, T.23 N , R.15 E Chilton Co.	K-Ar Whole Rock	400 ± 8 m.y.	Kish and Harper (1973); Kish (1982, Personal Communication)
Lay Dam Formation (secricite phyllite)	SE ¼ SE ¼ Sec. 24, T.23 N , R.14 E Chilton Co.	K-Ar Whole Rock	378 ± 8 m.y.	Kish and Harper (1973); Kish (1982, Personal Communication)
Lay Dam Formation (sericite phyllite)	NW ¼ Sec. 31, T.23 N , R.15 E Chilton Co.	K-Ar Whole Rock	412 ± 8 m.y.	Kish and Harper (1973); Kish (1982, Personal Communication)
Jemison Chert (sericite phyllite)	NW ¼ NW ¼ Sec. 7 T.22 N , R.15 E Chilton Co.	K-Ar Whole Rock	329 ± 7 m.y.	Kish and Harper (1973); Kish (1982, Personal Communication)
Hillabee Greenstone (meta-quartz dacite)	NW ¼ SW ¼ Sec. 12 T.21 S , R.6 E Clay Co.	K-Ar Hornblende	382 ± 14 m.y.	Russell (1978)
Hillabee Greenstone (meta-quartz dacite)	NW ¼ SW ¼ Sec. 12 T.21 S, R.6 E Clay Co.	Pb^{207}/Pb^{206} Zircon	444-458 m.y.	Russell (1978)
Jemison Chert-Erin Slate (graphite-sericite phyllite)	NW ¼ SE ¼ Sec. 17 T.24 N, R.18 E Coosa Co.	K-Ar Sericite	369 ± 18 m.y.	Wampler and others (1970)
Able Gap Group-Erin Slate	SE ¼ SW ¼ Sec. 18 T.21 S, R.6 E Clay Co.	Rb-Sr Whole Rock	379 m.y. (Sr^{87}/Sr^{86} = 0.7205)	R. Weimer (Unpub. data)

the degree of allochthony of the Appalachian crystalline rocks (Cook and others, 1980). Appalachian plate-tectonics models require that the eastern shelf of the proto–North American continent eventually became part of an active margin and was destroyed by deformation resulting either from continent-directed subduction or from other processes such as underthrusting beneath an island arc, microcontinent, or continent. There are apparently few data from southern Appalachian crystalline rocks that relate directly to stratigraphic and structural responses associated with the change of the continental margin to a convergent margin.

Attempts to correlate and compare the sequences of rocks within parts of the metamorphic terrain with the well-documented stratigraphic succession within the fold and thrust belt are hampered by a general lack of accurate knowledge of the age of the metamorphic sequences, as well as an apparent absence of rocks in the metamorphic sequences that are younger than early Paleozoic. In the southern Appalachian Blue Ridge belt, for example, only the late Precambrian to Cambrian clastic sequence is well represented. Rocks in the Murphy marble belt in north Georgia and North Carolina may be as young as Early Ordovician, but this correlation is uncertain.

The general lack of correlative sequences between the foreland and the crystalline terrain has resulted in a rather poor understanding of the nature of the transition between the foreland shelf and the more oceanward portion of the developing continental margin. The allochthonous nature of the orogen in the southern Appalachians would appear to imply that the location of this transition zone is covered beneath the crystalline thrust sheets, but it is also possible that the transitional sequence is partially preserved within some of the allochthonous masses that have been transported onto the foreland. Examination of the orogen in Alabama in the southernmost extent of the metamorphic belt provides insight into this problem.

As indicated above, in the southernmost Appalachians of Alabama the Talladega slate belt forms the frontal crystalline thrust sheet southeast of the foreland fold and thrust belt (Fig. 1). This belt contains a unique stratigraphic sequence including what appears to be an allochthonous portion of the lower Paleozoic foreland shelf sequence in the form of the Kahatchee Mountain and Sylacauga Marble Groups. This sequence of shelf clastic and carbonate rocks may represent the most complete and best preserved fragment of the ancient continental shelf now preserved within the metamorphic southern Appalachians. The shelf sequence appears to be overlain by a thick (4 to 5 km), dominantly clastic sequence, the Talladega Group, which is in turn overlain by the lower part of a major metavolcanic complex, the Hillabee Greenstone (Fig. 4). This is the only post-Middle Ordovician metasedimentary and metavolcanic sequence known to exist in the southern Appalachians. Recent research indicates that the metavolcanic complex may represent the distal portion of an Andean-type volcanic arc system that formed along the eastern edge of the proto-North American continent (Tull and Stow, 1980). The Talladega slate belt, therefore, contains stratigraphic elements evidently representative of a long history of passive margin shelf development, as well as rocks and structures associated with continental shelf collapse and subsequent volcanic and sedimentary lithologies that formed in response to subduction. The preservation of the members of this variable lithotectonic assemblage is significant in itself, but the occurrence of such a stratigraphic succession that may record the change from a divergent to a convergent plate boundary within a single crystalline thrust sheet is indeed unique.

The pre-Lay Dam Formation unconformity and the overlying diamictite facies marks a period of erosion, collapse, and subsidence of the continental shelf, associated with a time of major basement uplift. The relative position of the basement uplift(s) is not known because of the lack of current direction data from the Lay Dam Formation. The size and angularity of the clasts in the diamictites indicate transport distances that are much less than the strike length of the diamictite facies, and this implies a source approximately lateral to the basin margin. Whether or not this source was to the northwest or the southeast of the basin cannot be determined. If it were to the northwest, then the source region was subsequently overthrust by the Talladega slate belt nappe.

The cause of this continental shelf instability is uncertain at present. One hypothesis is that during the early stages of subduction along the ancient continental margin the shelf edge was uplifted, extended, and subsequently filled with foredeep material as it subsided. An alternative hypothesis would suggest that shelf uplift and collapse occurred within a compressional regime and that advancing basement-cored nappes shed debris into a subsiding foredeep. The coarse clastic deposition represented by the Lay Dam Formation and Butting Ram Quartzite was followed, however, by a period of regional Devonian chert deposition within the foreland and Talladega slate belt and not by the arrival in the Talladega slate belt of a series of basement-cored nappes. The regional chert deposition may have been synchronous with initiation of continental margin volcanism. This volcanic system apparently prograded onto the distal edge of the shelf in the vicinity of the Talladega slate belt and overrode part of the area of chert deposition.

This volcanic episode was probably a response to continent-directed subduction. Limited radiometric data from the Talladega slate belt and other parts of the Northern Alabama Piedmont, in conjunction with the stratigraphic data presented above, indicate that volcanism was followed rapidly by regional dynamothermal metamorphism and subsequent dynamic events.

ACKNOWLEDGMENTS

Studies in the Talladega slate belt that have contributed to the writing of this paper were supported by the University of Alabama Mineral Resources Institute, the Alabama Geological Survey (GSA 75-20 and GSA 80-2044), and the National Science Foundation (OSP 77-18691 and EAR-8109056). All forms of support are acknowledged and sincerely appreciated. My understanding of the geology of the Talladega slate belt has been greatly enhanced through field trips and discussions with various former graduate students at the University of Alabama, including Bill Carter, Terry Cook, Bert Hayes-Davis, Lamar Long, Bill Pendexter, David Sutley and Whitney Telle. A preliminary draft of this paper was reviewed by Denny Bearce, Charles Copeland, and Bill Thomas, and these reviewers have helped to improve the quality of the manuscript.

REFERENCES CITED

Adams, G. I., 1926, The crystalline rocks, *in* Adams, G. I., and others, Geology of Alabama: Alabama Geological Survey Special Report 14, p. 24–40.

Bearce, D. N., 1973, Geology of the Talladega metamorphic belt in Cleburne and Calhoun Counties, Alabama: American Journal of Science, v. 273, p. 742–752.

——1982, Lower Cambrian metasediments of the Appalachian Valley and Ridge province, Alabama: Possible relationship with adjacent rocks of the Talladega metamorphic belt: Geological Society of America Special Paper 191 (this volume).

Bird, J. M., and Dewey, J. F., 1970, Lithosphere plate-continental margin tectonics and the evolution of the Appalachian orogen: Geological Society of America Bulletin, v. 81, p. 1031–1060.

Butler, J. R., 1972, Age of Paleozoic regional metamorphism in the Carolinas, Georgia, and Tennessee, Southern Appalachians: American Journal of Science, v. 272, p. 319–333.

Butts, C., 1926, The Paleozoic rocks, *in* Adams, G. I., Butts, C., Stephenson, L. W., and Cooke, W., Geology of Alabama: Geological Survey of Alabama Special Report 14, p. 40–223.

——1940, Description of the Montevallo and Columbiana Quadrangles: U.S. Geological Survey Geological Atlas, Folio 226.

Carrington, T. J., 1973, Metamorphosed Paleozoic sedimentary rocks in Chilton, Shelby, and Talladega Counties, Alabama, *in* Carrington, T. J., ed., Talladega metamorphic front: Alabama Geological Society, 11th Annual Field Trip, Guidebook, p. 22–38.

Carrington, T. J., and Wigley, P. B., 1967, Redefinition of the Hillabee Schist, Alabama: Southeastern Geology, v. 8, p. 19–31.

Carter, W. W., 1981, Stratigraphic and structural relationships within the upper portion of the Talladega slate belt in the Pyriton area, Alabama: Geological Society of America Abstracts with Programs, v. 13, no. 1, p. 4.

Cook, F. A., Brown, L. D., and Oliver, J. E., 1980, The southern Appalachians and the growth of continents: Scientific American, v. 243, no. 4, p. 156–189.

Cook, T. A., 1981, Structural and stratigraphic problems associated with the southern termination of the Cheaha Quartzite, Clay, Coosa, and Talladega Counties, Alabama: Geological Society of America Abstracts with Programs, v. 13, no. 1, p. 5.

Crawford, T. J., and Medlin, J. H., 1973, The western Georgia Piedmont between the Cartersville and Brevard Fault zones: American Journal of Science, v. 273, p. 712–722.

Crickmay, G. W., 1936, Status of the Talladega series in southern Appalachian stratigraphy: Geological Society of America Bulletin, v. 47, p. 1371–1392.

Gastaldo, R. A., and Cook, R. B., 1981, A reinvestigation of paleobotanical evidence for the age of the Erin Shale, Clay County, Alabama: Geological Society of America Abstracts with Programs, v. 13, no. 1, p. 7.

Gilbert, O. E., Jr., 1973, The Sylacauga unconformity/Talladega fault: The Piedmont metamorphic front of Alabama as a stratigraphically controlled fault, *in* Carrington, T. J., ed., Talladega metamorphic front: Alabama Geological Society, 11th Annual Field Trip, Guidebook, p. 39–50.

Hatcher, R. D., Jr., 1972, Developmental model for the southern Appalachians, Geological Society of America Bulletin, v. 83, p. 1735–1760.

——1978, Tectonics of the western Piedmont and Blue Ridge, southern Appalachians: Review and speculation: American Journal of Science, v. 278, p. 276–304.

Hatcher, R. D., Jr., and Odom, A. L., 1980, Timing of thrusting in the southern Appalachians, U.S.A.: Model for orogeny? Journal of the Geological Society of London, v. 137, p. 321–327.

Hays, C. U., 1891, The overthrust faults of the southern Appalachians: Geological Society of America Bulletin, v. 2, p. 141–154.

Hurst, V. J., 1973, Geology of the southern Blue Ridge belt: American Journal of Science, v. 273, p. 643–670.

Jonas, A. I., 1932, Structure of the metamorphic belt of the southern Appalachians: American Journal of Science, v. 244, p. 238–243.

Kesler, T. L., 1950, Geology and mineral deposits of the Cartersville district, Georgia: U.S. Geological Survey Professional Paper, 224, 97 p.

Kish, S. A., and Harper, C. T., 1973, Potassium-argon geochronology of a portion of the southwestern Blue Ridge: Geological Society of America Abstracts with Programs, v. 5, no. 5, p. 409.

Long, A. L., 1981, Relationship between the structure and geochemistry of the copper deposits of the Hillabee Greenstone in the Millerville region, Clay County, Alabama [M.S. thesis]: Tuscaloosa, Alabama, The University of Alabama, 233 p.

McCalley, H., 1897, Report on the Valley Regions of Alabama, Part II, The Coosa Valley Region: Alabama Geological Survey, Special Report 9, 826 p.

McConnell, K. I., and Costello, J. O., 1981, The relationship between the Talladega Group and the Ocoee Supergroup near Cartersville, Georgia: Geological Society of America Abstracts with Programs, v. 13, no. 1, p. 30.

McLaughlin, R. E., and Hataway, D. J., 1973, Fossils in the Murphy Marble: Geological Society of America Abstracts with Programs, v. 5, p. 418–419.

Neathery, T. L., and Reynolds, J. W., 1973, Stratigraphy and metamorphism of the Wedowee Group, a reconnaissance: American Journal of Science, v. 273, p. 723–741.

Neathery, T. L., and others, 1975, Geologic profiles in the northern Alabama Piedmont, *in* Neathery, T. L., and Tull, J. F., eds., Geologic profiles of the northern Alabama Piedmont: Alabama Geologic Society, 13th Annual Field Trip, Guidebook, p. 1–8.

Odom, A. L., and Fullager, P. D., 1973, Geochronologic and tectonic relationships between the Inner Piedmont, Brevard Zone, and Blue Ridge belts, North Carolina: American Journal of Science, v. 273-A, p. 133–149.

Park, C. F., Jr., 1935, Notes on structure of the Erin Shale of Alabama: Journal of the Washington Academy of Sciences, v. 25, p. 276–279.

Pendexter, W. S., 1981, Stratigraphic relationships of the carbonate sequence in the Talladega slate belt in Chilton and Coosa Counties, Alabama: Geological Society of America Abstracts with Programs, v. 13, no. 1, p. 32.

Prouty, W. F., 1916, The crystalline and other marbles of Alabama: Alabama Geological Survey Bulletin, no. 18, 48 p.

——1922, Talladega slates of Alabama: Pan-American Geologist, v. 37, no. 5, p. 363–366.

——1923, Geology and mineral resources of Clay County: Geological Survey of Alabama, County Report no. 1, 190 p.

Rankin, D. W., 1976, Appalachian salients and recesses: Late Precambrian continental break-up and the opening of the Iapetus Ocean: Journal of Geophystical Research, v. 81, p. 5605–5619.

Rodgers, John, 1970, The tectonics of the Appalachians: New York, John Wiley and Sons, 271 p.

Rodgers, J., and Shaw, C. E., Jr., 1962, Age of the Talladega Slate of Alabama: Geological Society of America Special Paper 73, p. 226–227.

Russell, G. S., 1978, U/Pb, Rb-Sr, and K-Ar isotopic studies bearing on the tectonic development of the southernmost Appalachian Orogen, Alabama [Ph.D. dissertation]: Tallahassee, Florida, Florida State University, 197 p.

Shaw, C. E., Jr., 1970, Age and stratigraphic significance of the Talladega Slate: Evidence of pre-Middle Ordovician tectonism in central Alabama: Southeastern Geology, v. 11, p. 253–267.

Smith, E. A., 1888, Report of progress for the years 1884–1888: Alabama Geological Survey Report of Progress, 24 p.

——1903, Carboniferous fossils in Ocoee slates in Alabama: Science, v. 18, p. 244–246.

Stose, G. W., and Stose, A. J., 1944, The Chilhowee and Ocoee series of the southern Appalachians: American Journal of Science, v. 242, p. 369–390.

Sutley, D. E., 1977, A petrographic and paleoenvironmental study of the Jemison Chert, Chilton County, Alabama [M.S. thesis]: Tuscaloosa, Alabama, The University of Alabama, 74 p.

Telle, W. R., 1981, Depositional setting of the diamictite facies of the Lay Dam Formation, Talladega slate belt, Chilton County, Alabama: Geological Society of America Abstracts with Programs, v. 13, no. 1, p. 36.

Telle, W. R., Tull, J. F., Russell, C. W., 1979, Tectonic significance of the bouldery facies of the Lay Dam Formation, Talladega slate belt, Chilton County, Alabama: Geological Society of America Abstracts with Programs, v. 11, no. 4, p. 215.

Thomas, W. A., 1977, Evolution of Appalachian-Ouachita salients and recesses from reentrants and promontories in the continental margin: American Journal of Science, v. 277, p. 1233–1278.

Thomas, W. A., and Neathery, T. L., 1980, Tectonic framework of the Appalachian orogen in Alabama, *in* Frey, R. W., ed., Excursions in southeastern geology, Volume 2: American Geological Institute, p. 465–526.

Tull, J. F., 1977, The nature and significance of the Hollins Line, eastern Alabama Appalachians: Geological Society of America Abstracts with Programs, v. 9., p. 190.

——1978, Structural development of the Alabama Piedmont northwest of the Brevard Zone: American Journal of Science, v. 278, p. 442–460.

——1979, Stratigraphic and structural relationships of the eastern Talladega slate belt of Alabama, *in* Tull, J. F., and Stow, S. H., eds., The Hillabee metavolcanic complex and associated rock sequences: Alabama Geological Society, 17th Annual Field Trip, Guidebook, p. 3–13.

Tull, J. F., and Stow, S. H., 1980, The Hillabee Greenstone: A mafic volcanic complex in the Appalachian Piedmont of Alabama: Geological Society of America Bulletin, Part I, v. 91, p. 27–36.

Tull, J. F., and Thomas, W. A., 1980, Stratigraphic and structural transition between the external fold and thrust belt and the internal metamorphic belt in the Southern Appalachians: International Geological Congress, 26th, Paris, Abstracts, v. 1, p. 296.

Tull, J. F., and others, 1978, The Hillabee Greenstone: Stratigraphy, geochemistry, structure, mineralization, and theories of origin: University of Alabama Mineral Resources Institute Research Report No. 1, 100 p.

Wampler, J. M., Neathery, T. L., and Bentley, R. D., 1970, Age relations in the Alabama Piedmont, Alabama, *in* Bentley, R. D., and Neathery, T. L., eds., Geology of the Brevard fault zone: Alabama Geological Society, 8th Annual Field Trip, Guidebook, p. 81–90.

Warren, R. N., 1969, Geology of a portion of the Stump Hills in southern Shelby County and northern Chilton County, Alabama [M.S. thesis]: New Orleans, Louisiana, Tulane University.

Manuscript Accepted by the Society May 7, 1982

Printed in U.S.A.

Geological Society of America
Special Paper 191
1982

Relationship between Talladega belt rocks and Ocoee Supergroup rocks near Cartersville, Georgia

Keith I. McConnell
Georgia Geologic Survey
19 M.L. King, Jr. Drive
Atlanta, Georgia 30334

John O. Costello*
Georgia Geologic Survey
19 M.L. King, Jr. Drive
Atlanta, Georgia 30334

ABSTRACT

For many years, the question regarding what happens to the rocks of the Talladega belt in the vicinity of their apparent northeastern terminus near Cartersville has been the subject of controversy. This has coincided with the debate over the age and correlation of metasedimentary rocks that overlie the billion-year-old Corbin gneiss complex to the east of Cartersville. Both of these problems are interrelated, and the resolution of each is dependent on the other.

Stratigraphic relationships in the polydeformed rocks exposed in the Salem Church anticlinorium east of Cartersville indicate that the rocks unconformably overlying the Corbin gneiss complex are lithostratigraphic equivalents of the lowermost Ocoee Supergroup. These lithologies can be traced southwestward to the area east of Emerson where the Talladega belt has been presumed to end. Here, it is evident from studying the small- and large-scale structural features that folding has played an important role in the structural and stratigraphic complications that occur. Our mapping suggests that although part of the Ocoee Supergroup does disappear southwest of Cartersville because of folding, other parts of the Ocoee continue on to the southwest and into the Talladega belt.

In the Talladega belt of Alabama, rock units such as the Heflin Phyllite, Abel Gap Formation, and Lay Dam Formation are lithologically similar but may be much younger than parts of the lowermost Ocoee Supergroup sequence present in Georgia. Other rock units of the Talladega belt in Alabama also resemble parts of the Ocoee sequence, but they too are not directly relatable to the Ocoee. All long-range correlations can be considered only speculative until detailed mapping in western Georgia and eastern Alabama is completed. However, there is evidence to suggest that at least part of the Talladega belt is Precambrian in age and was deposited synchronously with the Ocoee Supergroup.

*Present Address: Earth Sciences and Resources Institute, Byrnes International Center, University of South Carolina, Columbia, South Carolina 29208.

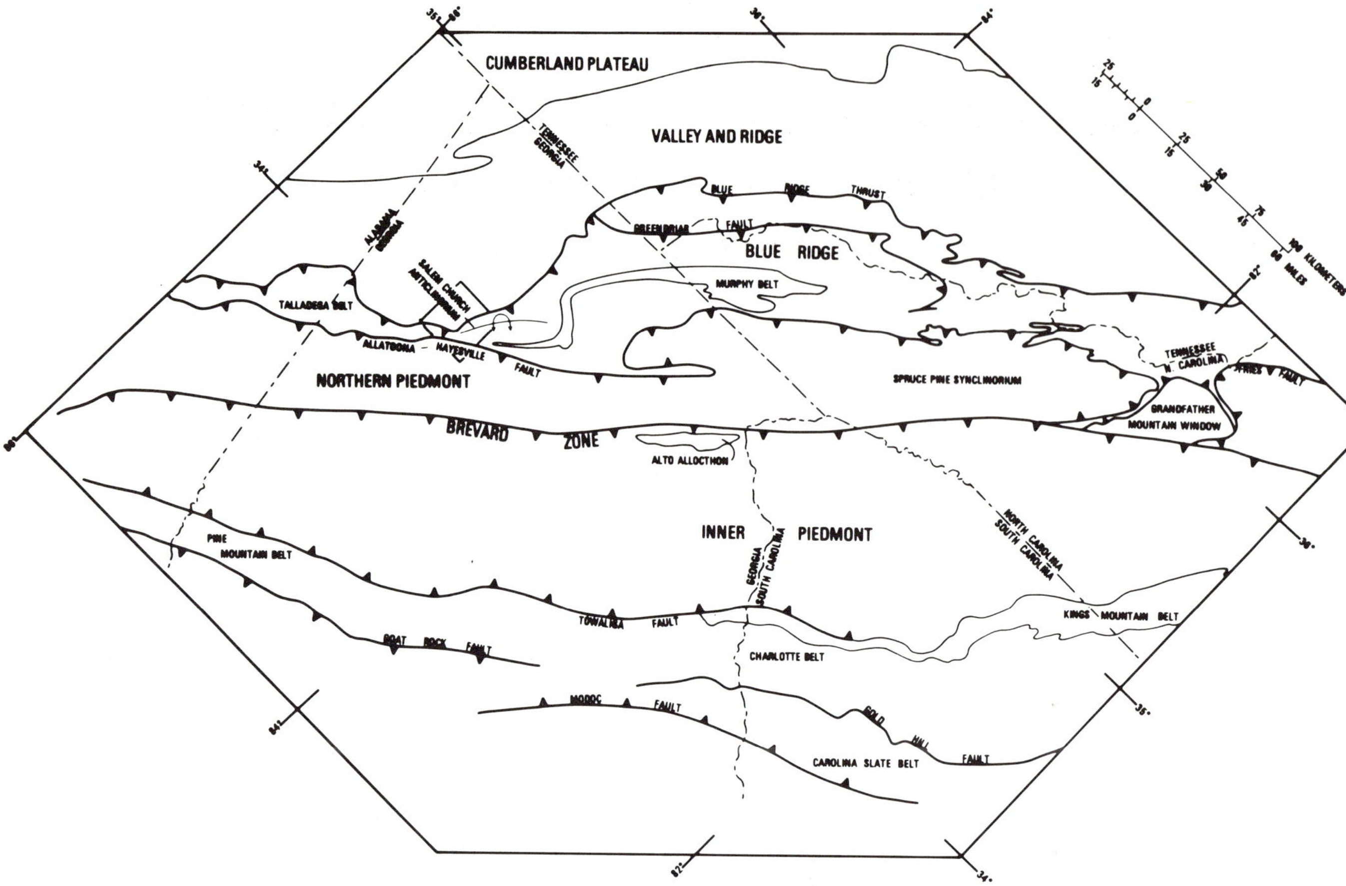

Figure 1. Generalized map of the major subdivisions and tectonic features of the southern Appalachians (modified after Williams, 1978; Hatcher and Odom, 1980). Study area is outlined.

INTRODUCTION

Ninety-three years after Smith (1888) first used the term "Talladega" to define rocks along Talladega Creek in Clay and Talladega Counties, Alabama, the exact limits of the belt as well as the age of units within it remain some of the most enigmatic problems in the southwestern part of the Appalachian orogen. Questions regarding the relationship of these rocks to Paleozoic lithologies in the Valley and Ridge to the north and other crystalline rocks to the southeast still plague interpretations regarding this area. This is particularly true for the rocks that have been included in the Talladega belt in Georgia (Fig. 1). Although some detailed mapping in the northeastern part of the Talladega in Georgia has been done (Crawford, 1976, 1977a, 1977b; McConnell and Costello, 1980), much of the belt remains unmapped and is still represented by broad bands of uniform color or pattern on geologic maps (Georgia Geologic Survey, 1976; Crawford, 1970). Lack of detailed information, particularly structural data, has limited authors of regional frameworks to speculation regarding the stratigraphic position of the Talladega. This speculation results not only from the lack of structural data but also from contradictory age interpretations given for various rock units in the Talladega. These interpretations range from middle to upper Paleozoic, based on fossils in the Jemison Chert and Erin Shale (Shaw, 1970, 1973; Gilbert, 1973; Gastaldo and Cook, 1981), to at least partially Precambrian, based on lithologic similarities and areal continuity with Ocoee Supergroup rocks exposed along trend with the Talladega in north-central Georgia (Hayes, 1900; Prouty, 1923; Stose and Stose, 1944; McConnell and Costello, 1980; Crawford and Cressler, 1981).

Disagreements over the age of rocks in the Talladega belt result primarily from confusion over the structural and stratigraphic relationships present at various localities within the belt. One such area occurs in Georgia in the area just south of Cartersville where Ocoee Supergroup rocks trend into the Talladega (Fig. 2). Although we can offer no comment in this report on problem areas in Alabama, we present what we believe is a sound interpretation of the structural and stratigraphic relationships between rocks of the Talladega belt and rocks of the Ocoee Supergroup at their apparent juncture near Cartersville.

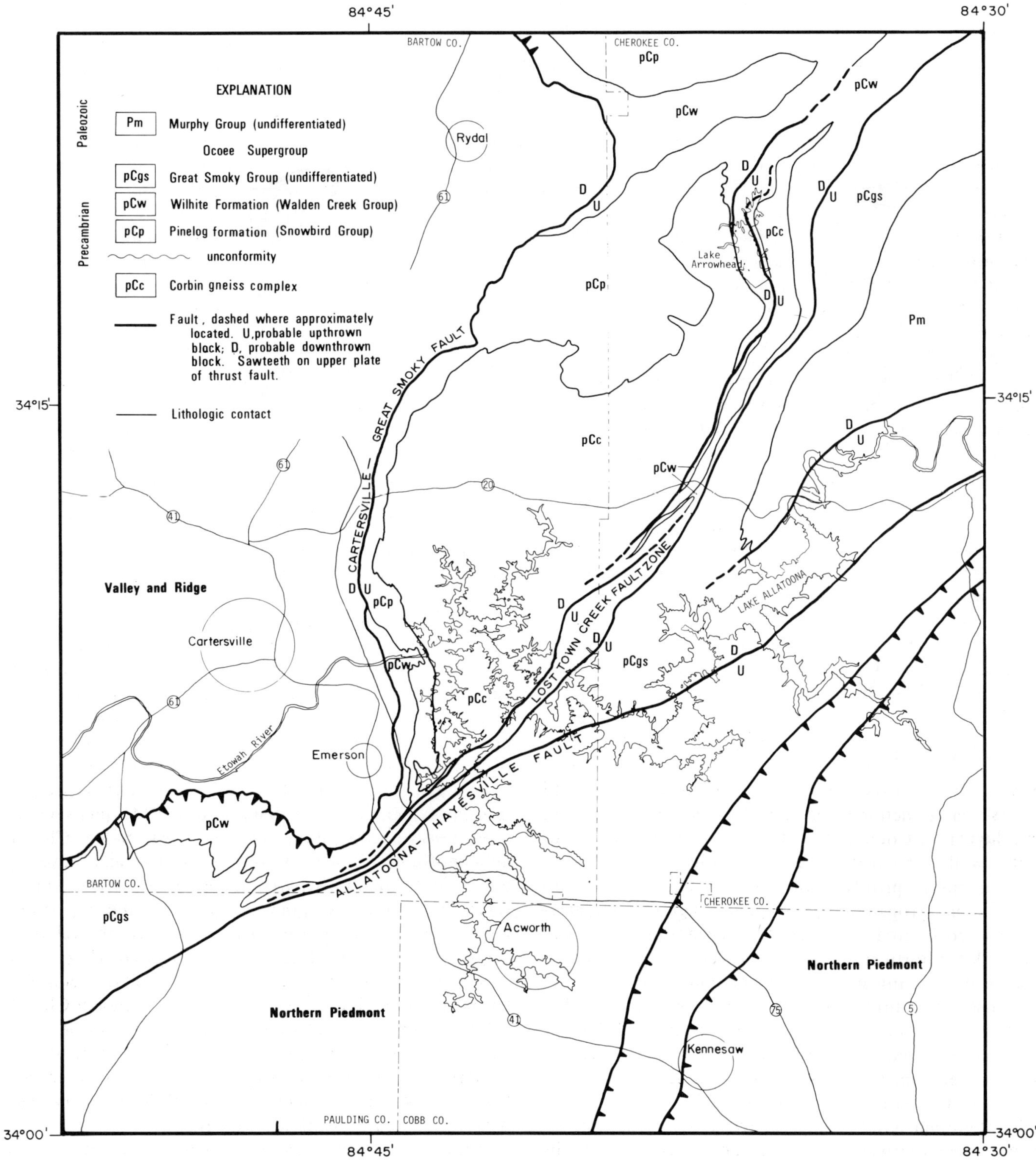

Figure 2. Generalized geologic map of the southwestern terminus of the Salem Church anticlinorium (modified after McConnell and Costello, 1980; Crawford, 1976; Cressler and others, 1979).

GEOLOGIC OVERVIEW OF THE CRYSTALLINE ROCKS EAST AND SOUTH OF CARTERSVILLE

This report deals with the area east and south of Cartersville, Georgia, (Fig. 2) which lies just north of the Emerson reentrant in the Cartersville–Great Smoky fault. The Cartersville–Great Smoky fault is the southern segment of the Blue Ridge thrust and extends from Tennessee into Georgia and Alabama (Fig. 1). This coincides with Hayes's (1900) Cartersville fault and the extension of the Great Smoky fault southward from Tennessee as previously suggested by Stose and Stose (1944). East of Cartersville, the Cartersville–Great Smoky fault has thrust rocks of the Ocoee Supergroup and Corbin gneiss complex (McConnell and Costello, 1980) over lower Paleozoic Shady Dolomite and Chilhowee Group.

The oldest rocks present in the study area are those that comprise the Corbin gneiss complex. The Corbin is exposed in the core of the northeast-trending Salem Church anticlinorium (Fig. 1) and represents the southwesternmost exposures of Grenville-age (Proterozoic Y) basement in the Blue Ridge geologic province. The Corbin complex is a mixed association of orthogneisses and paragneisses with compositions varying from granite to granodiorite, the most characteristic phase being coarse-grained quartz monzonitic gneiss with megacrysts of microcline in a plagioclase, biotite, garnet, and blue quartz groundmass.

Unconformably overlying the Corbin gneiss complex are rocks of the Pinelog and Wilhite formations (McConnell and Costello, 1980). Lithologic similarities suggest that the Pinelog and Wilhite formations are lithostratigraphic equivalents of the lower Ocoee Supergroup (i.e., the Snowbird and Walden Creek Groups that are exposed to the northeast in the Great Smoky Mountains of Tennessee; King and others, 1958). The Pinelog formation is composed of an interlayered sequence of metaconglomerate, metasandstone, and locally carbonaceous metasiltstone. Basal units that lie unconformably on the Corbin gneiss complex consist of well-sorted quartz-pebble metaconglomerate, metasandstone, and graphitic phyllite. These in turn are overlain by thickly bedded and graded metaconglomerates, cross-bedded metasandstones, graphitic phyllites, and thin lenses of immature, poorly sorted lithic metaconglomerates (diamictite). These lithologies grade upward into a relatively thin sequence of carbonaceous and noncarbonaceous, locally calcareous metasiltstone, metasandstone, and sandy marble of the Wilhite Formation (Fig. 2).

Stratigraphic relationships of the lower Ocoee Supergroup rocks are well defined on the western limb of the Salem Church anticlinorium; however, those on the eastern limb are obscured by intense faulting and transposition of the S_1 foliation by structures associated with F_2 folding. On the eastern limb, only thin slivers of the Pinelog formation are present, and the Wilhite Formation is locally in fault contact with the Corbin gneiss complex (Fig. 2). In this same area, units of the upper Ocoee Supergroup (Great Smoky Group) are in fault contact with the underlying Wilhite Formation. The Great Smoky Group is composed of a metamorphosed succession of conglomerates, sandstones, and siltstones which has been subdivided into three formations, from oldest to youngest: the Etowah formation, the Sweetwater Creek formation, and the Illinois Creek formation (McConnell and Costello, 1980). These rocks in turn grade upward into the rocks of the Murphy Group which are exposed in the northeast-trending Murphy syncline (Fig. 2).

Rocks of the Corbin gneiss complex contain evidence of Grenville-age (Proterozoic Y) granulite facies metamorphism (i.e., orthopyroxenes and Mg-rich garnets; Martin, 1974). The overlying metasediments show no effects of this episode of metamorphism and were probably deposited after the conclusion of this event. All rocks east of the Cartersville–Great Smoky fault do contain evidence of a later, Paleozoic-age metamorphic event. Metamorphic grade increases from lower greenschist facies near the Cartersville–Great Smoky fault to upper amphibolite facies in the Murphy syncline. From his $^{40}Ar/^{39}Ar$ work in western North Carolina, Dallmeyer (1975) concluded that metamorphism took place in the southern Blue Ridge approximately 480 $\pm$ 30 m.y. ago. However, other authors (Mose and Stransky, 1973) have proposed ages as young as 350 to 380 m.y. for peak of metamorphism in the Blue Ridge. Coincident with and subsequent to this metamorphic event, rocks east of the Great Smoky fault underwent polyphase deformation. First-generation isoclinal, overturned-to-recumbent F_1 folds (Fig. 3) have been refolded by coaxial, second-generation (F_2) folds, resulting in interference patterns as evidenced by the bending of the Murphy syncline around the northeastern terminus of the Salem Church anticlinorium (Fig. 1). Transposition of original layering during first-generation folding is commonly observed in the rocks of the Salem Church anticlinorium and is partially responsible for the confusion over the stratigraphic relationships in this area. Southeast-trending F_3 folds (Fig. 4) are superposed on earlier fold events, which results in a complex basin and dome interference pattern in the Salem Church anticlinorium. These F_3 folds also play a major role in the bending of Ocoee lithologies from a southerly trend to a westerly trend at Emerson (Fig. 2). Folding culminated with late open warps that trend northwest-southeast and have no penetrative fabric associated with them.

Rocks in the Salem Church anticlinorium and Murphy syncline have been overthrust from the southeast by rocks of the northern Piedmont (Fig. 2). This major zone of pre- and post-metamorphic thrusting has been termed "the Allatoona-Hayesville fault" (Hatcher and others, 1978; McConnell and Costello, 1980).

Figure 3. Photograph showing isoclinal recumbent folds (F_1) in the Pinelog formation being domed by second-generation (F_2) folds.

PREVIOUS INVESTIGATIONS

C. Willard Hayes and his associates first mapped the Cartersville area (unpublished Cartersville folio) and subsequently defined the Cartersville fault (Cartersville–Great Smoky fault in this report; Hayes, 1891, 1900). As originally drawn, the fault extended directly through the town of Cartersville and thrust rocks of the Ocoee "series" (presumed to be of Silurian age on the basis of Willis's and Keith's work in east Tennessee) over Paleozoic sediments (Hayes, 1891). In a subsequent report, Hayes (1900) relocated the fault a few miles to the east and redesignated the age of the Ocoee rocks as Algonkian. In his unpublished report on the Cartersville area, Hayes mapped rocks (his Wilhite slate and Pinelog conglomerate) that he believed to unconformably overlie the Corbin "granite" south and west of Emerson into Talladega belt rocks, which also included rocks of what he termed "Gilmer formation" (Fig. 5). Hayes's Gilmer formation is equivalent to the Great Smoky Group of this report (Table 1). Although Hayes lacked the sophisticated structural and tectonic theories now available, his map of the Cartersville area presented an accurate depiction of the lithologic distribution in the area and provided a remarkably perceptive stratigraphic correlation with the stratigraphy in the Ocoee to the northeast in the Great Smoky Mountains.

Somewhat later, Crickmay (1936) used the term Talladega "series" to include rocks of the known Talladega in Alabama and Georgia as well as those rocks present east of the Great Smoky fault and west of the eastern limb of the Murphy syncline (Fig. 1). He favored a Precambrian age for the Talladega "series." Stose and Stose (1944) believed that

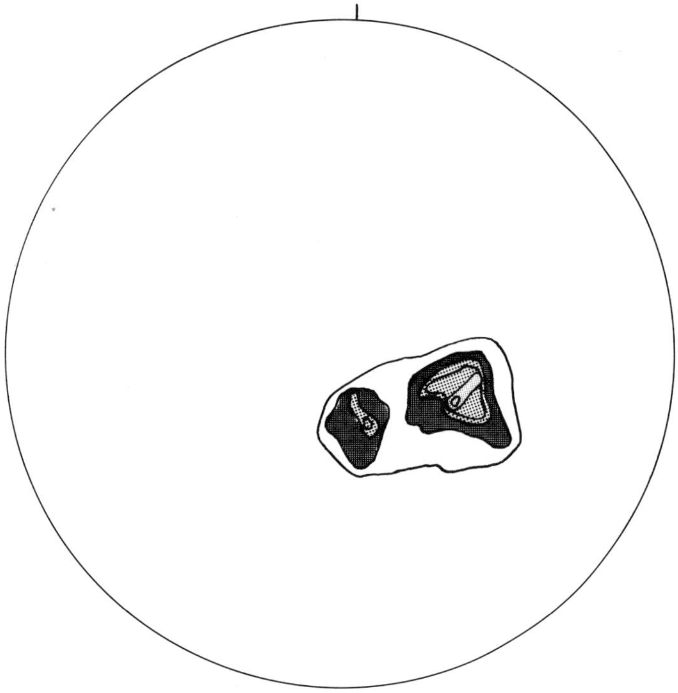

Figure 4. Fabric diagram depicting the trend and plunge of third-generation (F_3) folds (25 measurements). Contour intervals: 8%, 10%, 12%, 25%, 28%, 32%.

the Talladega "series" of Alabama was stratigraphically continuous with the Ocoee series of Hayes (1900) and were the first workers to suggest that the Snowbird "formation" was present around the Corbin gneiss complex. In 1950, Kesler suggested that the Cartersville fault did not exist and correlated rocks of the Talladega "series," as well as those that unconformably overlie the Corbin complex, with the Rome and Weisner. Crawford and Medlin (1970) indicated that the graphitic phyllites of the Talladega were equivalent to those of the Wedowee, Ashland, and Brevard belts. In a later report (Crawford and Medlin, 1973), they indicated that rocks of the Talledega "Group" represented an upwarping of rocks equivalent to those exposed to the southeast in the Austell-Frolona antiform. Cressler (1970), in his hydrologic report on Polk County, traced several rock units in the Talladega belt, which are lithologically similar to Ocoee lithologies, from Haralson County, Georgia, into Talladega County, Alabama. Hurst (1973) questioned the need for the Cartersville fault and suggested that the Talladega in Georgia was lying unconformably on lower Paleozoic sediments and was post-Early Ordovician in age.

Within the past six years, some efforts have been made to map parts of the Talladega belt in detail. Most of this work has been done in the area south and east of Emerson (Crawford, 1976, 1977a, 1977b; McConnell and Costello, 1980),

TABLE 1. CORRELATION CHART SHOWING THE RELATIONSHIP BETWEEN THE STRATIGRAPHY AND TERMINOLOGY USED BY HAYES, CRESSLER AND OTHERS, AND MCCONNELL AND COSTELLO

HAYES (1900)		CRESSLER AND OTHERS (1979)			MCCONNELL AND COSTELLO (1980)		
OCOEE "SERIES"	GILMER FORMATION	OCOEE SUPERGROUP	GREAT SMOKY GROUP	SMQ	OCOEE SUPERGROUP	GREAT SMOKY GROUP	ILLINOIS CREEK FORMATION
				CQMS			SWEETWATER CREEK FORMATION
				MSI			ETOWAH FORMATION
	WILHITE SLATE			pCop/GMS(?)			WILHITE FORMATION
	PINELOG CONGLOMERATE			pCo			PINELOG FORMATION
CORBIN "GRANITE"		CORBIN "GRANITE"			CORBIN GNEISS COMPLEX		

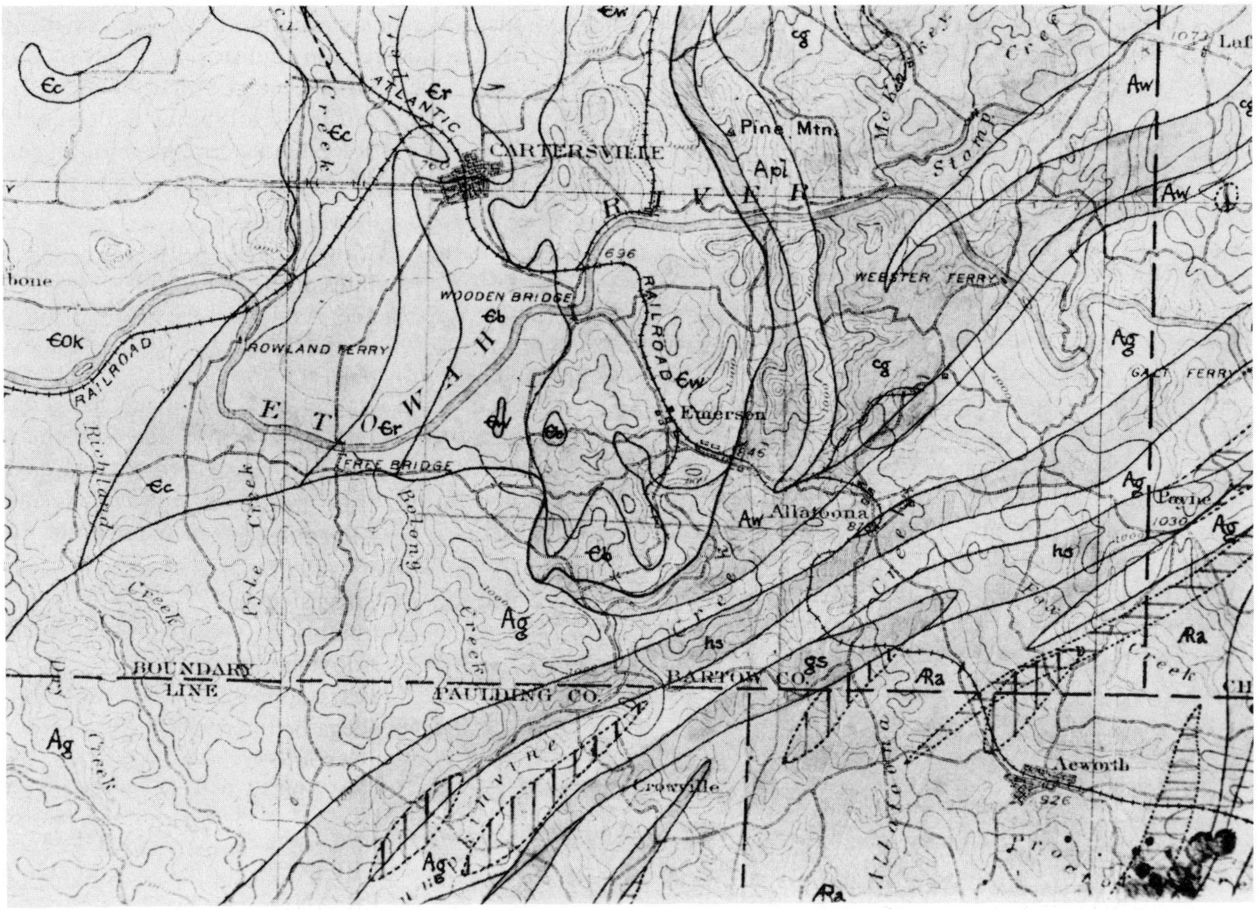

Figure 5. Reproduction of part of C. W. Hayes's Cartersville 30-minute sheet. AG = Gilmer formation; Aw = Wilhite slate; Apl = Pinelog conglomerate; and cg = Corbin granite.

although a small amount of work has been done in the western part of the belt (Weimer, 1976; McConnell, 1980, unpub. data). Cressler and others (1979) and Crawford and Cressler (1981) suggested that the Talladega "Group" and associated lithologies have overthrust the Great Smoky fault and the southwestern terminus of the Salem Church anticlinorium along a low-angle fault termed "the Emerson (Cartersville) fault." McConnell and Costello (1980) disagreed with this interpretation and suggested that rock units of the Talladega belt bend around the Emerson reentrant in the Cartersville–Great Smoky fault and merge with rocks of the Ocoee Supergroup.

CARTERSVILLE, GREAT SMOKY, AND EMERSON FAULTS

The presence or absence of the Cartersville–Great Smoky and Emerson faults east and south of Emerson is the crux of the controversy regarding the stratigraphic position of rocks of the Talladega belt in Georgia. In the Emerson area, rocks of the Talladega belt traced eastward from Alabama are thrust over the Ocoee lithologies, which unconformably overlie the Corbin gneiss complex, or are in stratigraphic conformity with them, or there is no Cartersville fault, and rocks of the Talladega are time-equivalents to the Paleozoic sequence exposed in the Valley and Ridge to the north and west. We discard the last interpretation because it has been shown convincingly that west of Emerson the Cartersville fault thrust Talladega rocks over Valley and Ridge rocks that range in age from Cambrian to Mississippian (Chowns, 1977; Cressler and others, 1979). Without question, therefore, faulting has played a major role in the tectonic development of this area, but the exact nature of faulting is still subject to interpretation as is the relationship between Talladega belt rocks and the Ocoee Supergroup.

Cressler and others (1979) and Crawford and Cressler (1981) have suggested that the Great Smoky fault, trending south from the Tennessee-Georgia line, has been overthrust from the south by the northeast-trending Emerson (Cartersville) fault. The trace of the Emerson fault follows that of the previously defined Cartersville fault from the Alabama-Georgia state line eastward to an area just south of Emerson where it is interpreted to have overthrust the Great Smoky fault. The Emerson is then suggested to continue to the northeast along the eastern limb of the Salem Church anticlinorium (Cressler and others, 1979). Overthrusting the Great Smoky fault has important implications regarding tectonism in the southern Appalachian orogen. It implies that a major northwest-directed compressional event occurred after the emplacement of the Blue Ridge thrust sheet. In addition, the presence of the Emerson fault in this area requires that only those rocks that stratigraphically underlie the Murphy Group rocks of the Murphy belt (in this report and in Crawford and Cressler [1981] believed to be lithostratigraphic equivalents of the Great Smoky Group) trend into the Talladega belt at Emerson. Because of these important implications, we believe that a close examination of the proposed Emerson fault is warranted. As defined, the Emerson fault is a nearly horizontal thrust that locally dips at a low angle (Cressler and others, 1979). Cressler and others (1979) indicated that this low angle of dip differentiates the Emerson from the Great Smoky fault which they indicated has a dip of from 40 to 45°. However, this low dip on the Emerson fault cannot be reconciled with observations of the fault zone on the eastern limb of the Salem Church anticlinorium where dips on the mylonitic foliation present in the Corbin gneiss complex are steep and in some cases near vertical. In addition, where the Emerson (Cartersville) fault is observed west of Emerson, it is essentially a brittle fault with no indication of ductile shearing. However, along the eastern limb of the Salem Church anticlinorium and the trace of the Emerson fault, the Corbin gneiss complex is characterized by a zone of intense mylonitization (Costello, 1978). Although some brittle faulting is evident near the eastern border of the Corbin, it is generally steeply dipping and can be shown readily to be related to the faulting out of overturned limbs of F_2 folds. The low-angle and brittle Emerson fault west of Emerson has been transformed into a high-angle zone of ductile shearing on the east side of the Corbin gneiss complex.

More doubt is cast on the presence of the Emerson fault in the area east of Emerson where it has been interpreted to have overthrust the Great Smoky fault (Cressler and others, 1979). At no place along this part of the fault is there any textural evidence in the phyllites and quartzites bounding it to suggest that any major faulting has occurred. No apparent offset of the S_1 schistosity where it gradually bends from an east-northeast trend west of Emerson to a north-northwest trend southeast of Cartersville (Fig. 6) is observed. The flexure of the S_1 schistosity, which crosses the proposed trace of the Emerson fault undisturbed, defines a conical fold pattern (Fig. 7) that has a trend of approximatley S60° E and plunges approximately 45°. Additional evidence for this folding event is present in the small-scale structures observed in this area (Fig. 4). Interference patterns exhibited by rock units in the Valley and Ridge to the north also suggest that this fold event has affected these rocks. It is doubtful that the flexure of the S_1 schistosity around the Emerson reentrant in the Cartersville–Great Smoky fault would remain undisturbed if cut by a major post-metamorphic fault. We suggest that the differences in the dip on the plane of the Great Smoky fault as observed southwest of and east of Emerson are the result of rotation of the fault plane by these syn- or post-emplacement folds (F_3).

Finally, similar lithologies and stratigraphic sequences exist on either side of the proposed Emerson fault. These lithologies can be traced southward across the trace of the interpreted Emerson fault into the Talladega. This relation is complicated by two factors: the termination of the Salem Church anticlinorium in this area and facies changes within the formations present here. Just east of Emerson, the Salem Church anticlinorium plunges beneath the surface as do the rocks of the Corbin gneiss complex and Pinelog formation. Calcareous, carbonaceous phyllites of the Wilhite Formation continue on to the southwest (Fig. 2). Although the graphite content and calcareous nature of the Wilhite varies along strike due to small-scale facies changes within the formation, the basic lithology (i.e., graphitic phyllite interlayered with sericite phyllite and minor conglomeratic meta-arkose) does not change.

In conclusion, we see no direct evidence to suggest that the Emerson fault, as drawn, exists in the rocks east of Emerson. While evidence of faulting is present along the east side of the Corbin gneiss complex, we believe that this represents an early phase of ductile shearing resulting from transposition and faulting out of the limbs of F_1 folds which has been overprinted by later, brittle faulting. This late faulting results from faulting out of the overturned limbs of F_2 folds as they were appressed against the massive Corbin gneiss complex. We have termed this zone of en-echelon faulting "the Lost Town Creek fault zone" (Fig. 2). In essence, the Cartersville–Great Smoky fault is the Cartersville fault that Hayes (1900) described over 80 years ago, and units that he observed to trend around the Emerson reentrant in the fault are accurately drawn (Fig. 5).

RELATIONSHIP BETWEEN TALLADEGA BELT ROCKS AND OCOEE SUPERGROUP ROCKS AT EMERSON

The confusion regarding the geology of the Emerson area and, in conjunction with that, the relationship of the Talladega belt rocks with Ocoee Supergroup rocks results from a complex interplay of faulting, folding, and stratigraphic environment. In order to understand this relationship, it is necessary to establish a reference point from which to expand from known Ocoee lithologies across the area east of Emerson into what has been defined as Talladega belt. Just such a reference point exists in the stratigraphic sequence present on the western limb of the Salem Church anticlinorium east of Cartersville. The transition from basement

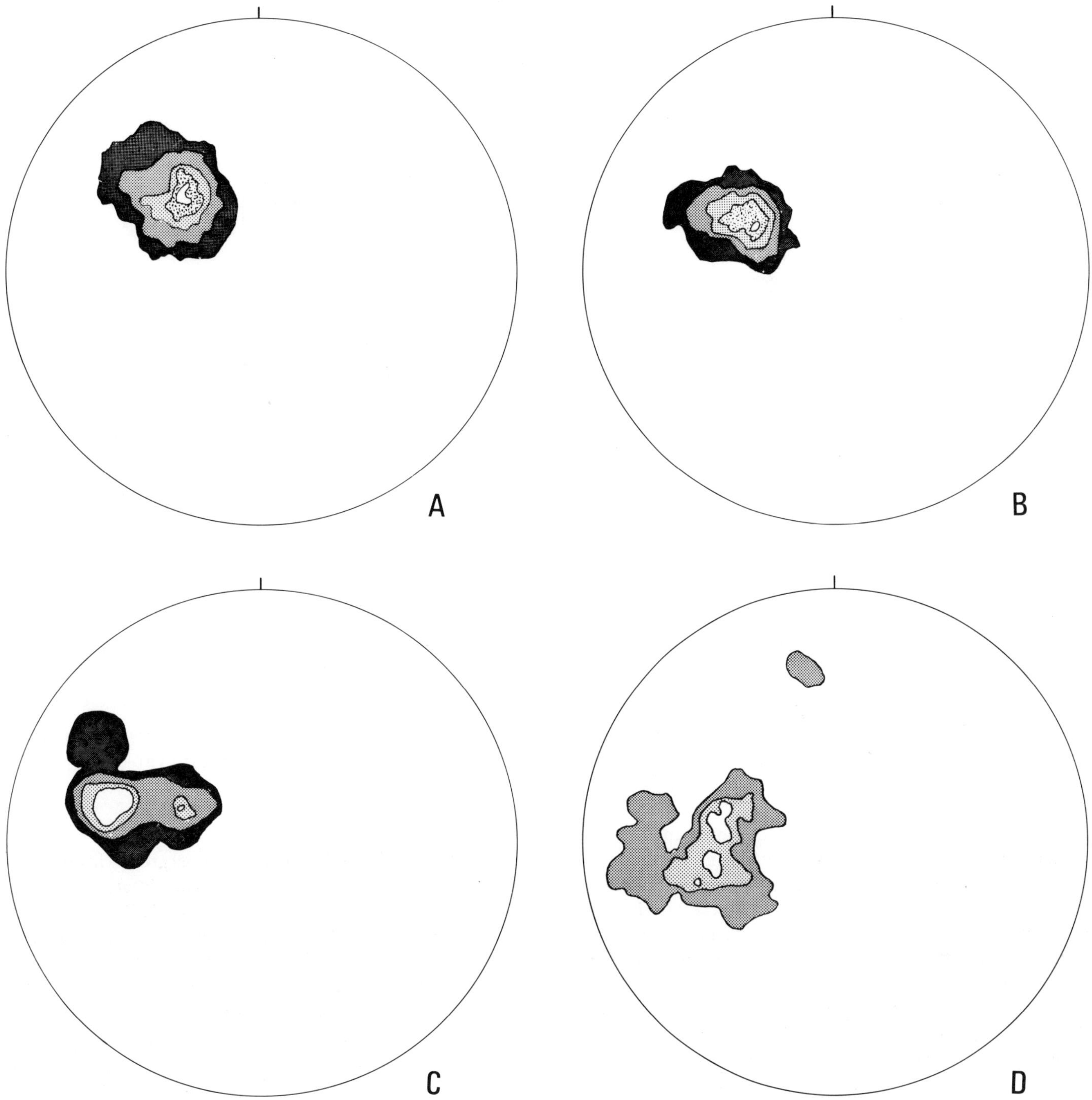

Figure 6. Fabric diagrams of poles to S_1 foliation showing the rotation of this surface around the Emerson reentrant in the Cartersville–Great Smoky fault. Diagram A = poles to S_1 foliation in rocks of the Talladega belt southwest of Emerson (74 measurements); B = poles to S_1 foliation in rocks of the Talladega belt on Interstate 75 (65 measurements); C = poles to S_1 foliation in rocks of the Talladega belt north of Interstate 75 (14 measurements); and D = poles to S_1 foliation in rocks of the Talladega belt/Ocoee Supergroup along Interstate 75 east of Emerson (34 measurements). See Figure 2 for locations.

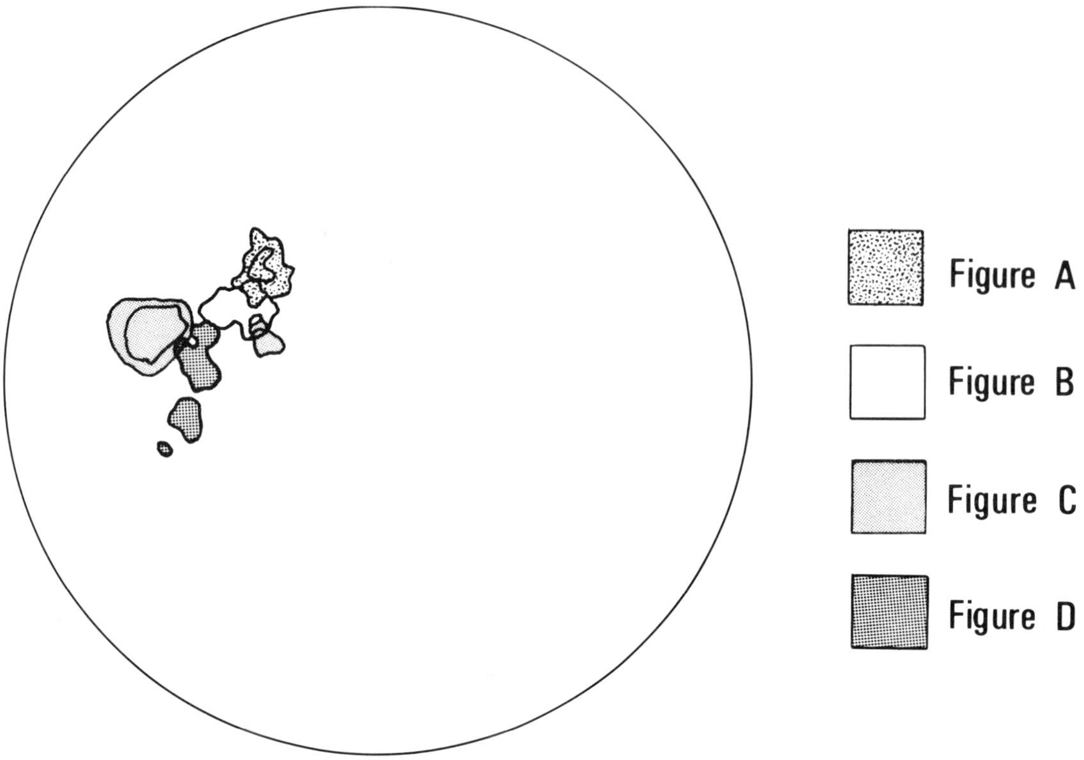

Figure 7. Fabric diagram of the highest concentrations from diagrams A through D in Figure 6 showing the rotation of S_1 foliation around the Emerson reentrant in the Cartersville–Great Smoky fault.

gneiss upward into an unconformable high-energy clastic sequence and then into a sequence that represents a lower energy, more restricted depositional environment is of paramount importance in efforts to decipher the relations east of Emerson. This same lithologic succession can be traced from the northeastern terminus of the Salem Church anticlinorium approximately 25 km to the northeast, southward to Interstate 75 southeast of Emerson. At the southeastern terminus of the Salem Church anticlinorium, the Corbin gneiss complex and Pinelog formation plunge beneath the surface, but the overlying Wilhite Formation continues on to the southwest uninterrupted and therefore forms part of the Talladega belt in this area. In conjunction with the plunging out of the massive lithologies like the Corbin gneiss and the quartzites of the Pinelog formation, faulting associated with transposition during F_2 folds becomes less apparent. Rocks of the Great Smoky Group, in fault contact with graphitic phyllites of the Wilhite Formation to the northeast, are believed to conformably overlie the Wilhite to the southwest of Emerson and therefore also compose part of the Talladega in this area (Fig. 2). All of this, with the exception of some revision of stratigraphic nomenclature and structural interpretations, is close to what Hayes proposed nearly 80 years ago. The Wilhite slate, Pinelog conglomerate, and Gilmer formation are stratigraphic equivalents with the Wilhite Formation, Pinelog formation and Great Smoky Group of this report (Table 1).

REGIONAL IMPLICATIONS

Undoubtedly, equating Ocoee Supergroup lithologies with parts of the Talladega belt poses some interesting questions regarding the age of the rocks in the Talladega belt of Georgia in contrast to the age proposed for the eastern part of the Talladega in Alabama. While the age of the Ocoee Supergroup is not exactly known, stratigraphic relationships in the Great Smoky Mountains indicate that it is older than the Lower Cambrian Chilhowee Group, but younger than the Grenville-age (Proterozoic Y) basement gneisses on which it lies unconformably (King and others, 1958). The same can be said of the Pinelog-Wilhite sequence in the Salem Church anticlinorium in that it is younger than the Grenville-age basement gneisses (Corbin gneiss complex) but older than the Chilhowee. Because of its equivalence with the Ocoee Supergroup, the northeastern part of the Talladega belt would have a similar relationship. Farther to the southwest, in the vicinity of Buchanan, Georgia, Weimer (1976) mapped what he believed to be lithostratigraphic equivalents to the uppermost Talladega in eastern Alabama as defined by Bearce (1973a; i.e., the Abel Gap Formation and Chulifinee Schist). However, recent mapping in the area by the senior author indicates that these rocks could just as well be lithostratigraphic equivalents of the Ocoee lithologies to the northeast. In Alabama, Bearce (1973b) originally defined the Heflin Phyllite as being older than the Able Gap Formation, which he suggested was equivalent to the lower Cambrian Weisner Formation. He also indicated that the Heflin was probably equivalent to the Lay Dam Formation found farther to the southwest. Parts of the Lay Dam Formation, the bouldery mudstone facies, are similar to the diamictite facies of the Pinelog formation. Both have boulders of Grenville-age granitic gneiss supported in a fine-grained siltstone matrix (Fig. 8). In addition,the carbonaceous Able Gap Formation is lithologically similar to the Wilhite Formation. Until recently, this correlation would seem to have been viable; however, Bearce (1981) now believes that the Abel Gap is equivalent to the Butting Ram Sandstone and is probably Devonian in age. This would preclude equating either the Abel Gap or Heflin Phyllite with the Ocoee units overlying the Corbin gneiss complex. However, recent mapping in the Borden Springs area (D. Bearce, June 1981, written commun.) indicates that there is an "Abel Gap–like" lithology present below the Heflin which is probably Early Cambrian or older in age. This unit may correspond to the Wilhite observed to the northeast near Emerson or perhaps one of the graphitic horizons in the Great Smoky group (McConnell and Costello, 1980). In addition, Tull (March 1981, oral communication) suggested that rocks below the Jumbo-Sylacauga carbonate sequence (i.e., the Talladega slate belt) may possibly be equivalent to Ocoee rocks observed near Emerson. However, this interpretation brings up some puzzling questions; namely, what part of the Ocoee sequence do rocks of the Talladega slate belt represent, and what happens to Murphy Group rocks that are known to stratigraphically overlie the Ocoee? Although carbonates of the Jumbo-Sylacauga sequence could be equated with the Murphy Marble, this correlation would be unlikely due to the absence in this area of the well-defined stratigraphic sequence above and below the Murphy Marble. Similar problems arise with other attempts at correlation between the Talladega belt of Georgia and that of Alabama, and any plausible interpretation of this relationship must await completion of detailed mapping in western Georgia and eastern Alabama.

Figure 8. Photograph showing the diamictite facies of the Pinelog formation.

CONCLUSIONS

Rocks that unconformably overlie the billion-year-old Corbin gneiss complex east of Cartersville are lithologically similar to rocks of the Ocoee Supergroup of the Great Smoky Mountains area. The lower Ocoee in this area has been separated into two units termed "the Pinelog and Wilhite formations" after Hayes's Pinelog conglomerate and Wilhite slate. Stratigraphically above the Wilhite Formation are rocks of the Great Smoky Group, which in turn are conformably overlain by rocks of the Murphy Group in the Murphy syncline. The Corbin gneiss complex, basal Ocoee Supergroup equivalents (Pinelog and Wilhite formations), and the Great Smoky Group, as well as their lithostratigraphic equivalents in the Talladega belt, have been thrust over the Lower Cambrian Chilhowee Group and Shady Dolomite along the Cartersville–Great Smoky fault. Lithostratigraphic units that were defined in the Salem Church anticlinorium east of Cartersville have been traced around the Emerson reentrant in the Cartersville–Great Smoky fault and into the Talladega belt. Assuming the age of the sediments unconformably overlying the Corbin gneiss is consistent with the Precambrian age given for the Ocoee in eastern Tennessee, then at least part of the Talladega belt in Georgia must be Precambrian in age. Reconnaissance mapping to the southwest indicates that these same units continue and may extend into Alabama.

ACKNOWLEDGMENTS

We express our appreciation for the constructive criticisms given in reviews by James F. Tull, Denny N. Bearce, and W. Robert Power. We also would like to posthumously recognize C. Willard Hayes for his work in the Cartersville area, most of which remains unpublished.

REFERENCES CITED

Bearce, D. N., 1973a, Geology of the Talladega metamorphic belt in Cleburne and Calhoun Counties, Alabama: American Journal of Science, v. 273, p. 742–745.

——1973b, Character of the Talladega belt in eastern Alabama: Alabama Geological Society, 11th Annual Field Trip, Guidebook, p. 10–21.

——1981, A closer look at the Talladega belt–Valley and Ridge province in eastern Alabama and western Georgia [abs.]: Geological Society of America Abstracts with Programs, Southeastern Section, v. 13, p. 2.

Chowns, T. M., 1977, Stratigraphy and economic geology of Cambrian and Ordovician rocks in Bartow and Polk Counties, Georgia: Georgia Geological Society, 12th Annual Meeting and Field Trip, Guidebook, 21 p.

Costello, J. O., 1978, Shear zones in the Corbin gneiss of Georgia: Georgia Geologic Survey Bulletin 93, p. 32–37.

Crawford, T. J., 1970, Geologic map of Carroll and Heard Counties, Georgia, *in* Hurst, V. J., and Long, S., 1971, Geochemical study of alluvium in the Chattahoochee-Flint area, Georgia: University of Georgia Institute of Community and Area Development, 52 p.

——1976, Geologic map of the Burnt Hickory Ridge quadrangle, Georgia: Georgia Geologic Survey Open-File Map.

——1977a, Geologic map of the Taylorsville quadrangle, Georgia: Georgia Geologic Survey Open-File Map.

——1977b, Geologic map of the Yorkville quadrangle, Georgia: Georgia Geologic Survey Open-File Map.

Crawford, T. J., and Cressler, C. W., 1981, The Talladega "Series"-Great Smoky-Emerson fault: Relationship in the Cartersville area, Georgia [abs.]: Geological Society of America Abstracts with Programs, Southeastern Section, v. 13, p. 5–6.

Crawford, T. J., and Medlin, J. H., 1970, Stratigraphic and structural features between the Cartersville and Brevard fault zones: Georgia Geological Society Guidebook 9, 35 p.

——1973, The western Georgia Piedmont between the Cartersville and Brevard fault zones: American Journal of Science, v. 273, p. 712–722.

Cressler, C. W., 1970, Geology and groundwater resources of Floyd and Polk Counties, Georgia: Georgia Geologic Survey Information Circular 39, 95 p.

Cressler, C. W., Blanchard, H. E., Jr., and Hester, W. G., 1979, Geohydrology of Bartow, Cherokee and Forysth Counties, Georgia: Georgia Geologic Survey Information Circular 50, 45 p.

Crickmay, G. W., 1936, Status of the Talladega Series in southern Appalachian stratigraphy: Geological Society of America Bulletin, v. 47, p. 1371–1392.

Dallmeyer, R. D., 1975, Incremental Ar^{40}/Ar^{39} ages of biotite and hornblende from retrograded basement gneisses of the southern Blue Ridge: Their bearing on the age of Paleozoic metamorphism: American Journal of Science, v. 275, p. 444–460.

Gastaldo, R. A., and Cook, R. B., 1981, A reinvestigation of paleobotanical evidence for the age of the Erin Shale, Clay County, Alabama [abs.]: Geological Society of America Abstracts with Programs, Southeastern Section, v. 13, p. 7.

Georgia Geologic Survey, 1976, Geologic map of Georgia: Georgia Geological Survey Map, 1:500,000.

Gilbert, O. E., Jr., 1973, The Sylacauga unconformity/Talladega fault: The Piedmont metamorphic front of Alabama as a stratigraphically controlled fault: Alabama Geological Society, 11th Annual Field Trip, Guidebook, p. 39–50.

Hatcher, R. D., Jr., and Odom, A. L., 1980, Timing of thrusting in the southern Appalachians, USA: Model for orogeny? Journal Geological Society of London, v. 137, p. 321–327.

Hatcher, R. D., Merschat, C. E., Milici, R. C., and Wiener, L. S., 1978, Structural transect in the southern Appalachians, Tennessee and North Carolina, *in* Field trips in the southern Appalachians: Tennessee Division of Geology, Report of Investigations 37, p. 5–52.

Hayes, C. W., 1891, The overthrust faults of the southern Appalachians: Geological Society of America Bulletin, v. 2, p. 141–154.

——1900, Geological relations of the iron-ores in the Cartersville district, Georgia: American Institute of Mining Engineers Transactions, v. 30, p. 403–419.

Hurst, V. J., 1973, Geology of the southern Blue Ridge belt: American Journal of Science, v. 273, p. 643–670.

Kesler, T. L., 1950, Geology and mineral deposits of the Cartersville District, Georgia: U.S. Geological Survey Professional Paper 224, 97 p.

King, P. B., Hadley, J. B., Newman, R. B., and Hamilton, W., 1958, Stratigraphy of Ocoee Series, Great Smoky Mountains, Tennessee and North Carolina: Geological Society of America Bulletin, v. 69, p. 947–966.

Martin, B. F., 1974, The petrology of the Corbin Gneiss [M.S. thesis]: Athens, Univeristy of Georgia, 133 p.

McConnell, K. I., and Costello, J. O., 1980, Guide to geology along a traverse through the Blue Ridge and Piedmont provinces of North Georgia, *in* Frey, R. W., ed., Excursions in southeastern geology: American Geological Institute, v. 1, p. 241–258.

Mose, D. G., and Stransky, T., 1973, Radiometric age determinations in the Southern Appalachian Blue Ridge: Geological Society of America Abstracts with Programs, Southeastern Section, v. 5, p. 422.

Prouty, W. F., 1923, Geology and mineral resources of Clay County, with special reference to the graphite industry: Alabama Geological Survey County Report 1, 190 p.

Shaw, C. W., 1970, Age and stratigraphic significance of the Talladega Slate: Evidence of pre-middle Ordovician tectonism in central Alabama: Southeastern Geology: v. 11, p. 253–267.

——1973, Stratigraphy, structure, and tectonics of rocks adjacent to the Piedmont physiographic front in Talladega County, Alabama: Alabama Geological Society, 11th Annual Field Trip, Guidebook, p. 56–65.

Smith, E. A., 1888, Report of progress for the years 1884–1888: Alabama Geological Survey Report of Progress, 24 p.

Stose, G. W., and Stose, A. J., 1944, The Chilhowee Group and Ocoee series of the southern Appalachians: American Journal of Science, v. 242, p. 367–390, 401–416.

Weimer, R. H., 1976, A geologic investigation of the Talladega Group in Polk and Haralson, Counties, Georgia [M.S. thesis]: Tallahassee, Florida State University, 76 p.

Williams, H., 1978, Tectonic-lithofacies map of the Appalachian orogen: St. John's Newfoundland, Canada, Memorial University of Newfoundland, scale 1:1,000,000.

MANUSCRIPT ACCEPTED BY THE SOCIETY MAY 7, 1982

Geological Society of America
Special Paper 191
1982

Talladega 'Series,' Great Smoky fault, and Emerson fault: Relationships in the Cartersville area, Georgia

Thomas J. Crawford
West Georgia College
Carrollton, Georgia 30118

Charles W. Cressler
U.S. Geological Survey
Doraville, Georgia 30360

ABSTRACT

Talladega "Series" rock units in Polk and Paulding Counties have been mapped through the Cartersville area along the southeastern fault boundary of the "Corbin granite complex," and correlated with rock units of the Great Smoky Group in Cherokee County. The Talladega "Series" in this area is bounded on the northwest by the Emerson (Cartersville) fault, along which the Talladega was thrust over (1) Valley and Ridge rocks ranging in age from Early Cambrian (Chilhowee Group) to Mississippian (Fort Payne Chert), (2) the Ocoee Supergroup, and (3) the "Corbin granite complex." The Emerson fault also overrode the Great Smoky fault, which separates Ocoee from Chilhowee rocks in the Cartersville area.

The thrust fault along the southeastern boundary of the Talladega "Series" is only one of numerous closely spaced imbricate faults within and southeast of the Talladega. The presence of closely spaced thrusts within the Talladega, the association of Hillabee Chlorite Schist–"Pumpkinvine Creek" lithologies with many of these faults, and the fact that detailed mapping has not been completed between the Rockmart-Yorkville area and the Alabama state line emphasize the need for caution in projecting correlations and interpretations across this area.

INTRODUCTION

The geology of the Cartersville area has been especially confounding for many years. Geologists have been particularly aware of the many interpretations concerning the areal distribution of lithologies, the age and correlation of rock units, the presence or absence of faults, and their nature and location in the Cartersville area. It was appropriate that the Georgia Geological Society, organized to promote the science of geology (particularly in Georgia), chose as its first annual field trip a consideration of the Cartersville fault problem (Bentley and others, 1966). Interpretations made up to 1966 were summarized in the Guidebook for that trip. Vernon J. Hurst presented later summaries in his papers, "The Piedmont in Georgia" (1970) and "Geology of the Southern Blue Ridge Belt" (1973).

It is not our intention here to review and summarize the literature concerning the Cartersville area; summaries are available in recent publications (McConnell and Costello, 1980, p. 243). The purpose of this short paper is to present interpretations based solely on our own detailed work in the area during the past 15 years. The data on which these interpretations are based have been published or made available in open file (see references cited). The interpretations and relationships apply only to the area shown in Figure 1. We do not imply that they are valid throughout the southeast nor, perhaps, throughout the extent of the referenced faults or rock units in Georgia.

TALLADEGA "SERIES"

The Talladega "Series" has been mapped in detail in the

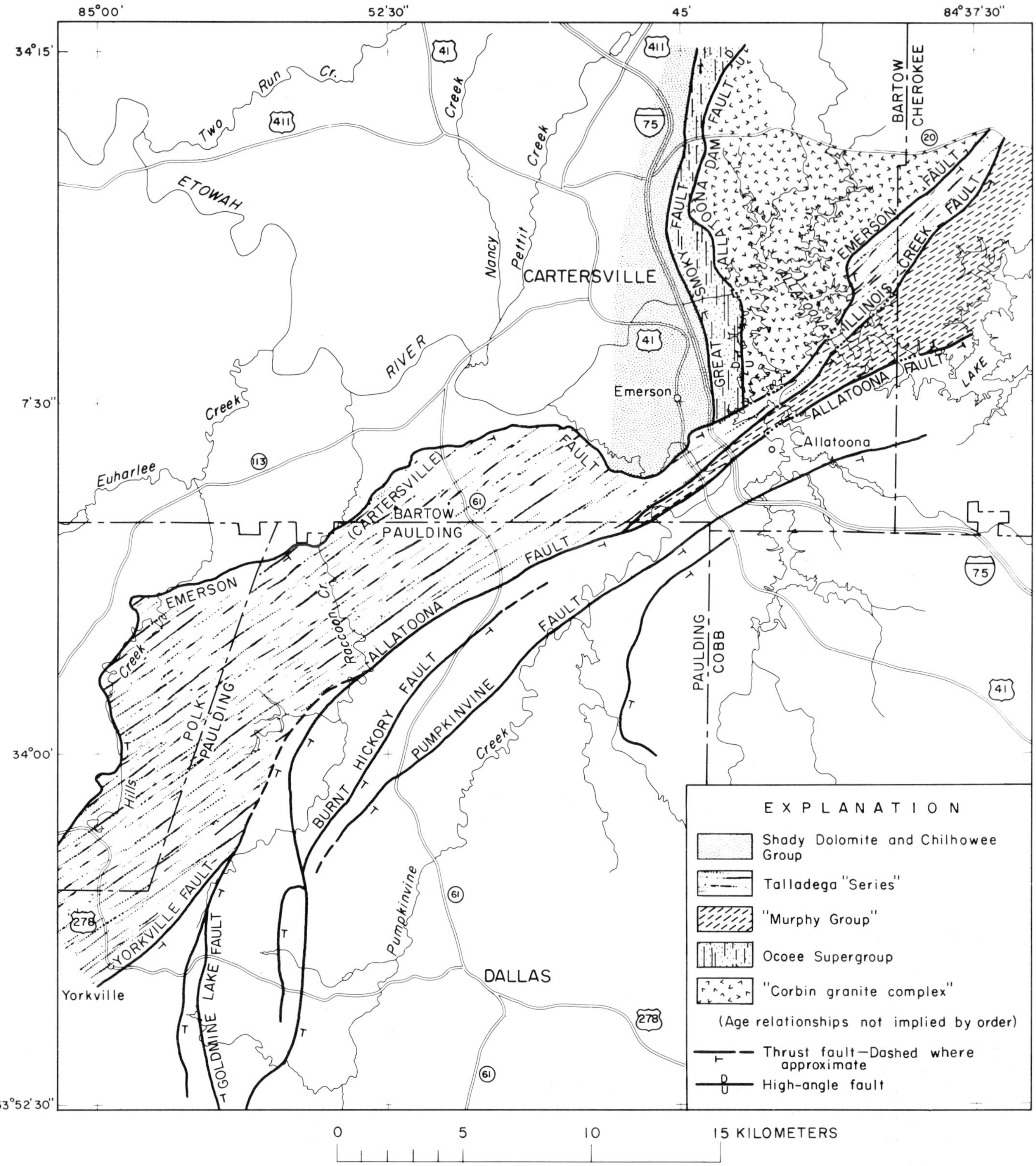

Figure 1. The Talladega "Series," Great Smoky fault, and Emerson fault: relationships in the Cartersville area, Georgia.

Burnt Hickory Ridge, Taylorsville, and Yorkville 7.5-minute quadrangles (Crawford, 1976, 1977a, and 1977b); also in part of the Acworth, the Allatoona Dam quadrangle, and part of the South Canton quadrangle. Reconnaissance mapping of Carroll, Haralson, and Paulding Counties was completed some years earlier (Hurst and Crawford, 1970; Crawford, 1971).

The Talladega "Series" in this area consists of three major lithologic units (minerals are listed in order of increasing abundance):

1. Schist-metagraywacke unit: Fine-grained (phyllitic) quartz-muscovite schist, and interlayered biotite-muscovite-quartz-feldspar metagraywacke. Both contain fine disseminated magnetite and ilmenite. The tiny black opaque minerals are conspicuous in the fine-grained schist; coarse platy ilmenite in residual vein quartz litter is characteristic of this unit. Layers of schist and metagraywacke are variable in thickness, on the order of 0.3 to 3.0 m.

2. Graphite schist unit: Fine- to medium-grained graphite-feldspar-quartz-muscovite schist, having moderate to good schistosity; graphite content is highly variable, but parts are very rich in graphite; includes thin quartzites and layers of micaceous graphitic metaconglomerate containing blue quartz pebbles up to 15 mm in long dimension.

3. Chloritic and calcareous schist-metagraywacke unit: feldspar-quartz-muscovite schist and muscovite-quartz-feldspar metagraywacke, both of which are chloritic and calcareous. The schist is medium-grained, very feldspathic and quartzose, and contains abundant small quartz pods. The metagraywacke is very feldspathic. Boundaries between the interlayered schist and metagraywacke are gradational and layering is poorly developed. An abundance of quartz pods gives the rock a "wavy" texture. This unit weathers brownish red to dark reddish brown.

In Bartow County near the Emerson (Cartersville) fault zone, there is a similar rock that contains large feldspar porphyroblasts and stretched quartz pebbles in a fine-grained quartz-mica-feldspar matrix. It is mylonitic in part and crops out only in a small area near the fault. This lithology is included here in the chloritic and calcareous schist-metagraywacke unit; however, it may be a fault slice, or may prove to be a distinctly different mappable unit.

MAJOR FAULTS AND RELATIONSHIPS

Parts of the Talladega "Series" are terminated both along strike and across strike by the Emerson (Cartersville) fault zone (Cressler and others, 1979), along which the Talladega was thrust over Valley and Ridge rocks (Fig. 1) ranging in age from Early Cambrian (Chilhowee Group) to Mississippian (Fort Payne Chert). In the vicinity of Emerson and to the west, the Chilhowee Group, Shady Dolomite, Rome Formation, Conasauga Formation, and Knox Group trend southward and have been overridden by the Talladega "Series" along the northeast-trending Emerson (Cartersville) fault zone.

East and northeast of Emerson, massive quartzites and phyllites of the Ocoee Supergroup have been faulted against the Cambrian Chilhowee Group and Shady Dolomite along the south-trending Great Smoky fault zone. The Ocoee Supergroup is bounded on the east by the south-trending Allatoona Dam fault, which has brought the "Corbin granite complex" against the Ocoee.

Southeast and east of Emerson, the distinctive lithologies and lithologic sequence of the Talladega "Series" have been mapped northeastward (Fig. 1) to where they have been thrust northwestward over the south-trending Great Smoky fault zone, the massive quartzites and phyllites of the Ocoee Supergroup, and the "Corbin granite complex."

Southeast of Emerson, at the junction of the Emerson (Cartersville) fault zone–Great Smoky fault zone, where the Talladega "Series" has been thrust over the Ocoee Supergroup, every effort was made to correlate northeast-trending Talladega units with south-trending Ocoee units. We could not make such a correlation. South-trending massive quartzites of the Ocoee Supergroup, which are distinctive ridge-formers east of Cartersville, do not continue beyond the Emerson–Great Smoky fault junction. The high-angle reverse faults and associated zones of intense brecciation that are common in the Chilhowee west of the Great Smoky thrust sheet also are terminated. Our detailed mapping has shown that the Ocoee Supergroup has been overridden by the Talladega "Series," which continues northeastward along the southeastern and eastern margins of the "Corbin granite complex" (Fig. 1). The Emerson (Cartersville) fault zone also continues northeastward and forms the boundary between the Talladega "Series" and the "Corbin granite complex" (Cressler and Crawford, 1976; Crawford, *in* Cressler and others, 1979). The lithologic units of the Talladega "Series" traced northeastward from the Emerson area correlate with lithologic units of the Great Smoky Group, which continue northeastward across Cherokee Conty.

The Talladega "Series" has been overridden from the southeast by the "Murphy Group" along what we refer to here as the Illinois Creek fault, previously unmapped (Fig. 1). The "Murphy Group," which approaches Emerson from the northeast, is bounded on the northwest by the Illinois Creek fault and on the southeast by the Allatoona fault (Hurst, 1973). The "Murphy Group" is completely cut out a few miles southwest of Emerson by the Allatoona fault. Detailed mapping has shown a series of imbricate faults, both to the northwest and southeast of the Allatoona fault, spaced only a few tens to a few hundred meters apart along which slices have been thrust to the northwest and west.

Southwest of Emerson, where the "Murphy Group" has been cut off by the Allatoona fault, the Talladega "Series" is in fault contact with a a sequence of lithologies characterized by mafic units—amphibolite, hornblende gneiss, and/or chlorite schist. Based on lithology and position, this sequence is, in part, correlative with the "Pumpkinvine Creek Formation" (Mc Connell and Costello, 1980, p. 4–6) and may be, in part, correlative with the Hillabee Chlorite Schist (greenstones) of Alabama (Tull and Stow, 1979, 1980). However, detailed mapping has not been completed between the Rockmart-Yorkville area and the Alabama state line, and direct correlation is still speculative. The closely spaced imbricate faults in this area considerably complicate the correlation of lithologic units by both cutting out and repeating parts of lithologic sequences. Projected and hasty correlations may greatly complicate future work in this area.

SUMMARY

The Cartersville area is one of complex geology (Kesler, 1950). It includes intensely folded sedimentary rocks; strongly deformed metamorphic rocks; igneous rocks that have intruded metasediments and then have been overlain unconformably by sediments which since have been metamorphosed; numerous overthrusts and high-angle reverse faults in both the sedimentary terrane and the igneous-metamorphic terrane; a lack of fossils in the metamorphic rocks; lithologies in the sedimentary areas that are very similar to those in the metamorphic areas; and similar lithologies in various thrust sheets, where continuity and position in a sequence are critical for interpretation.

We have been fortunate to work in and around the Cartersville area for long periods during the past 15 years, attempting to unravel the complexities piece by piece. Especially beneficial was the opportunity to proceed from detailed work outside the immediate Cartersville aea, where structural and stratigraphic relationships are less complex, and work into areas of more complex relationships in the vicinity of the intersection of the Great Smoky and Emerson faults. We have benefited from the efforts of previous workers, agreeing with some interpretations, disagreeing with others. As indicated above, there are numerous reasons for varied interpretations; continued detailed work will allow a closer approach to the true relationships.

ACKNOWLEDGMENTS

The authors acknowledge with appreciation the Georgia Department of Natural Resources, Georgia Geologic Survey, which supported Crawford's work during 1975–1977 in the Burnt Hickory Ridge, Taylorsville, and Yorkville quadrangles, as well as Cressler's work in the Cartersville area, through their cooperative program with the U.S. Geological Survey. Crawford's work in the Allatoona Dam area in 1978 was supported by the U.S. Geological Survey. Earlier work in Bartow, Haralson, and Paulding Counties was supported by the U.S. Department of Commerce and the Coosa Valley Area Planning and Development Commission. The work would not have been possible without the financial support of these agencies.

Also, we express our appreciation to Michael W. Higgins for his encouragement and interest in the area, and to both him and Carol J. Thurmond for reviewing the manuscript.

REFERENCES CITED

Bentley, Robert D., Fairley, William M., Fields, Herb H., Power, W. Robert, and Smith, James W., 1966, The Cartersville fault problem: Georgia Geological Society, Guidebook No. 4, Georgia Department of Mines, Mining and Geology, 38 p.

Crawford, Thomas J., 1971, Geologic map, Carroll and Heard Counties, Georgia, *in* Hurst, V. J., and Long, Sumner, Geochemical study of alluvium in the Chatahoochee-Flint area, Georgia: University of Georgia Institute of Community and Area Development.

——1976, Geologic map of the Burnt Hickory Ridge quadrangle, Georgia: Georgia Geologic Survey Open-File Report.

——1977a, Geologic map of the Taylorsville quadrangle, Georgia: Georgia Geologic Survey Open-File Report.

——1977b, Geologic map of the Yorkville quadrangle, Georgia: Georgia Geologic Survey Open-File Report.

Cressler, C. W., and Crawford, T. J., 1976, Cartersville fault found to be two intersecting thrusts, *in* Geological Survey Research 1976: U.S. Geological Survey Professional Paper 1000, 414 p.

Cressler, C. W., Blanchard, H. E., Jr., and Hester, W. G., 1979, Geohydrology of Bartow, Cherokee, and Forsyth Counties, Georgia: Georgia Department of Natural Resources Information Circular 50, 45 p.

Hurst, Vernon J., 1970, The Piedmont in Georgia, *in* Fisher, G. W., and others, ed., Studies of Appalachian geology, central and southern: John Wiley and Sons, Inc., p. 383–396.

——1973, Geology of the southern Blue Ridge belt: American Journal of Science, v. 273, p. 643–670.

Hurst, Vernon J., and Crawford, Thomas J., 1970, Sulfide deposits, Coosa Valley area, Georgia: Economic Development Administration, U.S. Department of Commerce, Washington, D.C., 190 p.

Kesler, T. L., 1950, Geology and mineral deposits of the Cartersville district, Georgia: U.S. Geological Survey Professional Paper 224, 97 p., 19pls.

McConnell, K. I., and Costello, J. O., 1980, Guide to geology along a traverse through the Blue Ridge and Piedmont provinces of north Georgia, *in* Frey, R. W., ed., Excursions in southeastern geology: Geological Society of America, Annual Meeting Guidebook, v. 1, p. 241–258.

Tull, James F., and Stow, Stephen H., 1979, the Hillabee metavolcanic complex and associated rock sequences: Alabama Geological Society, 17th Annual Field Trip, Guidebook, 64 p.

——1980, Structural and stratigraphic setting of arc volcanism in the Talladega slate belt of Alabama, *in* Frey, R. W., ed., Excursions in southeastern geology: Geological Society of America, Annual Meeting Guidebook, v. 2, p. 545–581.

Manuscript Accepted by the Society May 7, 1982

Printed in U.S.A.

Geological Society of America
Special Paper 191
1982

Lower Cambrian metasediments of the Appalachian Valley and Ridge province, Alabama; possible relationship with adjacent rocks of the Talladega metamorphic belt

Denny N. Bearce
Department of Geology
The University of Alabama in Birmingham
Birmingham, Alabama 35294

ABSTRACT

The Talladega belt in Alabama and Georgia is the northwesternmost belt of the Appalachian Piedmont metamorphic province. It contains low-rank metasediments and metavolcanics that have been thrust faulted onto Paleozoic sediments of the Valley and Ridge province via the Cartersville-Talladega fault system. The age of several formations in the southwestern part of the Talladega belt in Alabama has been determined to be Devonian, but controversy exists concerning the age of much of the rest of the belt. Another major problem has been the age and structure relationships of the Talladega belt to the Precambrian and Lower Cambrian rocks of the Blue Ridge province on strike with the Talladega belt to the northeast.

In the Borden Springs area, Cleburne County, Alabama, nappes of the Lower Cambrian Weisner and Shady Formations rest on younger Paleozoic rocks immediately northwest of the Talladega belt. A sequence composed mainly of slates and quartzites characterized by graded beds lies between the Talladega belt and the nappes of Weisner and Shady. Distinctive lithologies within this sequence are found also within the Talladega belt near Borden Springs and also near the southwest end of the belt in Alabama within metasediments immediately below the Jumbo Dolomite of the Sylacauga Marble Group. Although the slate-quartzite sequence has been interpreted in recent years as being Ordovician to Devonian, detailed mapping in the Borden Springs area indicates that it is correlative with the Early Cambrian Weisner and Shady, although somewhat different in sedimentary aspect from Weisner and Shady in nappes to the west. Therefore, the Talladega belt may contain rocks at least as old as Early Cambrian and may be at least partly equivalent in age to rocks of the Blue Ridge province.

The slate-quartzite sequence lies northwest, west, and southeast of an anticlinal region of younger Paleozoic sediments in western Georgia, over which it was thrust faulted. It thus forms an imbricated nappe sequence rooted, if at all, beneath the Piedmont province to the southeast.

INTRODUCTION

The Talladega belt in Alabama and Georgia is the northwesternmost belt of the Appalachian Piedmont metamorphic province. It contains sedimentary and volcanic rocks that have been metamorphosed to greenschist facies. Talladega belt rocks have been thrust faulted northwestward onto rocks of the Appalachian Valley and Ridge province. The Talladega belt is in turn overridden on the southeast and east sides by middle amphibolite facies metamorphic rocks of the Ashland Supergroup nappe and the Wedowee Group nappe (Tull, 1979, p. 3; Fig. 1, this paper).

Talladega belt rocks have presented major problems in age determination and in correlation with rocks of adjacent provinces. Lithologic ties have been made between Talladega belt rocks and fossiliferous rocks of the Valley and Ridge province at the southwest end of the belt (Carrington, 1973, p. 27–37). These correlations indicate a Devonian age for some Talladega belt units. Lithology-based correlations have been made from the southwest end of the belt northeastward to the Alabama-Georgia state line (Tull, 1979, p. 3–4, 59–60). The age of much of the Talladega belt remains in question, and various structural interpretations have suggested ages ranging from Precambrian to Pennsylvanian, in some cases for a single unit. Lack of clear-cut evidence for the age of much of the Talladega belt has also obscured the stratigraphic and structural relationships of the belt to the Appalachian Blue Ridge province, which adjoins the Talladega belt on strike to the northeast in northwestern Georgia and which contains Precambrian and Lower Cambrian rocks.

While lithologic ties have been made between the southwest end of the Talladega belt and the adjoining Valley and Ridge province, changes in stratigraphy northeastward in both the Talladega belt and the Valley and Ridge province have prevented similar ties being made at the northeast end of the Talladega belt. The degree of difference between the environments of deposition of Valley and Ridge province and northeastern Talladega belt rocks and the magnitude of displacement of Talladega belt rocks over those of the Valley and Ridge province in this region are both factors of uncertainty in an attempted correlation.

The present study has involved detailed mapping of a sequence of rocks in the vicinity of Borden Springs, Cleburne County, Alabama, which, while depicted on earlier

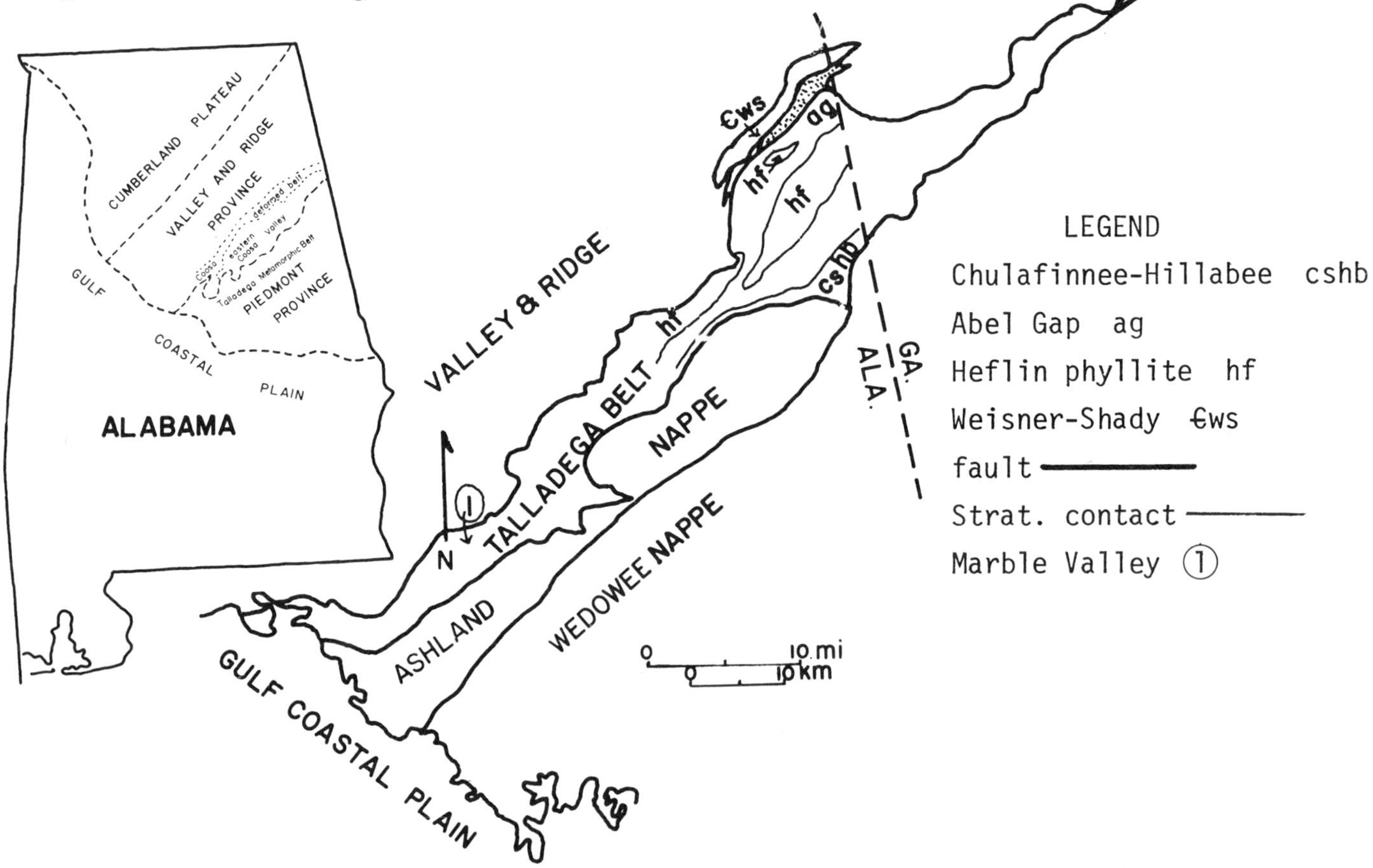

Figure 1. Regional maps locating study area. Slate-quartzite sequence stippled.

maps as lying in the Valley and Ridge province, has distinctive lithologies resembling those found in the adjoining Talladega belt. Earlier workers correlated this sequence with fossil-bearing Valley and Ridge province formations of various ages. The present study has resolved the age question and provided other evidence as well; correlation with adjacent rocks of the Talladega belt and with rocks of uncertain age at the southwestern end of the belt, 70 km distant, is suggested, and it appears possible that rocks of the same age exist in both the Talladega belt and the Blue Ridge province.

PRESENT INVESTIGATION

In southern Cherokee and northern Cleburne Counties, Alabama, nappes of Lower Cambrian Weisner and Shady Formations (Fig. 2, Indian Mountain, Augusta Mine Ridge) rest on a footwall of Valley and Ridge province formations ranging in age from Cambrian to Mississippian. These nappes form the northeast end of a series of allochthonous masses, all of Weisner and Shady and all apparently klippen, lying at most within a few kilometers of the Talladega metamorphic belt (Bearce, 1978; Fig. 1, this paper).

The rootless nature of the bodies of Weisner and Shady is evident from the lack of any stratigraphic continuity between the bodies and formations of known age surrounding them. However, in southern Cherokee and northern Cleburne Counties, the region between the nappes of Weisner and Shady and the Talladega belt (area approximately 15 km long and 1 to 5 km wide) is occupied by a sequence of slate and quartzite (Fig. 2) of recently disputed age. The sequence differs from the known Weisner and Shady to the northwest not only in being of slightly higher metamorphic grade and in showing evidence of more intense deformation but also in types of sedimentary structures. The Weisner of the klippen is characterized by cross-bedded quartzites; the quartzites in the sequence occupying the stippled area in Figure 1 are characterized by massive and graded beds. Slate within the sequence also displays grading of fine grained sand laminations.

Butts (1926), in compilation of the geologic map of Alabama, depicted the rocks of the study area as Weisner and Shady. Cressler (1970) mapped the easternmost extensions of the slate-quartzite sequence in western Polk County, Georgia, as Ordovician Rockmart Slate and Devonian Frog Mountain Sandstone, which are present immediately to the east in Georgia (Fig. 5). The writer, noting the problem that the middle Paleozoic and Lower Cambrian clastic sections in eastern Alabama and western Georgia are grossly similar in appearance, that the slate-quartzite sequence in question differs in general outcrop appearance from both known Weisner and Rockmart–Frog Mountain lithologies in adjoining areas, and that the slate-quartzite sequence bears strong resemblance to certain units in the adjoining Talladega belt, elected to map in detail the region occupied by this sequence in hopes of resolving the age dispute and establishing a tie between the Valley and Ridge province and the Talladega belt in eastern Alabama.

Figure 2 shows the distribution of the slate-quartzite sequence, which is thrust faulted to the northwest against Lower Cambrian Weisner Sandstone and Shady Dolomite (largely weathered to residual chert), and in turn is overthrust from the southeast by phyllites and quartzites of the Talladega belt along the Talladega (Cartersville) fault. It is faulted to the east against an extensive region of Cambrian-Ordovician Knox Group carbonate rocks. The sequence forms a northeast-trending belt, which is also in fault contact with Knox Group and younger rocks along the northwest edge of the region of Knox Group carbonate rocks for about 4 km east of the Alabama-Georgia state line.

The area occupied by the slate-quartzite sequence was mapped at a scale of 1:24,000. Reconnaissance was made of exposures of Rockmart Slate, Frog Mountain Sandstone, and other formations exposed to the east of the slate-quartzite sequence. Thin sections of the various lithologies in the slate-quartzite sequence and of the Rockmart Slate and Frog Mountain Sandstone were prepared and studied. The stratigraphy of the slate-quartzite sequence was determined from structural cross sections of the mapped area (Fig. 2).

STRATIGRAPHY OF SLATE-QUARTZITE SEQUENCE

Thickness evaluations of the various units in the slate-quartzite sequence are difficult because of limited exposure and tight to isoclinal folding. Coarse clastic units are apparently lenticular, making the stratigraphic sequence variable. Nevertheless, a general stratigraphic order is recognizable (Fig. 3).

A variable thickness of light olive–, light gray–, and light brown–weathering metaclaystone and metasiltstone forms the lower part of the sequence. In some areas this interval is faulted out and coarse-grained sandstone forms the basal unit (Fig. 2). In most areas, both along the faulted northwest margin of the slate-quartzite sequence and along thrust faults within the sequence, beds of coarse-grained, highly spherical, well-sorted, smokey quartz sand in a light gray-weathering clay matrix, ranging in thickness from a few millimeters to about 0.75 m, called "birdshot" sand, are present in this basal unit. "Birdshot" sand beds may be interlayered with black metasiltstone and metaclaystone as well as with beds up to 1 m thick of light and dark gray, coarse-grained, arkosic quartzite in the basal sequence. The "birdshot" sands are present within higher units of the stratigraphic sequence as well (Fig. 3).

Light gray, coarse-grained, feldspathic to arkosic quartzite of uniform texture, with well-rounded and highly

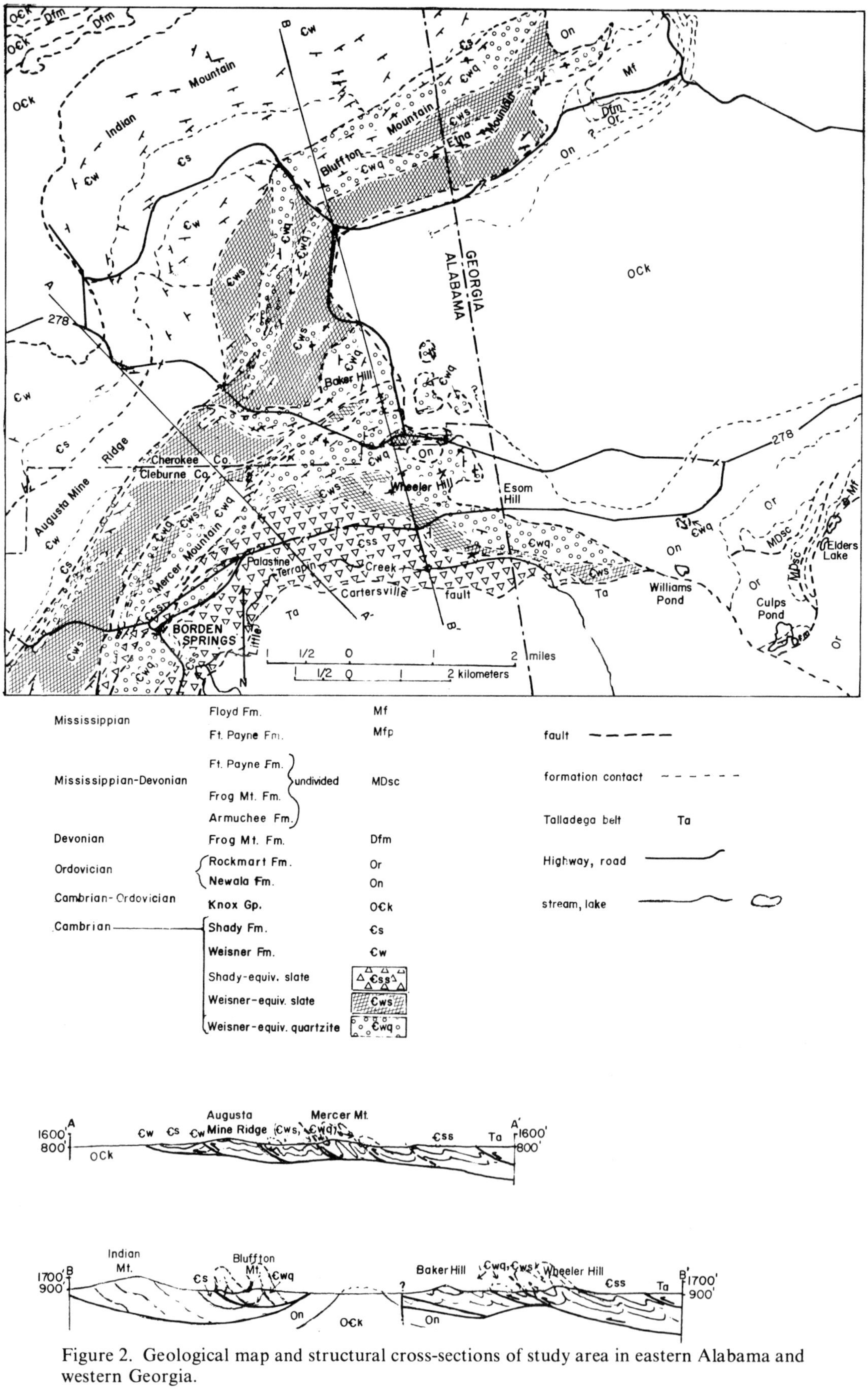

Figure 2. Geological map and structural cross-sections of study area in eastern Alabama and western Georgia.

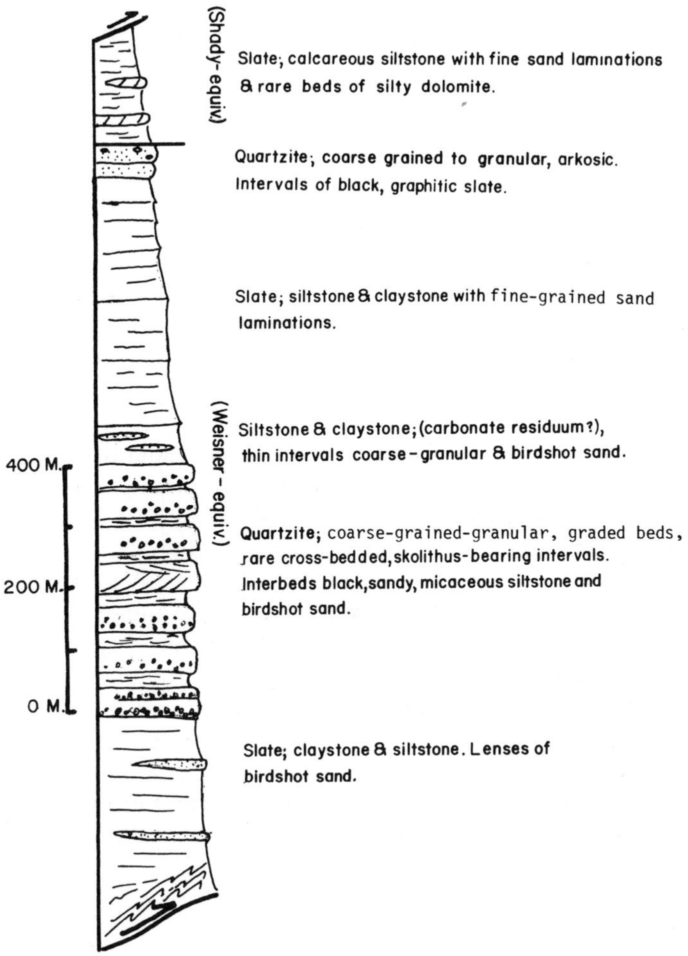

Figure 3. Slate-quartzite sequence stratigraphic section.

spherical, light gray quartz grains in massive-bedded ledges 2 to 3 m thick, overlies the metaclaystone-siltstone interval. On Baker Hill (Fig. 2) an interval of this quartzite has an exposed thickness of 43 m. In certain areas the massive-bedded, well-sorted quartzite is associated with poorly sorted, conglomeratic quartzite containing white quartz pebbles as long as 3 cm. This quartzite shows graded bedding on a scale ranging from a few centimeters to about 1 m. Good exposures of this quartzite with graded bedding are found on Etna Mountain, Georgia (Fig. 2). Metasandstone beds, generally argillaceous and medium to dark gray in color, are interbedded with the massive quartzite beds. These sandstone bodies are lenticular, as are the massive-bedded quartzites, and are interlayered and laminated with dark gray to black, carbonaceous, pyritic, finely laminated metasiltstone. The metasandstone bodies range from coarse- to very fine-grained and, together with interbedded siltstones, locally display cross-laminations, scour-and-fill structures, flame-tongue structures, and mud cracks. There is at least one unit of light gray, vitreous, medium- to coarse-grained, cross-bedded quartzite, locally displaying well-preserved *Skolithus* tubes, in the massive-bedded quartzite interval. Best exposures of this quartzite are seen on a ridge forming a northeast extension to Etna Mountain (Fig. 2). The writer has seen similar quartzites in eastern Alabama only in the Weisner Formation. Excellent exposures of *Skolithus*-bearing quartzite are present in the Weisner Formation on Indian Mountain (Fig. 2). The cross-bedded quartzite is probably only a few meters thick, but its similarity to the much thicker quartzite intervals of the Weisner Formation to the northwest of the slate-quartzite sequence is one line of evidence suggesting that the slate-quartzite sequence is at least in part Weisner-equivalent.

Locally, above the interval of various quartzites is an interval of yellow- to brown-weathering, platy to blocky, laminated metasiltstone and metaclaystone, which in highly weathered form resembles carbonate residum. This interval also contains light gray, friable, fine- to coarse-grained and granular, arkosic sandstone beds with numerous light gray-weathering pelite chips. Cross-bedding, graded bedding, and scour-and-fill structures are well preserved in this interval, which is near the structural base of the slate-quartzite sequence about 4 km west of Borden Springs on Highway 107 (Fig. 2). The entire interval is between 30 and 60 m thick.

The major part of the slate-quartzite sequence (Fig. 3), comprising much of the upper part of the section and also including intervals within the various quartzites in the lower parts of the section, is slate, derived from siltstone and claystone with sandstone laminations ranging in thickness from less than 1 mm to a few centimeters. The slate is dark gray to black where fresh, but typically is weathered to light olive-gray, pink, and tan. It displays a well-developed slatey cleavage. Sandstone laminations contain detrital mica and locally show graded bedding, small-scale flow rolls, scour-and-fill structures, and cross-laminations. The slate contains macroscopic tight to isoclinal folds that preclude accurate thickness measurements. Based on measurements in areas with fairly continuous outcrop control (cross-section A-A′, Fig. 2), the thickness of the upper slate interval may be between 400 and 800 m.

A second interval of quartzite, of the same general character as the first quartzite interval, overlies the slate interval. Interbeds of black, graphitic, pyritic, laminated metasiltstone are present in this quartzite interval. Some quartzite beds in this upper quartzite interval are calcareous. Rarely, clasts up to 10 cm long of dark gray silty limestone, as well as clasts of pelite and quartzite, in a granular, arkosic, quartzose matrix, occur in the upper quartzite interval. The upper quartzite interval is of variable thickness, ranging upward from 40 m.

The upper quartzite interval is overlain by dark gray metasiltstone with thin laminations of fine-grained metasandstone, which passes upward into weathered brown and yellow laminated silt and clay beds that appear to be carbonate residuum. The silt and clay beds are seen in road cuts;

on nearby slopes and knobs float blocks of ochre-brown, coarse-textured, limonitic chert are present. The silt and clay beds pass upward into medium to dark gray metasiltstone with thin light gray, fine-grained calcareous sand laminations and rare interbeds of medium gray silty dolostone up to 10 cm thick. This calcareous lithology has been seen only in the Little Terrapin Creek valley on the southeast side of the region occupied by the slate-quartzite sequence (Fig. 2). The metasiltstone has a phyllitic luster and was mapped previously as part of the Talladega belt. However, it is definitely in stratigraphic continuity with the slate-quartzite sequence. The residual chert float resembles that of the Shady Dolomite in Weisner-Shady nappes and klippen to the northwest, and the author interprets the calcareous siltstone and apparent carbonate residuum beds as being Shady-equivalent.

The top of the calcareous metasiltstone interval has been overthrust by the Cartersville fault and is concealed beneath graphitic phyllites and arkosic metasandstones of the Talladega belt (Fig. 2). Thickness of the sequence is difficult to estimate due to discontinuous exposure and to the certainty of isoclinal folds and the probability of imbrication within the interval. Thus, while the calcareous siltstone occupies a valley approximately 1 km wide and appears to dip fairly consistently to the southeast (Fig. 2), its stratigraphic thickness may be no more than 200 km (Fig. 2, cross-section A-A′).

RELATIONSHIP BETWEEN THE SLATE-QUARTZITE SEQUENCE AND THE TALLADEGA BELT

Metasediments of the Talladega belt adjoining the slate-quartzite sequence consist mainly of black phyllitic metasiltstone and metaclaystone with lesser amounts of medium to dark gray, pyritic, arkosic quartzite and metasandstone. These lithologies are nondistinctive and occur at many stratigraphic levels within the Talladega belt. However, one distinctive lithology, the "birdshot" sand, is interlayered with the black phyllites of the Talladega belt near Borden Springs. These rocks were originally mapped by the author (Bearce, 1973; Table 1, this paper) as Abel Gap Formation. However, there is now reason to believe they may be older. Recent mapping in Talladega and Coosa Counties, Alabama (Fig. 1, Marble Valley) has revealed the presence of sands resembling the "birdshot" sand (Pendexter, 1981, in prep. and personal commun.) within the Mount Zion Formation of Carrington (1973), also known as the Wash Creek Slate (Butts, 1940). The Wash Creek Slate underlies the Jumbo Dolomite, with gradational contact (Pendexter, 1981, in prep.), and is of uncertain age. The Jumbo Dolomite, also of uncertain age, is overlain unconformably by the Lay Dam Formation (Carrington, 1973, p. 29). The Lay Dam Formation, also of uncertain age, is overlain stratigraphically (nature of contact uncertain) by the Butting Ram Sandstone of Early Devonian age (Carrington, 1973; Tull,

TABLE I. CORRELATION OF VARIOUS TALLADEGA BELT SUBDIVISIONS

BUTTS (1926, 1940)	CARRINGTON, (1973)	BEARCE (1973)	TULL (1979)
S.W. TALLADEGA BELT	S.W. TALLADEGA BELT	N.E. TALLADEGA BELT	S.W. & N.E. TALLADEGA BELT
CARB. ERIN SLATE	CHILTON (UPPER, SE.) BLOCK		
			HILLABEE GREENSTONE
DEV. { YELLOWLEAF QUARTZ SCHIST	DEV. { YELLOWLEAF QUARTZ SCHIST &	CHULAFINNEE SCHIST	JEMISON CHERT
JEMISON CHERT	JEMISON CHERT		
BUTTING RAM SANDSTONE &	BUTTING RAM SANDSTONE	ABEL GAP FORMATION	DEV. { BUTTING RAM SANDSTONE
CHEAHA SANDSTONE	LAY DAM FORMATION	HEFLIN PHYLLITE	ERIN SLATE
JUMBO DOLOMITE	JUMBO DOLOMITE		LAY DAM FORMATION
QUARTZITE			CHEAHA SANDSTONE MBR
WASH CREEK SLATE	MT. ZION FORMATION		JUMBO DOLOMITE (ABSENT
FERRUGINOUS SANDSTONE MBR			N.E. TALLADEGA BELT)
BREWER PHYLLITE	METAMORPHIC FRONT		(WASH CREEK FORMATION)
CAMB. OR PREC.(?) { SYLACAUGA MARBLE	(LOWER, N.W.) BLOCK		
WAXAHATCHEE SLATE	CARB. { BREWER PHYLLITE		
SAWYER LIMESTONE MBR	SAWYER LIMESTONE		
	PARKWOOD FORMATION		
	FLOYD SHALE		

CAMBRIAN & ORDOVICIAN ROCKS

1979, p. 3). As the slate-quartzite sequence is correlative with the Weisner and Shady Formations, and potentially correlative in part with the Wash Creek Slate, then perhaps the Wash Creek Slate is Cambrian in age. Perhaps also the overlying Jumbo Dolomite is Shady-correlative. If these correlations prove correct, then deposition of the Talladega belt sequence was initiated in Early Cambrian time at the latest, possibly as sediments of a rifted margin prism, as were sediments of the Chilhowee Group in the Blue Ridge province to the northeast, in northwest Georgia (Thomas, 1977).

METAMORPHIC EFFECTS OF SLATE-QUARTZITE SEQUENCE

Prevalent mesoscopic metamorphic effects in the slate-quartzite sequence include an axial planar slaty cleavage, strongly expressed in metasiltstones and metaclaystones and reflected in the flattening of quartz granules in metasandstones. In thin sections of metasandstones and quartzites, quartz grains with strain laminae show tails and rims of crush quartz. Coarse white mica, probably of detrital origin, is oriented parallel to slaty cleavage. Some rocks contain small amounts of a red, pleochroic mica of low birefringence, possibly stilpnomelane. Quartzites contain epidote both in matrix between and filling cracks within quartz grains. In the vicinity of faults, cleavages are crenulated and have a slightly higher (phyllonitic) luster than do cleavages of slates elsewhere. Microscopically, all lithologies of the slate-quartzite sequence show extensive recrystallization of phyllosilicates to finely crystalline white mica of high birefringence, apparently sericite.

The slate-quartzite sequence appears to be of zeolite or lower greenschist facies, based on the general degree of recrystallization and the presence of epidote and possible stilpnomelane. The degree of slaty cleavage development and the extent of metamorphic recrystallization in the slate-quartzite sequence are noticeably higher than in the slates and quartzites of the Weisner Formation to the northwest. They are also higher than those of the Knox and younger formations, such as the Rockmart Slate, adjoining the slate-quartzite sequence to the east (Fig. 2).

STRUCTURES WITHIN THE SLATE-QUARTZITE SEQUENCE

The slate-quartzite sequence has been folded and imbricated. Two phases of folding are recognized. The first phase was accompanied by the development of axial planar slaty cleavage. The folds developed during this phase are flexural flow folds, typically overturned to the northwest with moderately to steeply inclined axial surfaces and hinges plunging gently southwest and northeast (Figs. 4A and 4C). The wavelengths of macroscopic first-phase folds range from a few meters to about 1.5 km (Fig. 2, Bluffton Mountain syncline). First-phase folds vary from open to isoclinal, depending on lithology. The second-phase folds consist of open flexural slip folds with steeply inclined axial surfaces and with hinges plunging moderately to the southeast and northwest (Fig. 4C). Figure 4B, a π diagram of poles to bedding, depicts a scattering of pole concentrations possibly reflecting multiple deformation. Assuming cylindrical folds, an attempt was made to define some possible great circle girdles utilizing the largest concentrations of hinges shown on Figure 4C. These concentrations are labeled on both diagrams, and the three great circles derived therefrom are shown on Figure 4B, where they are seen to pass through somewhat ambiguous partial girdles. A fourth great circle, trending east-west, is suggested by additional minor concentrations of poles at 90° and 270° on the perimeter of Figure 4B. This great circle implies a set of north-trending, nonplunging folds, which, while not suggested by hinge concentrations on Figure 4C, are present as macroscopic folds on Baker Hill and in the region southwest of Bluffton Mountain (Fig. 2).

Imbricate faults within the slate-quartzite sequence appear, on the basis of fault truncations, to be of the "breakback" variety and developed in a west-to-east sequence. There is an increasing degree of slaty cleavage development and of recrystallization of clay minerals to micas from fault block to fault block across the slate-quartzite sequence terrain. This progression in degree of penetrative deformation and recrystallization results in a marked contrast between rocks of the slate-quartzite sequence adjacent to the Cartersville fault and those of the Knox and younger formations adjacent to the Cartersville fault to the east (Fig. 2). While the Knox carbonate terrain–Talladega belt boundary (Cartersville fault) is easily recognized (Talladega belt phyllites juxtaposed on mid Paleozoic slates, sandstones, siltstones, and cherts), this fault is not so obvious crossing the slate-quartzite sequence. Here, Talladega phyllites are thrust faulted onto phyllitic calcareous metasiltstones (Shady-equivalent). If it were not for this distinct laminated calcareous metasiltstone, the Cartersville fault would juxtapose rocks of very similar clastic character and of similar structural fabric and metamorphic grade along the southeastern boundary of the slate-quartzite sequence. It might appear to be nothing more than another imbricate fault within the slate-quartzite sequence.

STRUCTURES BOUNDING THE SLATE-QUARTZITE SEQUENCE

The slate-quartzite sequence is thrust faulted against Weisner and Shady rocks to the northwest. This is evident from the presence of localized shearing and crumpling of strata along the Weisner-Shady/slate-quartzite sequence boundary. Figure 2 indicates that the Weisner and Shady of

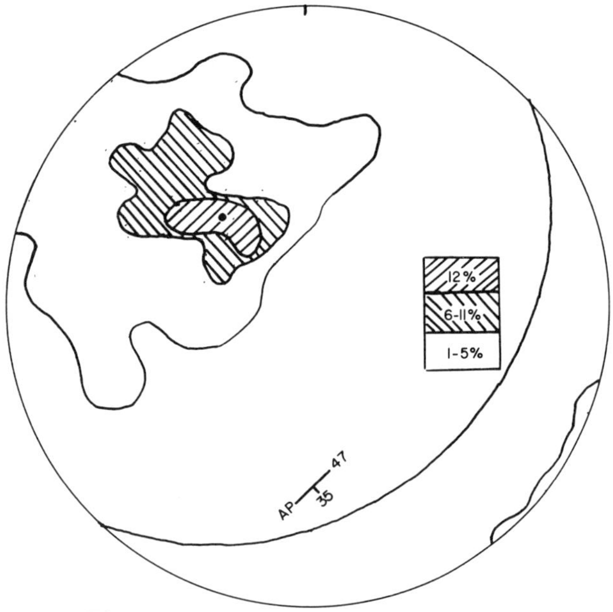

Figure 4A. Poles to 77 cleavages in slate-quartzite sequence.

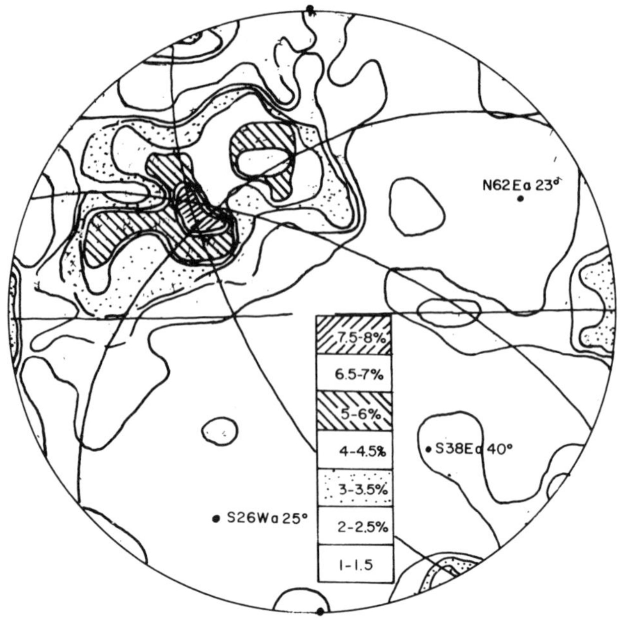

Figure 4B. Poles to 200 bedding planes in slate-quartzite sequence. Poles of fold hinge concentrations also shown from Figure 4C.

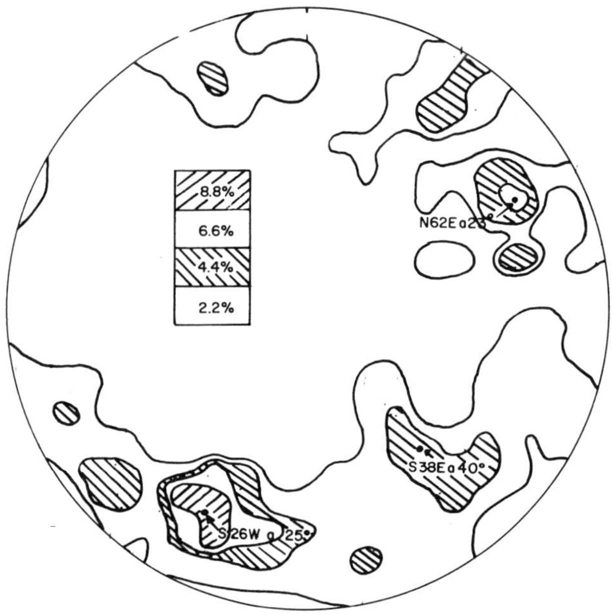

Figure 4C. Hinges of 45 mesoscopic folds in slate-quartzite sequence.

Indian Mountain and Augusta Mine Ridge are in turn thrust faulted onto younger Paleozoic rocks to the west and northwest. On the basis of structural and stratigraphic characteristics of bodies of Weisner and Shady elsewhere in Alabama, primarily their complete lack of stratigraphic contiguity with rocks surrounding them, these bodies have been interpreted to be klippen, resting with subhorizontal contact on surrounding rocks (Bearce, 1978). The Weisner and Shady of Indian Mountain and Augusta Mine Ridge are interpreted herein to be klippen also (Fig. 2, cross-section B-B′), on the basis of the large region of Knox and younger formations lying south of Indian Mountain and both south and east of the slate-quartzite sequence (Fig. 2).

Cressler (1970) interpreted the region of Knox and younger formations to be anticlinal in structure and to be in stratigraphic contiguity with the slate-quartzite sequence and the rocks of Indian Mountain to the northwest, which he tentatively correlated with the Ordovician Rockmart Slate and the Devonian Frog Mountain Sandstone (Fig. 5). The existance of such an anticline is supported by the pres-

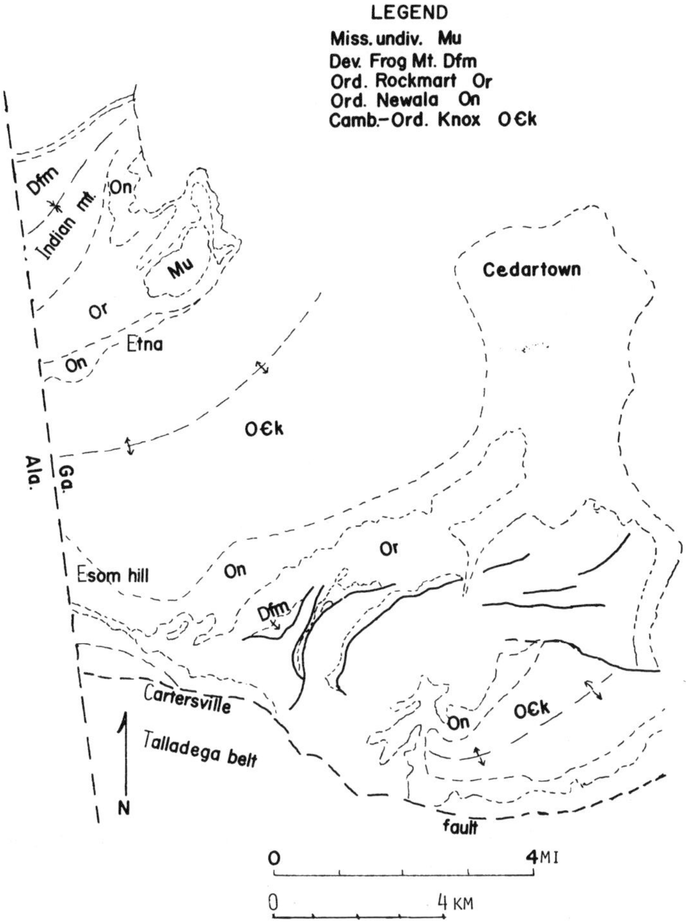

Figure 5. Geology of western Polk County, Georgia. Generalized from Cressler (1970).

ent study. In addition, there are parautochthonous bodies of Rockmart Slate and Frog Mountain Sandstone along the flanks of the anticline within a short distance of, or in contact with, the slate-quartzite sequence. However, in the years following Cressler's work, archaeocyathids were found in chert bodies on Indian Mountain (Bearce, 1977), establishing their age as Early Cambrian. The presence of identical chert associated with calcareous metasiltstones of the slate-quartzite sequence, as mentioned previously, provides the strongest evidence that the slate-quartzite sequence is also Early Cambrian, Weisner-Shady equivalent.

There is structural evidence as well that the contact between the slate-quartzite sequence and the region of Knox and younger formations is a tectonic contact. Along the east boundary of the slate-quartzite sequence, abutting Knox carbonates, limonite forms tabular, steeply inclined, slickensided masses up to 6 m thick. Phyllonitized slate and sheared quartzite lie on the west side of these limonite masses; brecciated, limonite-cemented Knox chert and sandstone lie on the east side. The slate-quartzite sequence forms a prong along the north flank of the anticline in Knox and younger formations. Exposures of Frog Mountain Sandstone adjacent to the southern edge of the prong are sheared, crenulated, and phyllonitized. Slate within the prong near the southern contact is likewise crenulated and phyllonitized.

DEFORMATION PHASES OF THE SLATE-QUARTZITE SEQUENCE

The slate-quartzite sequence was subjected to an initial deformation phase that led to development of flexural flow folds, mentioned previously, with an axial planar cleavage. This cleavage is manifested by flattening of quartz grains and recrystallization of clay minerals to finely crystalline mica. Although this cleavage was locally gently folded in a later fold phase, it is fairly consistent in attitude, as Figure 4A shows. A single maximum of cleavage poles reflects axial surfaces striking northeast and dipping moderately to the southeast. This is close to the main concentration of bedding poles in Figure 4B, reflecting the fact that first-phase folds in the slate-quartzite sequence tend to be isoclinal (Fig. 2, section A-A′, folds on Mercer Mountain; section B-B′, folds on Bluffton Mountain and Wheeler Hill).

The initial fold phase, then, was accompanied by metamorphism. Present evidence suggests that this was an Acadian (Devonian-Mississippian) deformational event. Tull (1979, p. 13) reports radiometric ages for the Hillabee Chlorite Schist, a metavolcanic unit of the Talladega belt, of about 382 m.y. Near the study area, the Middle Ordovician Rockmart Slate displays a well-developed slaty cleavage, but siltstones of the overlying Devonian Frog Mountain Sandstone do not. This suggests that at the time of deformation the Frog Mountain Sandstone was not yet deposited or was buried to insufficient depth for cleavage development.

The second fold phase, manifested in folding of cleavage and localized development of southeast and northwest plunging, open, flexural slip folds, is of uncertain age.

Thrust faulting has moved the slate-quartzite sequence over rocks as young as Late Mississippian (Fig. 2, Floyd Formation), and so terminated no earlier than Late Mississippian. As mentioned previously, the thrust faults developed in a west-to-east sequence, including the Cartersville fault. The Weisner-Shady sequence of Indian Mountain and Augusta Mine Ridge and the equivalent slate-quartzite sequence were emplaced as a major nappe, rootless, or rooted beneath the Talladega belt, containing open to isoclinal folds, by means of a subhorizontal thrust fault in Late Mississippian time or later. This nappe was then imbricated and overridden by the Talladega belt rocks during and after Late Mississippian time.

Development of the anticlinal structure in Knox and younger formations east and south of the slate-quartzite sequence is inferred to have followed nappe emplacement, warping the nappe as well. Subsequent erosion removed the rocks of the nappe from the crest of the anticline but left the prong of slate-quartzite sequence along its northwest side (Fig. 2, cross-section B-B′).

REGIONAL ASPECTS

The Weisner and Shady Formations Form klippen in the Valley and Ridge province immediately west of the Talladega belt scattered along almost the entire length of the exposed belt in Alabama. The slate-quartzite sequence is exposed in the Valley and Ridge province only in the study area, although its equivalents appear to be present as the Wash Creek Slate in the southwest part of the Talladega belt scattered along almost the entire length of the exposed belt. Sedimentary structures are better preserved in the slate-quartzite sequence than in the Wash Creek Slate and most other units of the Talladega belt due to the greater degree of shearing and recrystallization of rocks within the Talladega belt. Thus, the slate-quartzite sequence provides a clearer view of the transitional marine environment that existed along the southeastern edge of the continent in Early Cambrian time.

Mack (1980, p. 497–517) has included Early Cambrian clastic rocks in Alabama and Georgia in the Chilhowee Group (in Alabama, the Weisner Formation of Butts, 1926, p. 61, 64). Mack subdivided these rocks into four formations, the Cochran, Nichols, Wilson Ridge and Weisner Formation in ascending order, in part fluviatile and in part shallow marine. The author cannot, with certainty, correlate any specific part of the slate-quartzite sequence with any specific one of Mack's four formations. However, some of the sedimentary characteristics Mack has described are seen in the slate-quartzite sequence. The cross-bedded *Skolithus*-bearing quartzite that Mack finds in his Wilson Ridge and Weisner Formations, deposited presumably in a shallow water, high energy marine environment, are present, but rare, in the slate-quartzite sequence. The dark gray cross-laminated metasiltstone with lenses (possibly channel deposits) of coarse-grained metasandstone and quartzite, flame structures, and mud cracks in the slate-quartzite sequence are similar to lithologies Mack described in his Wilson Ridge Formation as shallow marine deposits. Graded metasandstones and metaconglomerates of the slate-quartzite sequence resemble Mack's description of the Cochran Formation. The close association of graded beds, pelite clasts, and scour-and-fill structures with shallow-water features such as mud cracks and *Skolithus* suggests a fluviatile environment with prograding deltas containing coarse channel sands. Slumps down oversteepened leading faces of the channel deposits could create small localized turbidity currents, resulting in graded beds.

Rodgers (1968), Thomas (1977), and Thomas and Neathery (1980) have discussed the evolution of the eastern margin of North America during the late Precambrian and Paleozoic. Figure 6 incorporates their ideas and illustrates a possible sequence of events leading to the present general stratigraphic and structural picture in eastern Alabama and western Georgia.

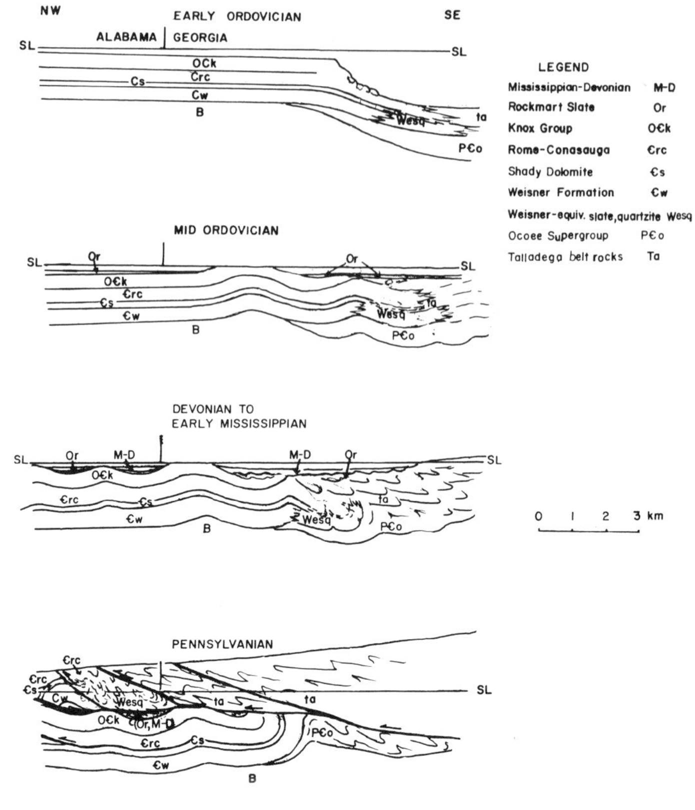

Figure 6. Possible sequence of events deforming slate-quartzite sequence and associated rocks.

From late Precambrian to Early Cambrian time the proto-Atlantic Ocean (Iapetus Sea) was opening, and a passive margin existed on the southeast coast of North America. During Early Cambrian Weisner time, a prism of fluviatile sediments, derived from Precambrian basement highlands to the west (Mack, 1980, p. 497–517), prograded eastward as a series of deltas into a shallow sea. Waves and longshore currents formed a series of bars from channel deposits to the west near shore. These bars are characterized by cross-bedded, *Skolithus*-bearing quartzites. Other channel sands, building eastward below wave base and below the influence of surface longshore currents, developed graded beds through the slumping of delta fronts.

The supply of coarse clastics diminished later in Early Cambrian time, and a carbonate bank developed along the passive margin, its seaward border in Alabama and Georgia lying somewhere beneath rocks of the present-day northern Piedmont (Thomas, 1977, p. 1242) or farther to the east. The Early Cambrian Shady Dolomite, the lowest unit of the carbonate bank, transgressed westward over the Weisner clastics. By this time the sea to the east of the carbonate bank had probably deepened considerably for Shady bank carbonates change eastward (in the study area) to calcareous siltstone. The presence of intermediate metaigneous intrusives in the Alabama Piedmont south of the study area suggests that this sea may have been a back-arc basin with the possible advent of subduction farther to the east in Middle or Late Cambrian time (Deininger, 1975, p. 58–59). The sea continued to transgress westward onto the continent after Shady time (Thomas and Neathery, 1980, p. 467) and remained through Early Ordovician time. During this time span a carbonate sequence with intervals of generally fine-grained, craton-derived clastic sediment was deposited on the carbonate bank.

Middle Ordovician time saw a reversal of the paleoslope with the onset of the Taconic Orogeny (Atkins and Higgins, 1980, p. 34). While there is no evidence of taconic metamorphism in the Talladega belt, late Precambrian and early Paleozoic sediments underlying the eastern edge of the carbonate bank were probably warped. An extensive terrain underlain by Knox and earlier formations was exposed to erosion providing clastic debris incorporated in the Middle Ordovician Rockmart Slate.

During Devonian to Mississippian time, late Precambrian to middle Paleozoic sediments of the Talladega belt and eastern Valley and Ridge province were folded and in part metamorphosed. The slaty cleavage of the Weisner-Shady-equivalent slate-quartzite sequence and of the Middle Ordovician Rockmart Slate developed during this Acadian deformational event.

A second phase of deformation of uncertain age refolded Acadian folds in Rockmart and older formations. The carbonate bank was arched into a series of anticlines and synclines that influenced Mississippian sedimentation as far west as Birmingham, Alabama (Thomas, 1972, p. 67).

Low-angle thrust faulting, occurring during and/or after Late Mississippian time, emplaced nappes of Weisner, Shady, and equivalent rocks onto a terrain of folded, thrust-faulted Knox and younger formations. Late-stage folding, either accompanying or occurring shortly after nappe emplacement, arched both nappes and footwall into a series of broad anticlines and synclines. Subsequent erosion reduced the nappes in areas outside the study area to isolated klippen and produced a half-window through an imbricated nappe on the east side of the study area.

ACKNOWLEDGMENTS

The author would like to express his appreciation for the constructive criticisms offered in reviews by James F. Tull and William A. Thomas.

REFERENCES CITED

Atkins, R. L., and Higgins, M. W., 1980, Superimposed folding and its bearing on geologic history of the Atlanta, Georgia, areas, *in* Frey, R. W. ed., Excursions in southeastern geology: American Geological Institute, v. 1, p. 19–35.

Bearce, D. N., 1973, Geology of the Talladega metamorphic belt in Cleburne and Calhoun Counties, Alabama: American Journal of Science, v. 273, p. 742–754.

——1977, Stratigraphic problems of the eastern Coosa Valley, *in* Walker, K. R., and Roeder, D., eds., Cambrian and Devonian stratigraphic problems of eastern Alabama: Alabama Geological Society, 15th Annual Field Trip, Guidebook, p. 37–53.

——1978, Structures of the eastern Coosa Valley, Alabama: American Journal of Science, v. 278, p. 4621–476.

Butts, C., 1926, The Paleozoic rocks, *in* Adams, G. I., Butts, C., Stevenson, L. W., and Cooke, W., eds., Geology of Alabama: Alabama Geological Survey Special Report 14, p. 40–223.

——1940, Description of the Montevallo and Columbiana quadrangles, Alabama: U.S. Geological Survey Atlas Folio 226, 20 p.

Carrington, T.J., 1973, Metamorphosed Paleozoic sedimentary rocks in Chilton, Shelby, and Talladega Counties, Alabama, *in* Carrington, T. J., ed., Talladega metamorphic front: Alabama Geological Society, 11th Annual Field Trip, Guidebook, p. 22–38.

Cressler, C. W., 1970, Geology and ground-water resources of Floyd and Polk Counties, Georgia: Geological Survey of Georgia Information Circular 39, 95 p.

Deininger, R. W., 1975, Granitic rocks in the northern Alabama Piedmont, *in* Neathery, T. L., and Tull, J. F., eds., Geologic profiles of the northern Alabama Piedmont: Alabama Geological Society, 13th Annual Field Trip, Guidebook, p. 49–62.

Mack, G. H., 1980, Stratigraphy and depositional environments of the Chilhowee group (Cambrian) in Georgia and Alabama: American Journal of Science, v. 280, p. 497–517.

Rodgers, J., 1968, The eastern edge of the North American continent during the Cambrian and Early Ordovician, *in* Zen, E-an, White, W. S., Hadley, J. B., and Thompson, J. B., Jr., eds., Studies of Appalachian geology: Northern and maritime: New York, Interscience, p. 141–149.

Thomas, W. A., 1972, Mississippian Stratigraphy of Alabama: Geological Survey of Alabama, Monograph 12, 121 p.

——1977, Evolution of Appalachian-Ouachita salients and recesses from re-entrants and promontories in the continental margin: American Journal of Science, v. 277, p. 1240–1244.

Thomas, W. A., and Neathery, T. L., 1980, Tectonic framework of the Appalachian orogen in Alabama, *in* Frey, R. W., ed., Excursions in southeastern geology: American Geological Institute, v. II, p. 465–526.

Tull, J. F., 1979, stratigraphic and structural relationships of the eastern Talladega slate belt of Alabama, *in* Tull, J. F., and Stow, S. H., eds., The Hillabee metavolcanic complex and associated rock sequences: Alabama Geological Society, 17th Annual Field Trip, Guidebook, p. 3–23.

Manuscript Accepted by the Society May 7, 1982

Printed in U.S.A.

Geological Society of America
Special Paper 191
1982

Stratigraphy and structure of the central Talladega slate belt, Alabama Appalachians

Terry A. Cook
Department of Geology and Geography
University of Alabama
University, Alabama 35286

ABSTRACT

The Talladega slate belt in eastern Alabama represents a crystalline thrust sheet composed of low-grade metasediments. Three major lithologic sequences comprise the Talladega slate belt: (1) the Kahatchee Mountain Group, (2) the Sylacauga Marble Group, and (3) the Talladega Group. The contact relationships between the Sylacauga Marble Group and the Talladega Group indicate that the phyllites and slates of the Talladega Group rest unconformably on marbles of the Sylacauga Marble Group. Previous workers have identified specific stratigraphic sequences within the Talladega Group both in the north-central portion of the Talladega slate belt and in the southern portion of the belt. Little work has been carried out in the south-central portion of the Talladega slate belt, a region intermediate between areas to the northeast and southwest where the regional stratigraphy has been defined. To the northwest in Cleburne and Clay Counties, Alabama, the Talladega Group has been broken down into the Heflin Phyllite, the Able Gap Formation, and the Chulafinnee Schist. To the southwest in Chilton County, Alabama, similar units have been mapped as the Lay Dam Formation, the Butting Ram Sandstone, and the Jemison Chert. These units have not been mapped through this intermediate south-central region of the Talladega slate belt because of the absence by faulting of a major sandstone unit, the Cheaha Quartzite, which has been used for regional correlation. Another prominent unit, the Jemison Chert, which outcrops to the southeast of the Cheaha Quartzite, continues across this region and was used to correlate the regional stratigraphy from the northeast with that in the southwest. Detailed mapping has shown that a small slice of paper-thin quartzites of the Jemison Chert interval has overridden the Cheaha Quartzite. The geometric relationships between these two units, the differing petrologic character of these ridge-forming lithologies, the duplication of the Jemison Chert interval, and the emplacement of this imbricate slice of Jemison, in addition to structural fabric data, suggest that this termination of the Cheaha Quartzite is fault related.

INTRODUCTION

The Talladega slate belt in eastern Alabama and western Georgia represents a northeast-trending crystalline thrust sheet consisting of low-grade metasediments and metavolcanic rocks that are bounded both above and below by distinctively different lithologic sequences. To the northwest is the unmetamorphosed lower Paleozoic sedimentary sequence of the Appalachian foreland, and to the southeast are structural blocks consisting of amphibolite facies metasediments and metavolcanics of the Ashland Supergroup and the Wedowee Group (Tull, 1978). The Talladega slate belt near Sylacauga, Alabama, contains three major rock sequences: (1) a lower clastic sequence, the Kahatchee Mountain Group; (2) a carbonate sequence, the Sylacauga Marble Group; and (3) an upper clastic sequence, the Talladega Group (Fig. 1).

The area described in this study is located southeast of Sylacauga in Clay, Coosa, and Talladega Counties, Alabama. It is situated in the central portion of the Talladega slate belt in Alabama and includes the upper clastic sequence of the belt, the Talladega Group (Tull and Stow, 1979; Fig. 2, this paper). This area was chosen for study because of its central location between regions in which the stratigraphy has been worked out in more detail and because a regionally extensive ridge-forming unit, the Cheaha Quartzite, which is a key to correlations in the belt, appears to terminate in this area.

The major units involved in this study are the clastic lithologies of the Talladega Group, including the Lay Dam Formation with its Cheaha Quartzite Member and Erin Slate Member (Erin Shale of Butts, 1926), the Butting Ram Sandstone, and the fossiliferous Lower to Middle Devonian Jemison Chert interval (Butts, 1926; Carrington, 1973). Also included in the study area is the metavolcanic Hillabee Greenstone (Tull and Stow, 1980). With the exception of the Cheaha Quartzite, which appears to terminate in southwestern Clay County, and the Butting Ram Sandstone, which

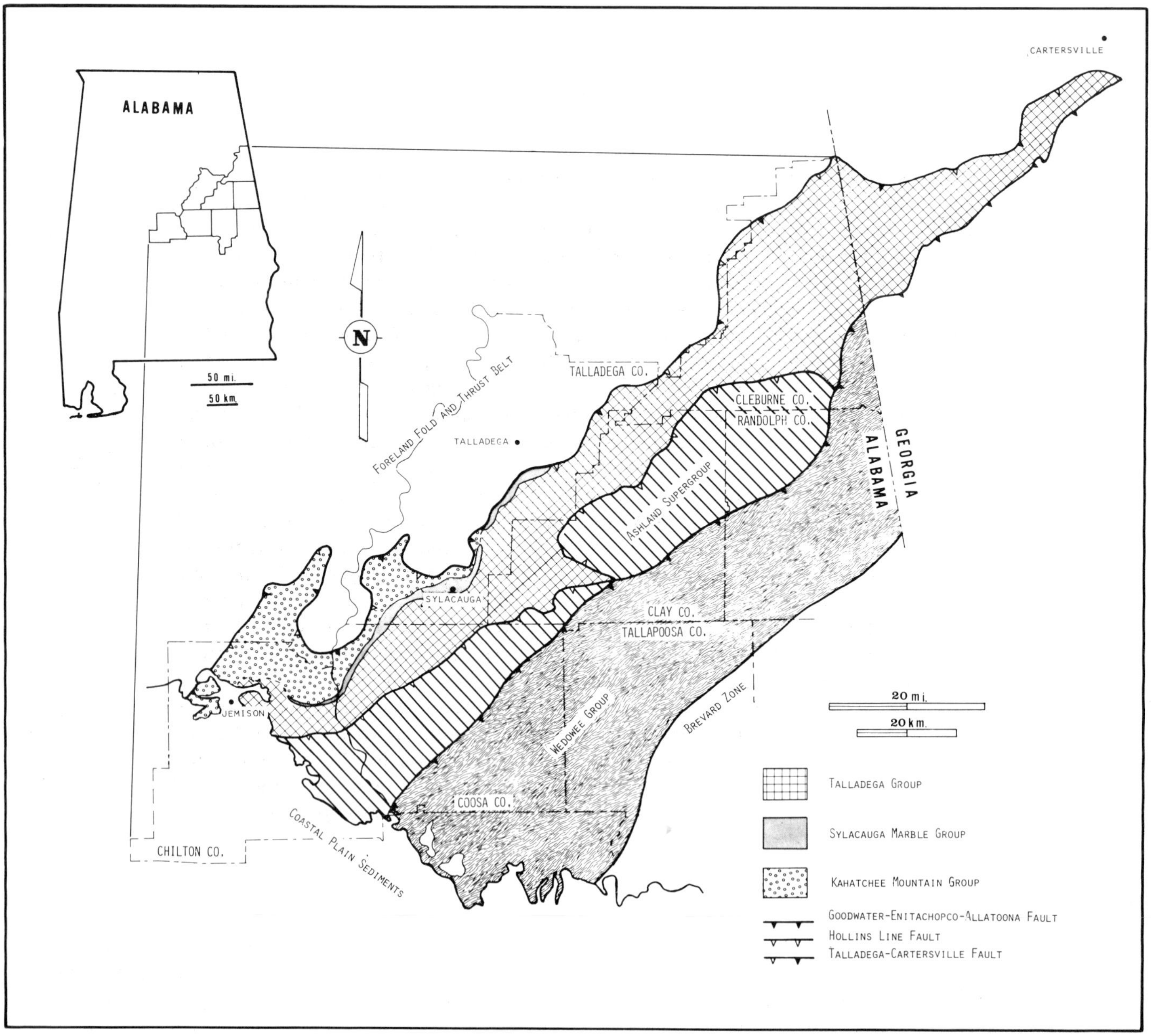

Figure 1. Regional tectonic setting of the Talladega slate belt, Alabama.

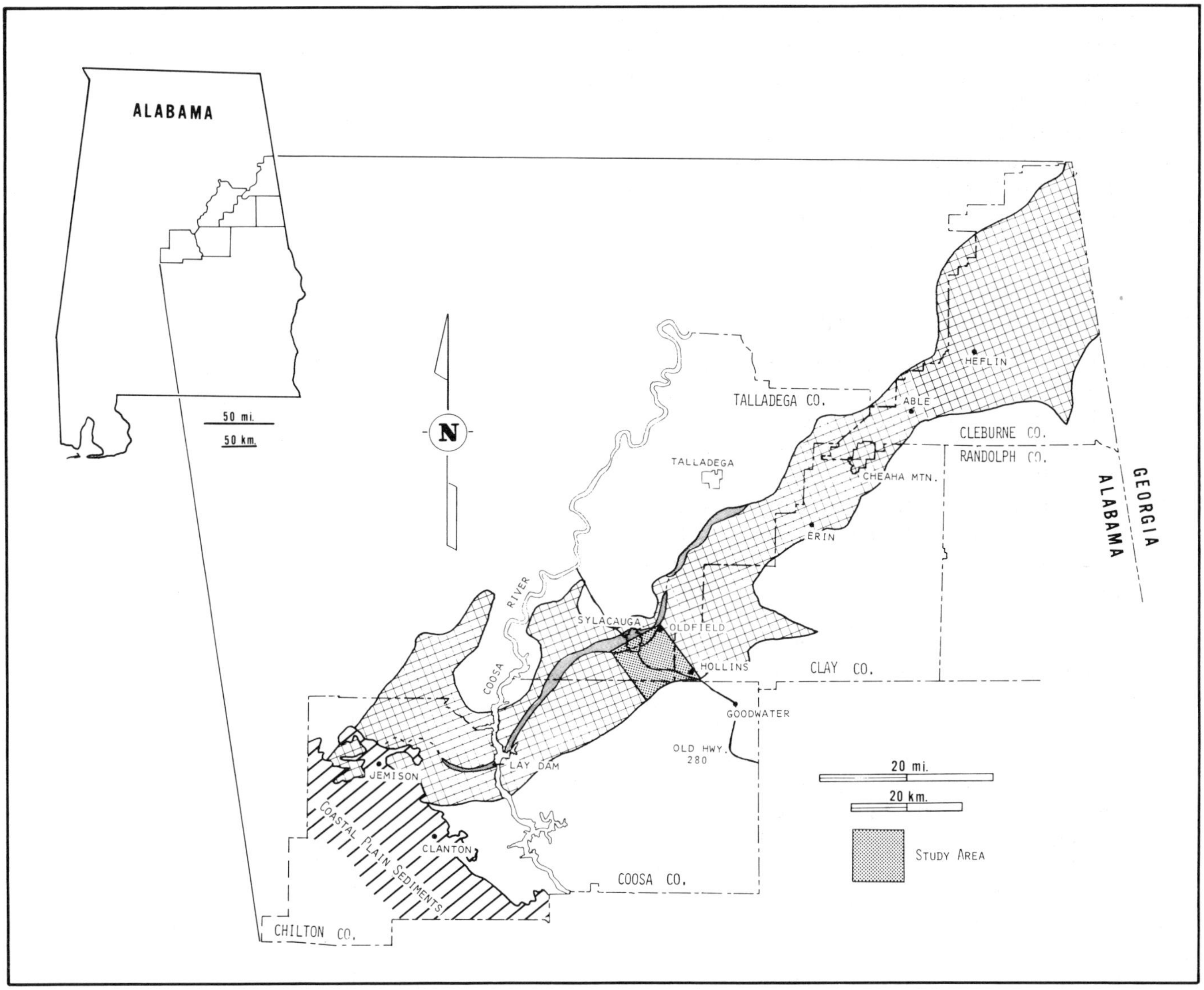

Figure 2. Location map of study area in south-central portion of the Talladega slate belt.

may only occur discontinuously in the study area, the above units appear to be continuous through the study area. Correlations of these units throughout the Talladega slate belt are based on distinctive lithologic features and on stratigraphic position. However, problems arise in the central portion of the belt because of the apparent termination of the Cheaha Quartzite. The Talladega Mountains, which are formed by the quartzite unit, end abruptly in southwestern Clay County. It appears that the termination of the Cheaha Quartzite is coincident with that of the mountain range. A less prominent ridge continues along strike to the southwest of the Cheaha Quartzite, but before this ridge joins the Cheaha ridge, it curves toward the northeast and continues in a position to the southeast of the Cheaha. This less prominent ridge is composed of dense, fine-grained quartzites and phyllitic quartzites that are in a regional stratigraphic position equivalent to the Jemison Chert interval, directly below the Hillabee Greenstone.

Southwest of this area in Chilton County the regional stratigraphy has been defined by Carrington, and the terminology used in this study is modified from that of Carrington (1973), as well as that of Butts (1926). To the northeast of this area in Cleburne County the regional stratigraphy has been defined in most detail by Bearce (1973), and correlation of terminology used in this report with that used by Bearce is shown in Figure 3. Bearce included the clastic rocks below the Cheaha Quartzite within his Heflin Phyllite and included the Cheaha Quartzite and Erin Slate in his Able Gap Formation. The Jemison Chert interval appears to be in an equivalent stratigraphic position to Bearce's Chulafinnee Schist.

The present study was undertaken in an effort to provide

detailed stratigraphic and structural data for the region southeast of Sylacauga, Alabama, which is the key in correlation of the regionally significant metaclastic sequence referred to as the Talladega Group. The absence of detailed work in this region has prevented accurate correlation of the reasonably well established Talladega Group stratigraphy to the west of the Coosa River in Chilton County with sequences to the northeast that have been studied in Clay and Cleburne Counties, Alabama.

STRATIGRAPHIC RELATIONSHIPS

Lying below the upper clastic sequence of the Talladega slate belt is the Sylacauga Marble Group, which consists of fine-grained, white to cream-white marble, intercalated with thin layers of green phyllite. Prouty (1916) mapped these marbles in a continuous belt, about 35 miles long, in Talladega and northern Coosa Counties. This marble belt generally forms a broad valley with small local ridges or knolls held up by cherty dolomitic facies within the marble. Butts (1926) reported that most of the marbles within the Sylacauga Marble Group consist of very pure calcium carbonate (96% to 99% $CaCO_3$), but Prouty (1916) mapped distinct beds of dolomite within the Sylacauga marble. The nature of the contact of the Sylacauga Marble Group with the phyllites of the overlying Lay Dam Formation has been in debate for a number of years. Prouty (1916) interpreted this contact to be a thrust fault, but later investigations by Butts (1926) led to the interpretation that the Sylacauga Marble occurred as a single unit within the Talladega slate belt. Investigations by Shaw (1970) led to his interpretation that the phyllites and slates of the Lay Dam Formation unconformably overlie the Sylacauga marbles, which he correlated with the Cambrian and Ordovician carbonate strata in the adjacent foreland sequence to the northwest. Field work in the vicinity of Sylacauga carried on by this writer also indicates an unconformable relationship between the carbonate rocks of the Sylacauga Marble Group and the overlying clastic rocks of the Lay Dam Formation. The best exposures of this contact are at the Hickman Quarry (Prouty, 1916), also known as the Crooked Creek locality (Shaw, 1970), just east of Oldfield in southeast Talladega County (Fig. 4), and at the Herd Quarry (Prouty, 1916), just to the northeast of the Hickman Quarry (Fig. 4). Both exposures show a very irregular marble surface with both marbles and phyllites showing no deformational features that would be associated with faulting along their contact (Fig. 5).

The Lay Dam Formation, which overlies the Sylacauga Marble sequence, is the oldest unit of the Talladega Group and is the most extensive and diverse of all the units investigated. The Lay Dam Formation was defined by Carrington (1973) in Chilton County as occurring stratigraphically between the Jumbo Dolomite (of the Sylacauga Marble Group) and the Butting Ram Sandstone. Southeast of Sylacauga the Lay Dam consists dominantly of silver to greenish-gray, slightly calcareous sericite phyllites and slates interbedded with numerous coarse-grained sandstone bodies. These coarser clastic units are generally made up of arkosic metaconglomerates or coarse-grained arkosic metasandstones, and are generally dark greenish-gray in color. The sandstone bodies are small and isolated and cannot be traced for any appreciable distance and appear to grade into the finer-grained phyllites that surround these coarser-grained bodies. Griffin (1951) originally noted that these lithologies were intercalated with the phyllites below the Cheaha Quartzite. Most metasandstone units are very

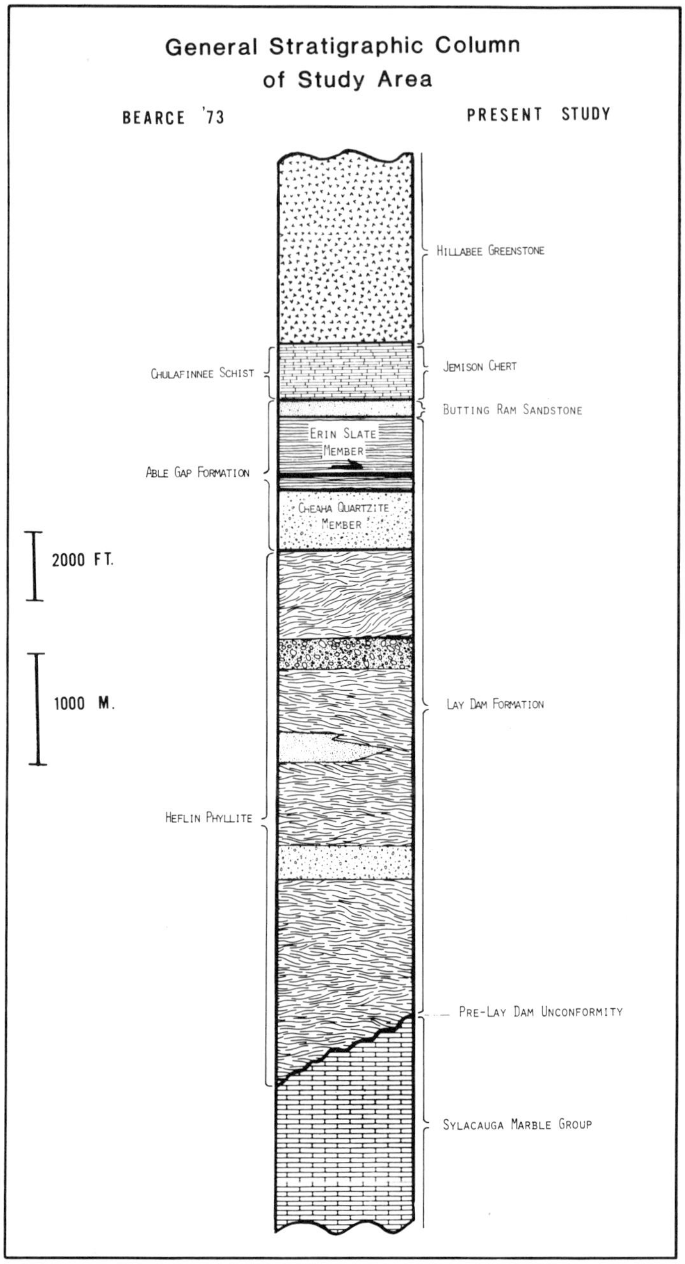

Figure 3. Generalized stratigraphic column illustrating nomenclature used in this study and by previous workers.

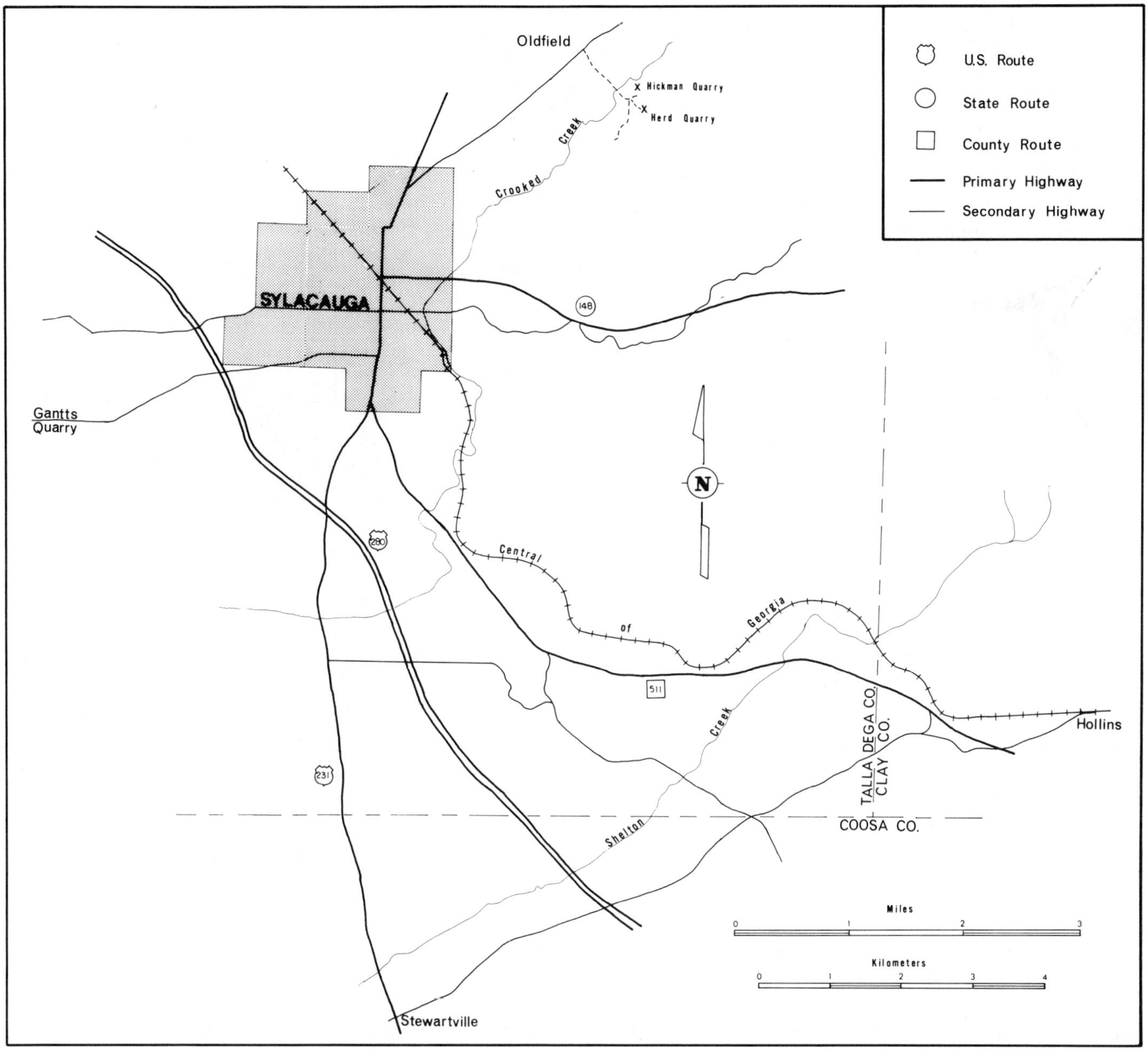

Figure 4. Location map of prominent cultural features and localities.

immature, poorly sorted, and contain abundant feldspar and quartz grains and detrital rock fragments; these rocks grade into metadiamictites. Rock fragments in the diamictites are sheared and flattened and include metashale and metacarbonate cobbles up to 20 cm long. The diamictite facies found in this study area is not as extensively developed as the facies found to the southwest in Chilton County (Telle, 1981) but probably represents distal lobes from the main bodies of diamictite farther to the southwest. Isolated diamictite bodies have been traced as far north as Alabama Highway 148 (Fig. 4). The approximate thickness of the Lay Dam Formation in the study area is 4,600 m.

The lower Cheaha Quartzite contact appears to be conformable with the Lay Dam Formation units below, and therefore the Cheaha Quartzite probably represents a member of the Lay Dam Formation. This contact, observed only at a few localities, is representated by interlayering of maroon sandy phyllites, typical of the lower portion of the Lay Dam Formation, with coarse-grained quartzites of the Cheaha Quartzite Member. The Cheaha Quartzite is composed of a generally light gray to white, coarse- to medium-grained quartzite, locally conglomeratic. Generally, very few feldspar grains are observed, and smoky quartz is dominant in both the conglomeratic and coarse-grained meta-

Figure 5. Outcrop of unconformity between phyllites of the Lay Dam Formation (upper part of photo) and marbles of the Sylacauga Marble Group (lower part of photo). Relief on unconformity is about 3 m.

sandstone portions of the unit. The Cheaha Quartzite exhibits an overall fining upwards sequence, grading from coarse-grained sandstones and conglomerates at the base up to medium-grained, light gray to white quartzites at the top of the unit. This overall fining upwards of the Cheaha Quartzite was recognized by Butts (1926) in his report of the Talladega Series. Primary facing criteria, such as cross-bedding and graded bedding, seen in the study area and in adjacent areas to the northeast all indicate that the Cheaha is upright. It is believed that the Cheaha represents a thicker, more extensively developed sandstone body similar to others below it within the Lay Dam Formation. The average thickness of the Cheaha Quartzite in the northeastern portion of the study area is 440 m.

The Cheaha Quartzite ends abruptly to the south against a fault within the Talladega slate belt in extreme southern Clay County. The reappearance of the Cheaha Quartzite from below this fault along strike to the southwest is significant. Within this region the Cheaha Quartzite appears to thin to the southwest in northern Coosa County. At U.S. Highway 280 a small ridge exposes smoky quartz conglomerates and coarse- to medium-grained quartzites, which are very similar lithologically to the Cheaha Quartzite to the northeast (Fig. 6).

This quartzite is expressed topographically as a smaller ridge than the ridge formed by the Cheaha Quartzite to the northeast, but it is thought to be equivalent to the Cheaha. This unit represents a much thinner (240 m) and somewhat finer-grained interval than the Cheaha to the northeast. The dominant lithology forming the smaller ridge is a gray, thin- to medium-bedded, fine- to medium-grained quartzite. Locally, coarse-grained conglomerates predominate.

The age of the Cheaha Quartzite at this time is uncertain. Smith (1903) suggested that the Cheaha was Pennsylvanian in age because it was interbedded with the overlying Erin Slate. At that time he mapped the Erin Slate as a Pennsylvanian sequence, based upon its reported content of plant fossils.

Recent mapping by this author indicates that the Cheaha Quartzite is a member of the Lay Dam Formation. Southwest of the study area in Chilton County the Lay Dam Formation occurs stratigraphically beneath the Lower to Middle Devonian Jemison Chert interval (Carrington, 1973; Tull, 1979). Therefore, the Cheaha Quartzite lies stratigraphically below the Jemison Chert interval and is inferred to be Lower to Middle Devonian or older. Although Smith (1903) interpreted the black slates of the Erin as Pennsylvanian in age, mapping by the author has indicated interbedded relationships between the Erin and the Jemison Chert, which suggest a Lower to Middle Devonian or older age for the Erin Slate. The Butting Ram Sandstone to the southwest in Chilton County is also reported to contain Devonian invertebrate fossils (Carrington, 1973), and where present in the study area the Butting Ram Sandstone occurs between the Jemison interval and the Erin Slate.

The Erin Slate consists of black graphitic sericite phyl-

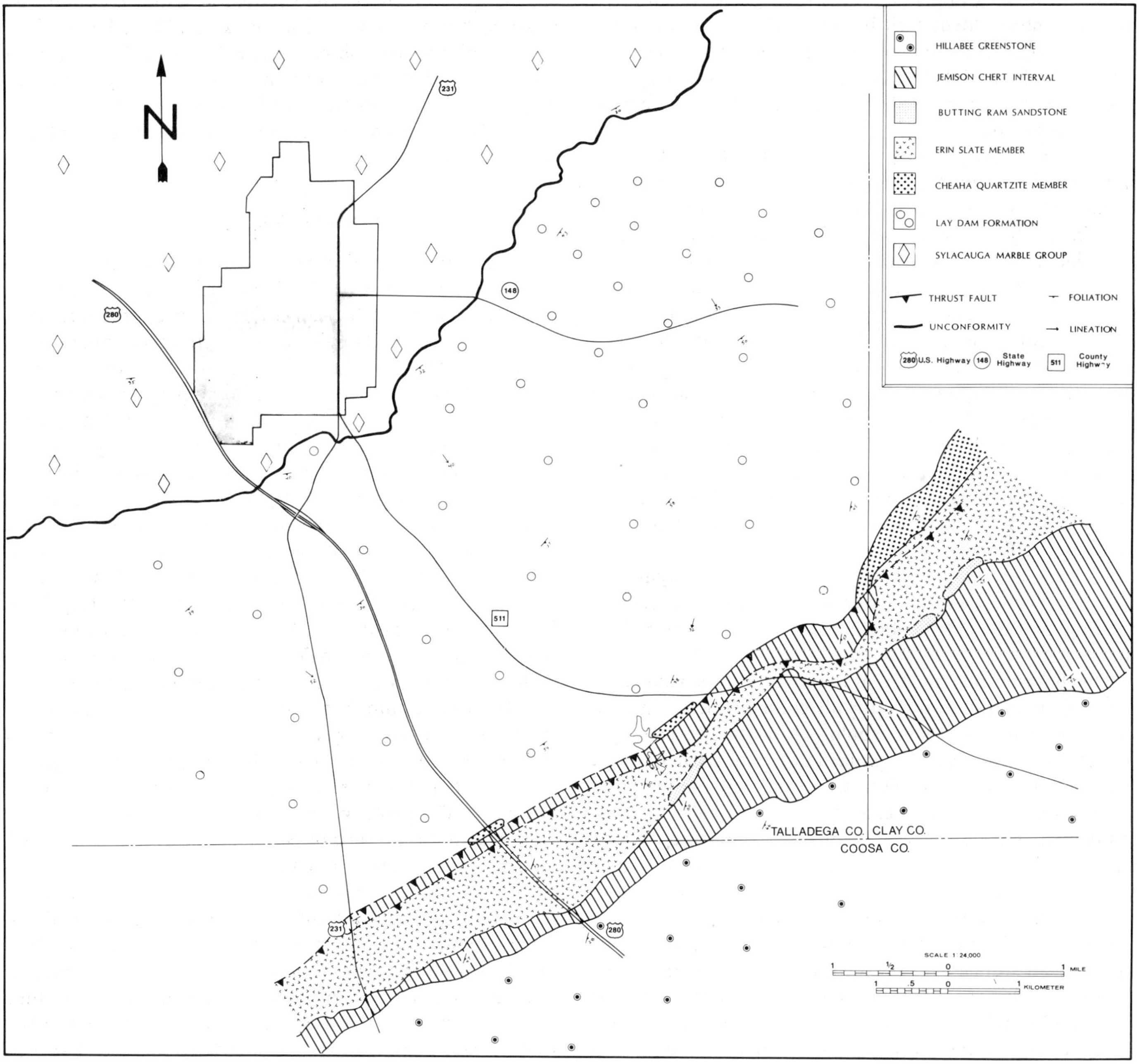

Figure 6. Geologic map of study area.

lites. The Erin found within the study area is quite similar to that at the type locality of the Erin in northwestern Clay County, some 33 km to the northeast. Prouty (1923) noted the presence of black carbonaceous slates to the east of Talladega Mountain generally throughout the entire length of the Talladega slate belt in Clay County. Prouty's map of Clay County (1923) shows several areas of graphitic slates in western Clay County (Fig. 2). The Erin Slate also appears to be a member of the Lay Dam Formation because of the similarity of lithologies and also the contact between the Cheaha Quartzite and the Erin Slate, both northeast and southwest of the region where the Cheaha has been concealed by faulting, appears to be stratigraphically conformable and interbedded. Griffin (1951) defined the Cheaha Quartzite–Erin Slate boundary at the point where the phyllites of the Erin Slate become the dominant lithology. The Erin Slate averages over 700 m in thickness throughout most of the study area.

At some localities lying just above the Erin Slate and immediately below the Jemison Chert interval is a thin green, chloritic, arkosic metasandstone unit that may be correlative with the Butting Ram Sandstone of Butts (1926). The thickness of this metasandstone is variable; in some localities it may be absent, while in other localities it may be as much as 50 m thick. The Butting Ram Sandstone is very discontinuous but is generally traceable across the study area. Carrington (1973) reported invertebrate fossils from this unit to the southwest in Chilton County, which assigns a Devonian age to the Butting Ram Sandstone.

Lying above the Butting Ram Sandstone, or in contact with the Erin Slate where the Butting Ram is absent, is the Jemison Chert interval. The contact of the Erin Slate and Jemison Chert interval can be observed at several localities and is represented by a series of interbedded white, paper-thin quartzites intercalated with black graphitic phyllites of the Erin Slate. This contact zone is important in demonstrating that a stratigraphic contact exists between these units.

The Jemison Chert interval found within the study area generally occurs as two slightly different distinctive lithologies. The typical Jemison lithology is a white to pale light gray, very dense, very fine-grained quartzite. This lithology lies at the base of the Jemison Chert interval and underlies the prominent ridge that can be traced through the entire study area. The other major Jemison lithology is a white to silver gray, slightly graphitic, paper-thin phyllitic quartzite that is found in the upper part of the Jemison Chert interval. This phyllitic quartzite is complexly folded, preventing an accurate thickness determination of the Jemison Chert interval. The Jemison interval lithology occurs in an arcuate thrust slice that, as mentioned above, locally overrides the Cheaha Quartzite. These quartzites of the Jemison interval have undergone intense deformation that has resulted in shearing, flattening and polygonization of quartz grains, and total recrystallization. None of the lower units within the study area appear to show this intensity of deformation. The average thickness of the Jemison Chert interval in the upper plate is 415 m in the study area.

The Hillabee Greenstone lies above the Jemison Chert interval (Tull and Stow, 1979). This study did not include the entire Hillabee sequence but only involves investigations along the Jemison-Hillabee contact. This contact is represented by an interbedded sequence of white paper-thin quartzite intervals with generally a maximum thickness of about 15 cm, thinning upward, and dark green foliated and laminated mafic phyllite, representing metamorphosed basaltic ash beds. This interbedded sequence is not continuously exposed but appears to be as thick as 150 m. Based upon stratigraphic relationships of the Hillabee to the underlying fossiliferous Jemison Chert and upon K-Ar age dates on magmatic hornblende from the Hillabee, a Devonian age for the Hillabee Greenstone complex is suggested (Tull and Stow, 1979).

STRUCTURAL RELATIONSHIPS

Structural and stratigraphic relationships between the units forming the two prominent ridges in this area, the Cheaha Quartzite ridge and the Jemison Chert ridge, indicate that faulting is the cause of the southwest termination of the Cheaha Quartzite, as well as the cause of duplication of the Jemison stratigraphy. Detailed mapping has shown that a light gray, slightly graphitic paper-thin quartzite (Jemison) is faulted onto thick, coarse-grained to conglomeratic quartzites at the south end of the Cheaha ridge. This fine-grained quartzite generally follows the same trend as the main Jemison ridge to the southeast but appears to be a thrust duplicated slice of the Jemison Chert interval, which has overridden the Cheaha Quartzite and is in fault contact with the stratigraphically lower portion of the Lay Dam Formation for a distance of 8.8 km (Fig. 6). The thrust fault that overrides the Cheaha Quartzite appears to occur mostly within the Erin Slate. The regional dip of the units in the structural upper block (those units to the southeast of the imbricate thrust) averages between 55° and 60° to the southeast, whereas the units below this imbricate thrust generally dip at a shallower angle around 30° to 35° to the southeast. The plane of the fault where measured generally dips in a southeastward direction at about 30°. These geometric relationships would allow this imbricate thrust fault to cut through a portion of the Jemison Chert interval at some depth with the fault bringing up a portion of the Jemison and emplacing it over the Cheaha Quartzite (Fig. 7). The suggestion that this segment of the Jemison interval is fault bounded is supported by the fact that the unit terminates both to the northeast and the southwest in the study area. Faulting is also suggested by a marked change in orientation of planar and linear structural features across

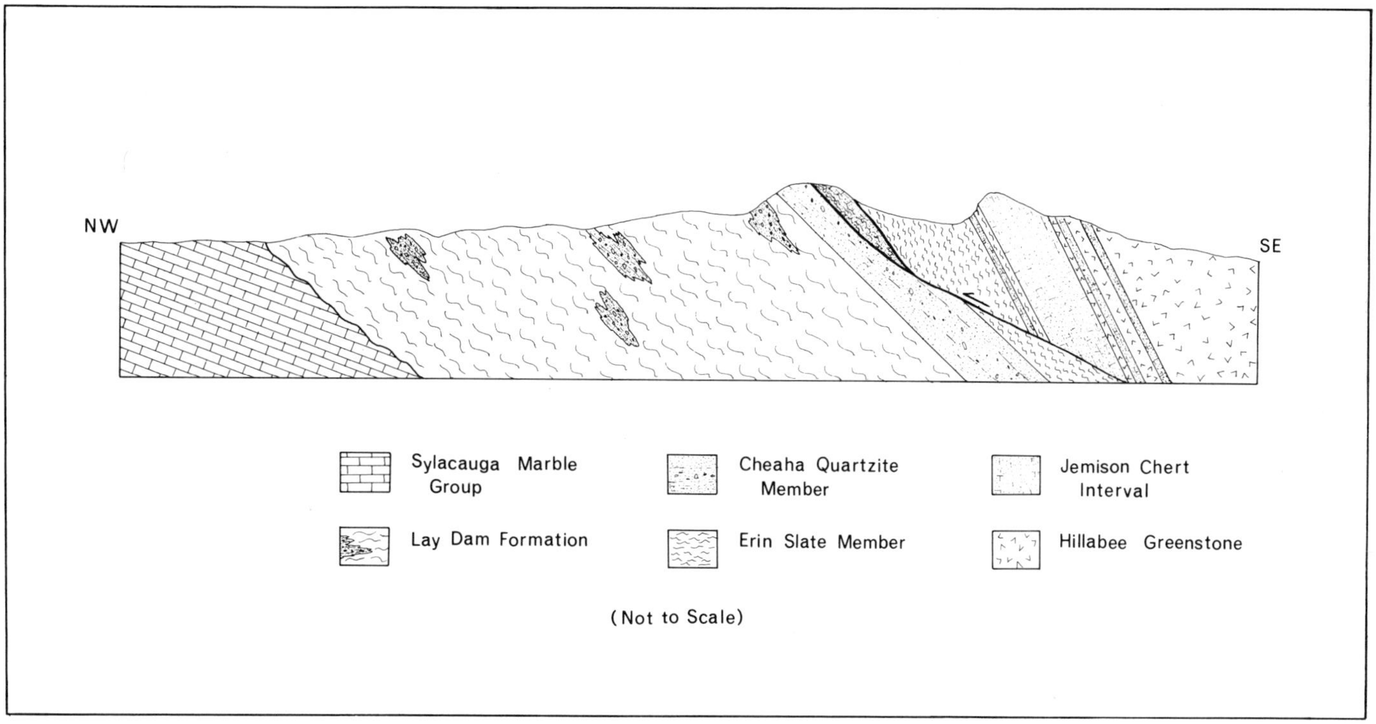

Figure 7. Schematic cross-section showing imbricate thrust fault cutting the Jemison Chert interval and emplacing slice of Jemison over Cheaha Quartzite (not to scale).

the contacts of the imbricate slice. Three subareas were defined within the study area, and structural data from these domains are illustrated in Figure 8. Subarea I includes lithologies located structurally below the thrust slice including a portion of the Erin Slate, the Cheaha Quartzite, and the Lay Dam Formation below the Cheaha. Structural data from the Sylacauga Marble sequence were not included in Subarea I. Subarea II consists of the thrusted slice of the Jemison Chert interval, and subarea III contains lithologies structurally above the thrust slice, including the structural upper portion of the Erin Slate, the Butting Ram Sandstone, the Jemison Chert and the Hillabee Greenstone (Fig. 9). Foliation orientations throughout the study area exhibit the same general strike, but a slight change in dip occurs between the different subareas. S_1 foliations in subarea III generally dip steeper than foliations from subarea I (Fig. 8). Stereogram plots of lineations from the various subareas show a diversity, which also may indicate structural discontinuity between these blocks (Fig. 8).

This imbricate faulting appears to have taken place after the major episode of deformation and is therefore faulting a previously deformed region. This would account for the apparent reverse situation of younger rocks being thrusted over older rocks. This smaller imbricate thrust appears to occur in an equivalent stratigraphic position and to affect similar lithologies as the High Falls Branch Fault (Carter, 1981), which occurs to the northeast of the study area. This fault appears to have developed as a splay related to faulting along the Hollins Line fault. Similar imbricate thrusting has occurred elsewhere along this stratigraphic level in the Talladega slate belt (Tull and Stow, 1979, p. 49). The stratigraphic position of this fault within the study area may be related to a facies change within the Cheaha Quartzite, which created mechanical anisotropies within the sequence.

DEPOSITIONAL ENVIRONMENTS

The Sylacauga Marble Group represents a possible Cambrian-Ordovician passive shelf sequence that probably formed on the eastern edge of the proto-North American continent. This sequence includes units probably as young as the Early Ordovician Knox Group (Shaw, 1970). During post-Early Ordovician time a major tectonic change took place forming the pre-Lay Dam Formation unconformity and was followed by deposition of the Talladega Group sediments. Shaw (1970) referred to the unconformity as evidence of pre-Middle Ordovician deformation or tectonism. This tectonism may have changed the passive divergent carbonate shelf margin into an active convergent continental margin. A major basin was created in which both coarse- and fine-grained sediments of the Lay Dam Formation were deposited. Within this sequence numerous large diamictite flows were developed to the southwest in Chilton County (Telle, 1981). The distal portions of these debris flows are recognized as the diamictite bodies that are traceable into the study area in Talladega and Clay Coun-

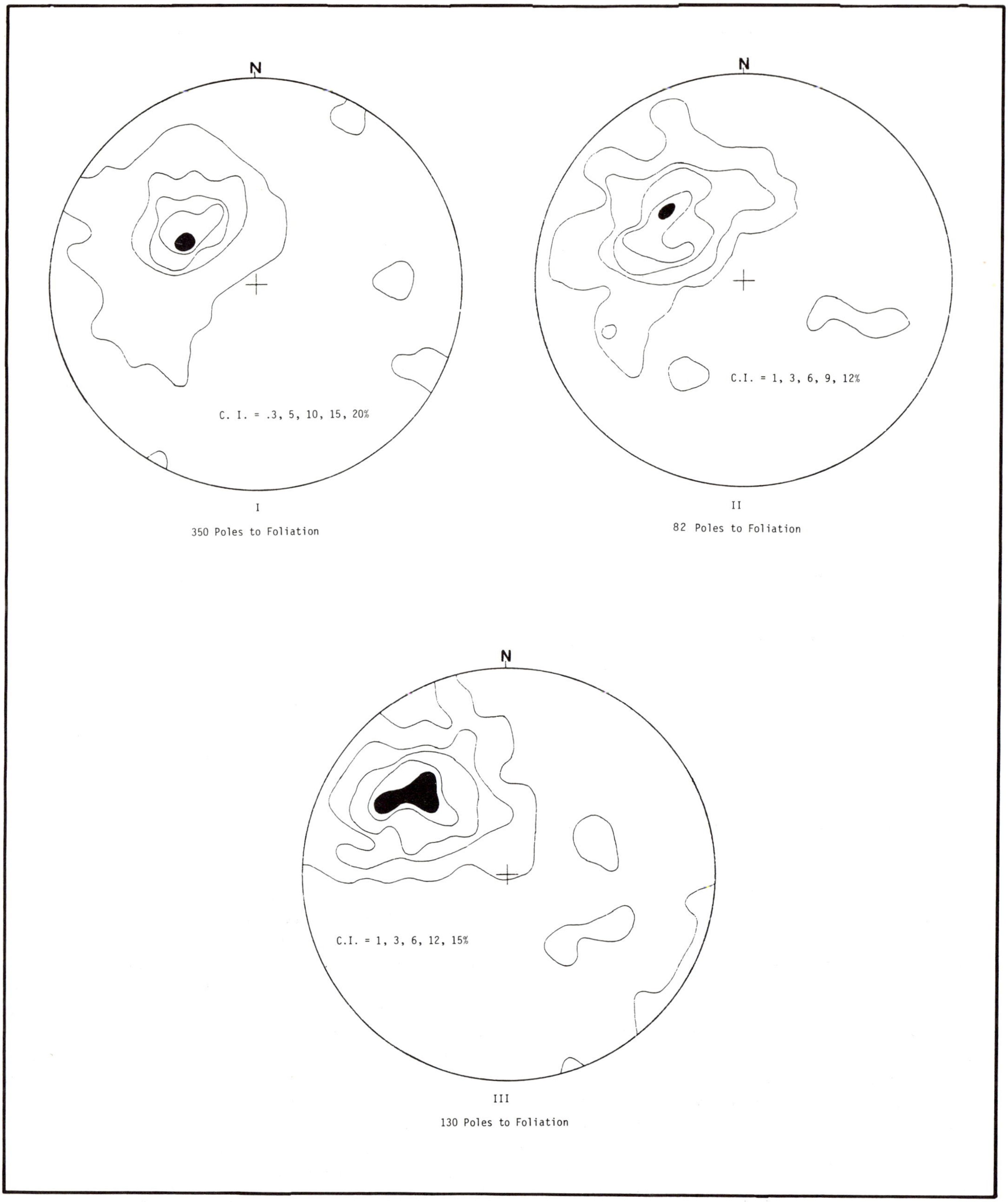

Figure 8 (This and facing page). Lower hemispere equal area projections of structural data from subareas illustrated in Figure 9. Data illustrated include: foliation (S_1 cleavage surfaces), L_1 — Lineations (cleavage–compositional layering intersections, mineral lineations). C.I. = contour intervals.

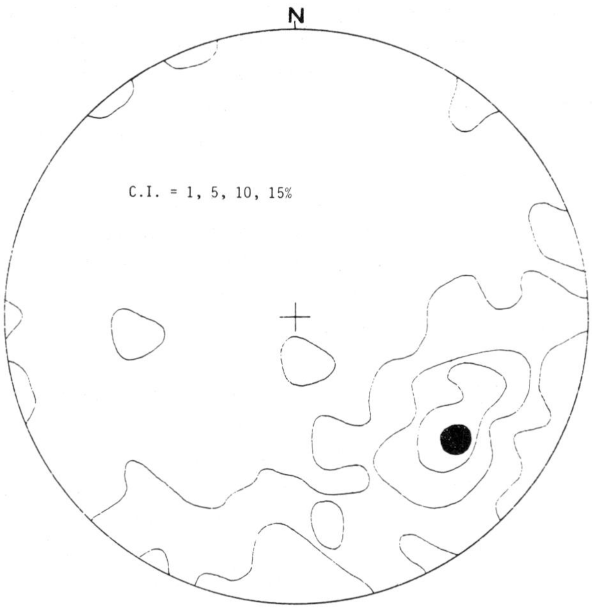

I

69 L_1 - Lineations

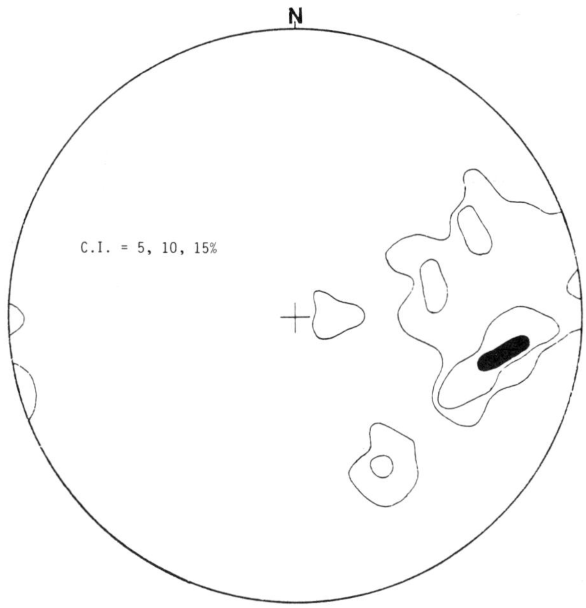

II

19 L_1 - Lineations

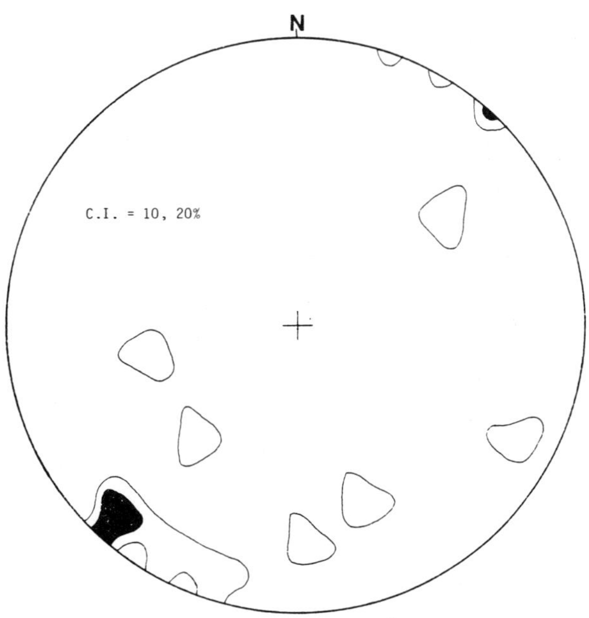

III

10 L_1 - Lineations

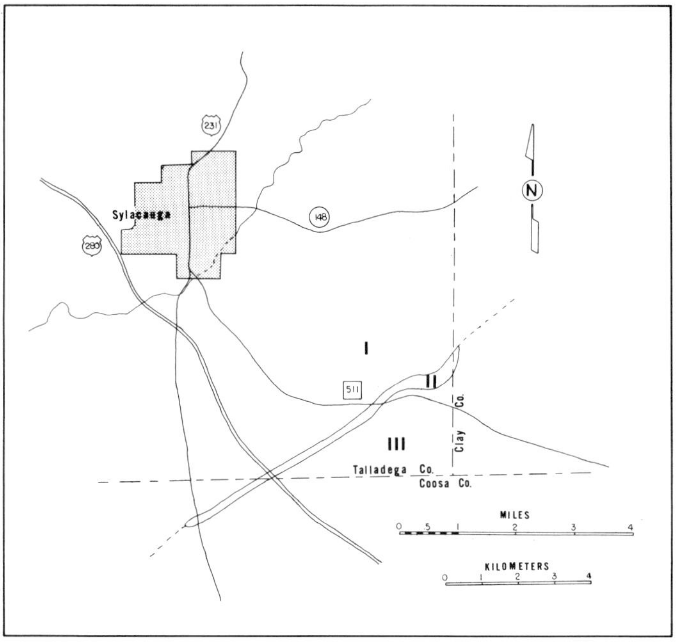

Figure 9. Location map of three subareas studied for structural analysis.

ties. Finer-grained laminated slates of the Lay Dam Formation may represent more distal turbidite facies. The Cheaha Quartzite may represent a regressive event involving the development of a beach environment. The Erin Slate appears to represent a period of lagoonal conditions that prograded over the Cheaha. In this model the isolated and discontinuous Butting Ram Sandstone may represent tidal channel deposits. The depositional environment of the Jemison Chert in the study area is uncertain. Total recrystallization and deformation of the Jemison in the study area has made intepretation of the original lithology difficult. The interbedded nature of the Jemison Chert and the Hillabee Greenstone indicates that volcanic activity was closely associated with the deposition of at least the upper part of the Jemison interval. Volcanism may have served as a possible source of the silica which was either deposited as a volcanogenic chert or was available for total silicification of an originally fine-grained sandstone.

CONCLUSIONS

Mapping the geology of the Talladega slate belt in the frontal crystalline thrust block of the southern Appalachians of Alabama has demonstrated that a regional stratigraphy can be recognized throughout much of the belt. The key to defining the stratigraphy of the Talladega Group, as with any sequence, is the identification and tracing of particular distinctive lithologic units. The Cheaha Quartzite represents an important lithologic unit that is distinct and traceable throughout the north-central portion of the Talladega slate belt of Alabama from Cleburne County into northeast Coosa County. Mapping and correlation of the stratigraphy in the south-central portion of the belt with that to the north has been hampered by what appeared to be the southern termination of the Cheaha Quartzite in Clay County. This study has involved mapping this stratigraphy in the south-central Talladega slate belt and has succeeded in establishing a stratigraphy that correleates with that already described by other writers in other parts of the belt. From this study, it appears that the Cheaha Quartzite continues to the southwest into Coosa County as a much thinner and somewhat finer-grained sequence. Localized faulting has imbricated the Jemison Chert interval and has truncated a portion of the Cheaha Quartzite in southwestern Clay County. Similar imbrication of this sequence has been noted at other localities along strike. Diamictites of the Lay Dam Formation may be traced from Chilton County into the study area, and they lie stratigraphically below the Cheaha Quartzite. The Jemison Chert interval and the Hillabee Greenstone both continue into the study area from Chilton County. These two units overlie a black graphitic slate that may be traced to the northeast into the Erin Slate. Also it is noted that all these units occur stratigraphically above the Sylacauga Marble and the pre–Lay Dam Formation unconformity. Major lithologic unit contacts, although subsequently complicated by faulting, were probably originally stratigraphic and commonly gradational. The Erin Slate Member, Butting Ram Sandstone, and Jemison Chert interval all show gradational stratigraphic relations. The Butting Ram Sandstone and Jemison Chert interval are reported to contain Devonian fossils in Chilton County. This suggests that the section lying below the Jemison interval must be Devonian or older.

ACKNOWLEDGMENTS

Field expenses, laboratory facilities, and technical assistance for this study were provided in part by the Geological Survey of Alabama, under direction of Thomas J. Joiner, State Geologist, and through graduate assistantships provided by the Department of Geology and Geography, the University of Alabama.

The author wishes to acknowledge the following individuals for their assistance during the preparation of this manuscript: Charles W. Copeland, Jr., Geological Survey of Alabama; W. Gary Hooks, the University of Alabama; and Denny N. Bearce, the University of Alabama in Birmingham.

The author would like to express a special appreciation to James F. Tull for his advice and critical review of this study in the field and assistance in preparation of this manuscript.

REFERENCES CITED

Bearce, D. N., 1973, Geology of the Talladega metamorphic belt in Cleburne and Calhoun Counties, Alabama: American Journal of Science, v. 273, p. 742–752.

Butts, C., 1926, The Paleozoic rocks, *in* Adams, G. I., Butts, C., Stephenson, L. W., and Cooke, W., Geology of Alabama: Geological Survey of Alabama Special Report 14, p. 40–223.

Carrington, T. J., 1973, Metamorphosed Paleozoic sedimentary rocks in Chilton, Shelby, and Talladega Counties, Alabama, *in* Carrington, T. J., ed., Talladega metamorphic front: Alabama Geological Society, 11th Annual Field Trip, Guidebook, p. 22–38.

Carter, W. W., 1981, Stratigraphic and structural relationships within the upper portion of the Talladega slate belt in the Pyriton area, Alabama [abs.]: Geological Society of America Abstracts with Programs, v. 13, no. 1, p. 4.

Griffen, R. H., 1951, Structure and petrography of the Hillabee Sill and associated metamorphics of Alabama: Geological Survey of Alabama Bulletin 63, 74 p.

Prouty, W. F., 1916, Preliminary report on the crystalline and other marbles of Alabama: Geological survey of Alabama Bulletin 18, 202 p.

——1923, Geology and mineral resources of Clay County: Geological Survey of Alabama, County Report 1, 190 p.

Shaw, C. E., Jr., 1970, Age and stratigraphic relations of the Talladega Slate: Evidence of pre-Middle Ordovician tectonism in central Alabama: Southeastern Geology, v. 4, p. 253–267

Smith, E. A., 1903, Carboniferous fossils in Ocoee slates in Alabama: Science, v. 18, p. 244–246.

Telle, W. R., 1981, Depositional setting of the diamictite facies of the Lay Dam Formation, Talladega slate belt, Chilton County, Alabama [abs.]: Geological Society of America Abstracts with Programs, v. 13, no. 1, p. 36.

Tull, J. F., 1978, Structural development of the Alabama Piedmont northwest of the Brevard Zone: American Journal of Science, v. 278, p. 442–460.

Tull, J. F., and Stow, S. H., eds., 1979, The Hillabee metavolcanic complex and associated rock sequences: Alabama Geological Society, 17th Annual Field Trip, Guidebook, 66 p.

Tull, J. F., and Stow, S. H., 1980, The Hillabee Greenstone: A mafic volcanic complex in the Appalachian Piedmont of Alabama: Geological Society of Alabama: Geological Society of America Bulletin, v. 91, p. 27–36.

MANUSCRIPT ACCEPTED BY THE SOCIETY MAY 7, 1982

Printed in U.S.A.

Geological Society of America
Special Paper 191
1982

Stratigraphic relationships of the carbonate sequence in the Talladega slate belt, Chilton and Coosa Counties, Alabama

William S. Pendexter
Department of Geology
University of Alabama
University, Alabama 35486

ABSTRACT

A major carbonate sequence occurs within the lower part of the Talladega slate belt in Chilton, Coosa, and Talladega Counties. The carbonate units are overlain by a major regional unconformity known as the pre-Lay Dam Formation unconformity. The carbonate sequence below the unconformity is represented in different areas by the Jumbo Dolomite, the Marble Valley carbonates, and the Sylacauga marbles. At the type location in Chilton County, the Jumbo is a 67-m-thick, predominantly thickly bedded dolostone. The contact with the underlying slates of the Wash Creek Slate (Mount Zion Formation) is an interlayered zone of dolostone and fine, commonly graphitic, clastic rock. This zone grades upward into a dolostone that contains a few pelitic layers. Near the base,the Jumbo contains intraclasts and recrystallized fragments up to 6 cm in length. Rounded quartz grains are disseminated in the lower section of the Jumbo. Near the middle of the massively bedded dolostone is a layer that contains intraclasts of massive and laminated carbonate as much as 12 cm in length. Just below the unconformity in the type section, the upper part of the Jumbo contains laminations of fine-grained clastic rock. Along strike to the northeast and southwest the unconformity appears to have erosional relief. Less than 1 km west of the type location the unconformity truncates the carbonate sequence completely. To the northeast the pre-Lay Dam Formation unconformity appears to rise in the section in the Marble Valley carbonates and the Sylacauga marbles, exposing a very thick carbonate sequence. Stratigraphic and structural relationships of the Jumbo, Marble Valley, and Sylacauga marbles are not yet resolved. Initial data indicate that the Jumbo occurs stratigraphically below and to the northwest of the Marble Valley carbonates, suggesting that the Jumbo is the oldest carbonate unit in the sequence.

INTRODUCTION

The Talladega slate belt consists of an allochthonous sequence of metamorphosed clastic, carbonate, and volcanic rocks in the southernmost Appalachians of Alabama and Georgia. A distinctive sequence of carbonate rocks, the Sylacauga Marble Group (Tull, 1982), in the lower (northwest) part of the Talladega slate belt, may be traced for at least 100 km along strike. The Sylacauga Marble Group is believed to be a trailing margin carbonate sequence that developed during the lower Paleozoic. In the southwestern part of the outcrop belt, this carbonate sequence is known as the Jumbo Dolomite (Butts, 1926). This carbonate sequence has been recognized at several locations including Jumbo, Marble Valley, and Sylacauga (Fig. 1). The continuity, contact relationships, and stratigraphy of the carbonate sequence are in debate. One purpose of this study was to determine the stratigraphic and structural relationships of the carbonate sequence and thereby to provide constraints on its age and relationships to the other rocks within the

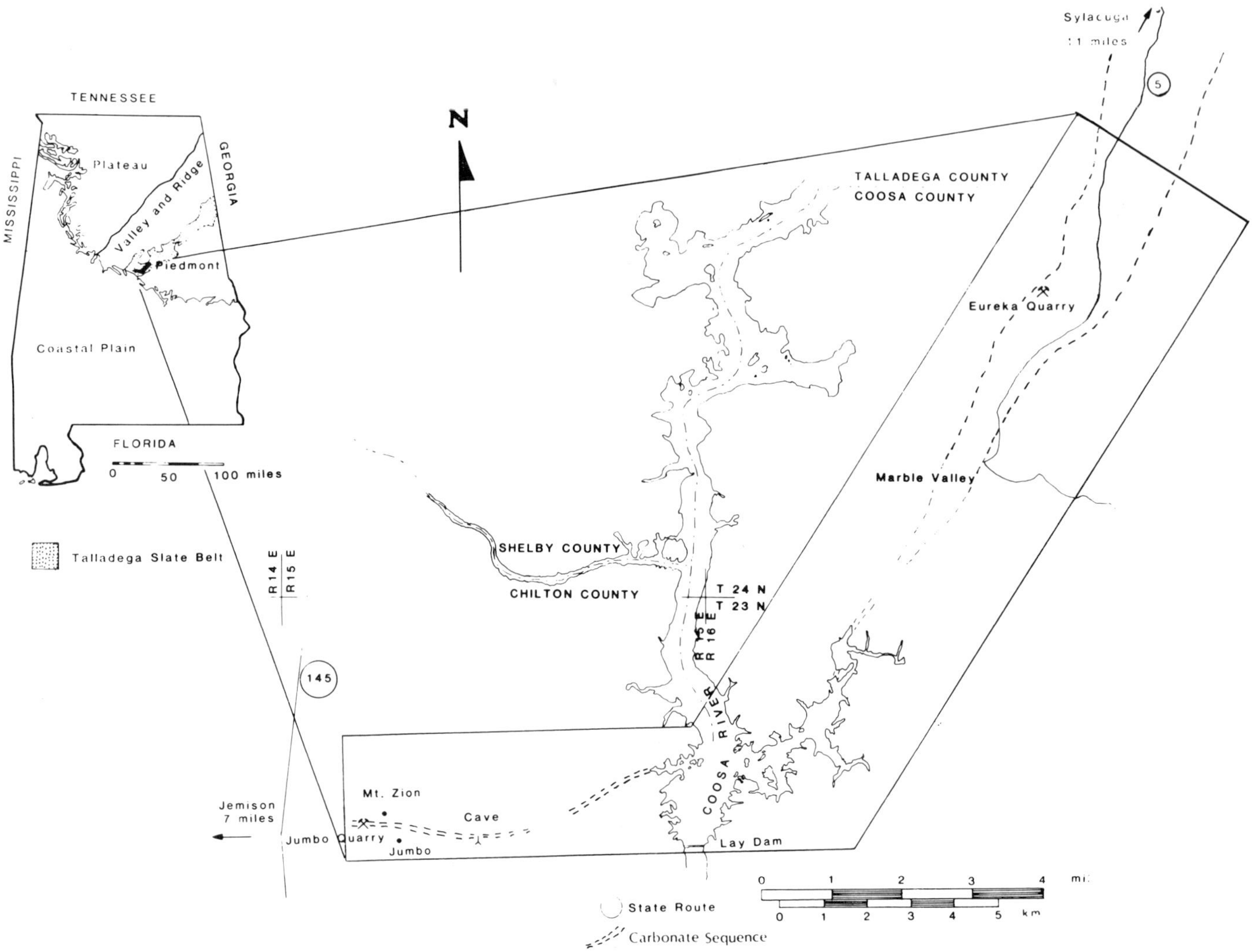

Figure 1. Location map illustrating the trend of the carbonate sequence within the study area.

Talladega slate belt and the Appalachian Foreland sequence to the northwest.

The lower (northwestern) part of the Talladega slate belt in Coosa, Chilton, and Talladega Counties, Alabama, consists of lower greenschist facies phyllites, quartzite, and marbles. The Talladega slate belt (Fig. 2), is bounded on the northwest by the Talladega-Cartersville fault system (McCalley, 1897; Butts, 1926) and the Paleozoic sedimentary rocks of the foreland fold and thrust belt. The southeastern border of the Talladega slate belt is the Hollins Line fault (Tull, 1978) along which rocks of amphibolite facies have been thrust over the rocks of the Talladega slate belt.

The portion of the Talladega slate belt that is the subject of this study includes two predominantly fine-grained clastic units that bracket the carbonate sequence. The Sylacauga Marble Group is underlain by the Wash Creek Slate (Butts, 1940) and is overlain by the Lay Dam Formation (Carrington, 1973).

STRATIGRAPHY

Lower Clastic Unit

Below the marble sequence in the study area is a sequence of fine-grained phyllitic rocks that are commonly graphitic. These rocks range in color from dark gray to black in graphitic units and from yellow, silver, and maroon to tan in nongraphitic units. This sequence was originally referred to by Butts (1940) as the Wash Creek Slate. Later, Carrington (1973) referred to a black slate unit below the Jumbo Dolomite in Chilton County as the Mount Zion Formation. The Wash Creek Slate (Mount Zion Formation) is composed of clay and silt-size grains of quartz, sericite, chlorite, and graphite, and is locally pyritiferous. The phyllites are locally very thinly laminated but are commonly very homogenous. The thickness of the Wash Creek Slate has not been determined within the study area because of a lack of a definitive lower boundary in this area. Butts (1940), how-

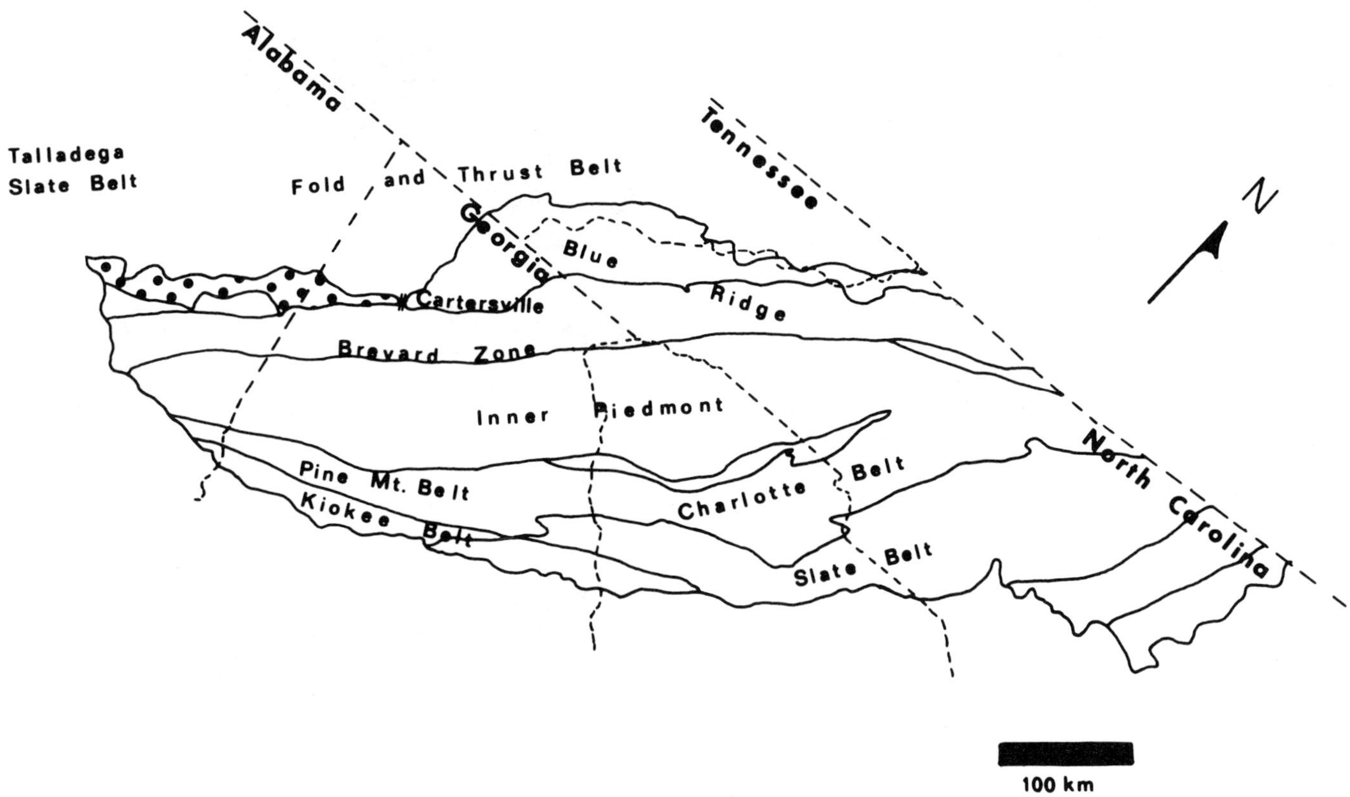

Figure 2. Regional setting of the Talladega slate belt in Alabama and Georgia.

ever, determined the lower boundary to be a stratigraphic contact with the underlying Brewer Phyllite, which is a purplish-brown sandy phyllite, in contrast to the fine-grained black, yellow, silver, or maroon Wash Creek Slate. Carrington (1973), in contrast, suggested that the base of the Mount Zion Formation, which he did not equate with the Wash Creek Slate of Butts (1940) in its entirety, is a major fault: the Chilton fault.

Approximately 30 m below the top of the Wash Creek Slate in Coosa County the average grain size of the phyllitic lithology increases upward and the more typical Wash Creek phyllites grade into a 10-m-thick sandy phyllite or silty metasandstone. This sandy phyllite is directly overlain by a 5- to 10-m-thick, coarse-grained, ridge forming quartzite that crops out on the prominent ridges to the northwest, which are 30 to 45 m higher than the adjacent carbonate valley. This quartzite may be traced for at least 4 km within the study area and is composed of elongated, coarse-grained, smokey quartz grains in a matrix of finely recrystallized quartz and minor sericite and chlorite. This quartzite has sharp contacts with the overlying and underlying phyllites. This unit probably correlates with the "conglomeratic sandstone member" of the Wash Creek Slate of Butts (1940), which also is found in the upper Wash Creek Slate to the west of the study area in Chilton County.

Jumbo Dolomite

The Wash Creek Slate is stratigraphically overlain by the regionally developed carbonate sequence. The Jumbo Dolomite (Butts, 1926) has been mapped from just west of the Jumbo quarry to approximately 1 km north of Paint Creek (Fig. 1). Farther to the northeast the carbonate sequence is represented by the Marble Valley carbonates. Outside the study area, farther to the northeast, the carbonates may be traced into the Sylacauga Marbles (Shaw, 1970).

The contact between the Wash Creek Slate and the overlying carbonate rocks is a gradational contact as much as 5 m in thickness. This gradational contact consists of interlayered, fine-grained pelitic and carbonate rocks with an increasing proportion of carbonate laminae upward. The interlayered pelitic and carbonate laminae have sharp, well-defined contacts with one another. The carbonate layers are composed of fine-grained dolomite, and the pelitic layers, typical of the upper portion of the Wash Creek Slate, contain quartz, sericite, chlorite, and graphite. Thin pelitic layers persist up through the carbonate where the dolomite is the predominant rock component. West of the Coosa River (Fig. 1) the Jumbo Dolomite is a fine- to medium-grained dolostone, thinly bedded at the base with an increasing bed thickness upward.

The lower 12 m of the Jumbo contains several layers which contain numerous intraclasts and coarsely recrystallized fragments. These intraclasts and recrystallized fragments are seen at the Jumbo Quarry and on the east bank of the Coosa River in Coosa County (Fig. 1). The intraclasts in this layer range from 1 to 2 cm to 6 cm in length (Fig. 3). The recrystallized fragments are somewhat smaller with a maximum length of 3 cm. There seems to be no sorting of these fragments by size or type, and all exhibit random orientation. The recrystallized fragments may be fossil shell fragments, but no morphological features remain to substantiate this possibility.

The lower part of the Jumbo Dolomite also contains disseminated, well-rounded, very fine to fine-grained quartz sand suspended in a matrix of carbonate. The grains of quartz are commonly fractured. The quartz sand-bearing carbonate is in irregularly shaped pods surrounded by carbonate matrix that is barren of quartz grains.

Above the lower 12 m, the Jumbo is a massively bedded, medium- to fine-grained, gray dolostone. Near the middle of the Jumbo, at the type locality, one bed contains large intraclasts that include both laminated and massive carbonate rocks (Fig. 4). These intraclasts range from 2 to 12 cm in length and show no perferred orientation or signs of sorting. Some of these clasts have been secondarily silicified. The thin pelitic layers in the Jumbo are more abundant in the uppermost 8 m at the type locality, and bedding thickness decreases upward in this uppermost part of the Jumbo.

Figure 3. Slabs from the lower 12 m of the Jumbo Dolomite showing intraclasts and recrystallized fragments.

Figure 4. Slab from the middle of the Jumbo Dolomite showing massive and laminated intraclasts.

Marble Valley Carbonates

Exposures of the Marble Valley carbonates are limited to a few stream beds and two abandoned quarries. It is difficult to establish the internal stratigraphy because of this lack of exposure. The Marble Valley carbonates are in gradational contact with the underlying Wash Creek Slate. The lower part of the carbonate is dark gray in color, while it is very light gray to white in color above the basal 20 m. This light-colored carbonate contains few pelitic layers and is thickly bedded. A chert-bearing unit occurs about 200 m above the base of the Marble Valley carbonates. The chert-bearing unit forms discontinuous hills in Marble Valley which rise up to 25 m above the valley floor. This chert ranges in color from white to tan to orange and brown and commonly contains pores partially filled with remobilized quartz. Prouty (1916) also recognized chert-bearing units in the Marble Valley–Sylacauga belt and equated some of these chert-bearing units with portions of the Cambro-Ordovician Knox Dolomite of the foreland fold and thrust belt. Above the chert-bearing unit in the Marble Valley area is a monotonous sequence of thickly bedded white to light gray calcite marble. The Jumbo Dolomite is probably equivalent only to the lower portion of the Marble Valley carbonates and not to the entire section. The chert-bearing unit appears to be stratigraphically above the Jumbo equivalent in the Marble Valley region.

The contact between the carbonate sequence in the study area and the overlying Lay Dam Formation is a major regional unconformity known as the pre–Lay Dam Formation unconformity (Tull and Stow, 1980). This unconformity appears to have erosional relief that is expressed over relatively short distances. For example, at the type locality, the Jumbo Quarry, the marble sequence is 67 m thick, while at the "cave" locality (Fig. 1) 2.4 km to the east, the marble is only 14.6 m thick (Fig. 5) This erosional surface truncates the Jumbo Dolomite completely less than 2 km to the west of the type locality. At other locations east of the type locality the Jumbo also appears to be absent in the section because the erosional surface has cut through the entire carbonate sequence, and the rocks of the Lay Dam Formation lie directly upon the Wash Creek Slate.

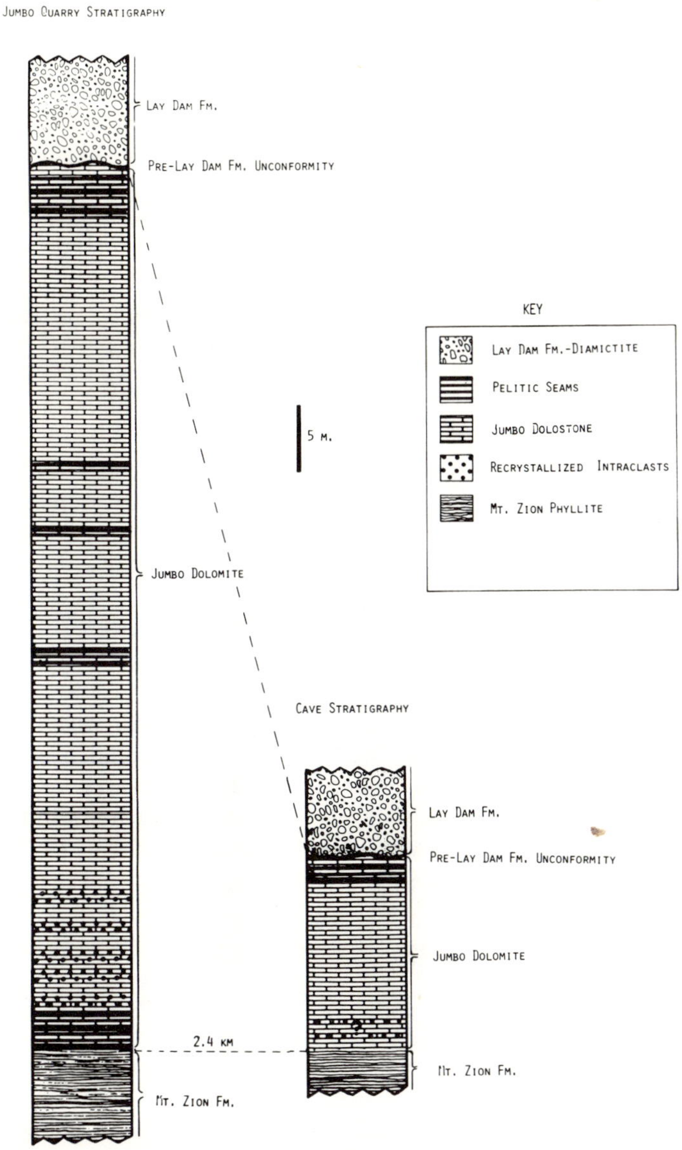

Figure 5. Stratigraphic columns illustrating the difference in thickness of the Jumbo Dolomite over a short distance.

The carbonate sequence expands to a total thickness of more than 900 m in Marble Valley (Fig. 6), and to the northeast near Sylacauga, the sequence has a thickness of 1,700 m (Shaw, 1970).

Upper Clastic Unit

The uppermost unit within the study area is the Lay Dam Formation. This formation is composed of a diamictite facies interlayered with finely laminated sericite phyllite and arkosic sandstones and ranges up to 4.5 km thick in Chilton County (Tull, 1979). The diamictite facies contains large clasts (1.5 m) of Precambrian age granitic basement rocks (Telle and others, 1979), as well as clasts of chert, carbonate, pelite, and quartzite, in a matrix of sericite, chlorite, and

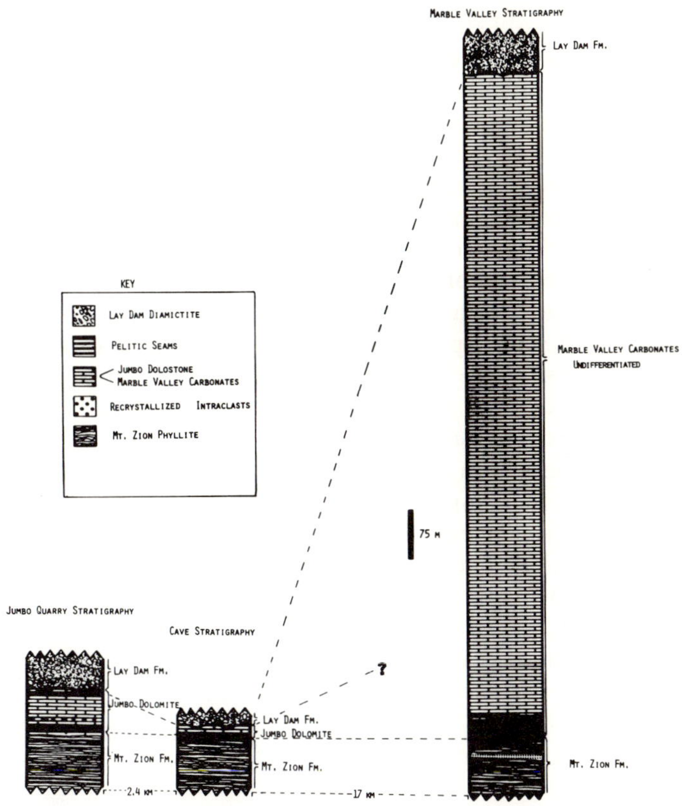

Figure 6. Stratigraphic columns showing the expansion of the carbonate sequence toward the northeast.

detrital grains of quartz, feldspar, and rock fragments. The diamictite facies appears to form lens-shaped units within the Lay Dam Formation.

DISCUSSION OF AGE RELATIONSHIPS

The age of the carbonate sequence in the Talladega slate belt has never been satisfactorily determined. Carrington (1972) reported the presence of fossil spores in the Jumbo Dolomite that indicated a Devonian age for the Jumbo. These spores were discarded before a definitive analysis could be made, and further study did not reveal additional fossil evidence for the age of the Jumbo. Other than the report of fossil spores from the Jumbo Dolomite, no fossils have been reported from the entire carbonate sequence. The chert unit in Marble Valley contains features that appear to be organic. These "fossils" have just recently been found and are currently being examined. If these features do prove to be fossils and can be confined to a narrow enough time range, then much light may be shed on the age and stratigraphic relationships of the lower part of the Talladega slate belt.

Carrington (1973) suggested that a fault (Chilton fault) emplaced the Jumbo and the underlying Mount Zion Formation over a sequence which he correlated with the Mississippian Floyd Shale. Carrington based his meta-Floyd correlation on relationships that he felt suggested that the meta-Floyd could be traced from an unmetamorphosed condition across the Cartersville-Talladega fault (zone), in Chilton and Shelby Counties, through pelitic rocks of increasing metamorphic grade.

Butts (1926) believed that the lower (northwestern) contact of the carbonate sequence was a fault contact. He mapped carbonate rocks in the Talladega slate belt north of the calcite marble sequence as windows into the carbonate rocks of the subjacent foreland fold and thrust belt. Prouty (1916) believed that the carbonate sequence was fault bounded both above and below. He did not extend the carbonate sequence as far south as Marble Valley.

The Wash Creek Slate was originally believed by Butts (1940) to be Precambrian to Paleozoic in age. Shaw (1970) suggested that the clastic units below the carbonate sequence are equivalent to the Lower Cambrian Weisner Formation.

Early to Middle Devonian macrofossils in the Jemison Chert in Chilton County (Butts, 1926), stratigraphically above the Jumbo Dolomite, provide an upper age limit to the lower part of the Talladega slate belt. The Jemison Chert, which is stratigraphically higher than the carbonate sequence, indicates that the carbonate sequence is pre-Early to Middle Devonian.

Potassium-argon age dates from the Wash Creek Slate and Lay Dam Formation in Chilton County (Kish and Harper, 1973a, 1973b) suggest that a regional metamorphic event affected the Talladega slate belt during the Early Devonian. These data also place an upper limit on the age of the sequence. Because of these dates and from the stratigraphic continuity that is believed to exist within the Talladega slate belt, the carbonate sequence in the Talladega slate belt is believed to be equivalent to part of the Cambro-Ordovician carbonate sequence in the subjacent foreland fold and thrust belt.

The rocks of the Wash Creek Slate, Sylacauga Marble Group, and the Lay Dam Formation exhibit multiple deformation features. All of the pelitic rocks contain a pervasive slaty cleavage. At least two post-foliation folding events, which are represented by mesoscopic and crenulation folds, have been recognized in the study area. Mesoscopic fabric data are illustrated in Figure 7 and show similar groupings of poles to S_1 both above and below the carbonates. These data support the hypothesis that the contacts above and below the carbonate sequence are stratigraphic, implying that no tectonic movement has taken place after the foliation was formed.

Primary sedimentary structures indicate that the stratigraphic sequence is regionally upright, but locally the section has been overturned by tight folding. The conclusion that the carbonate sequence is bounded by stratigraphic contacts differs from those of Prouty (1916) and Butts (1926), who suggested tectonic contacts either above, below,

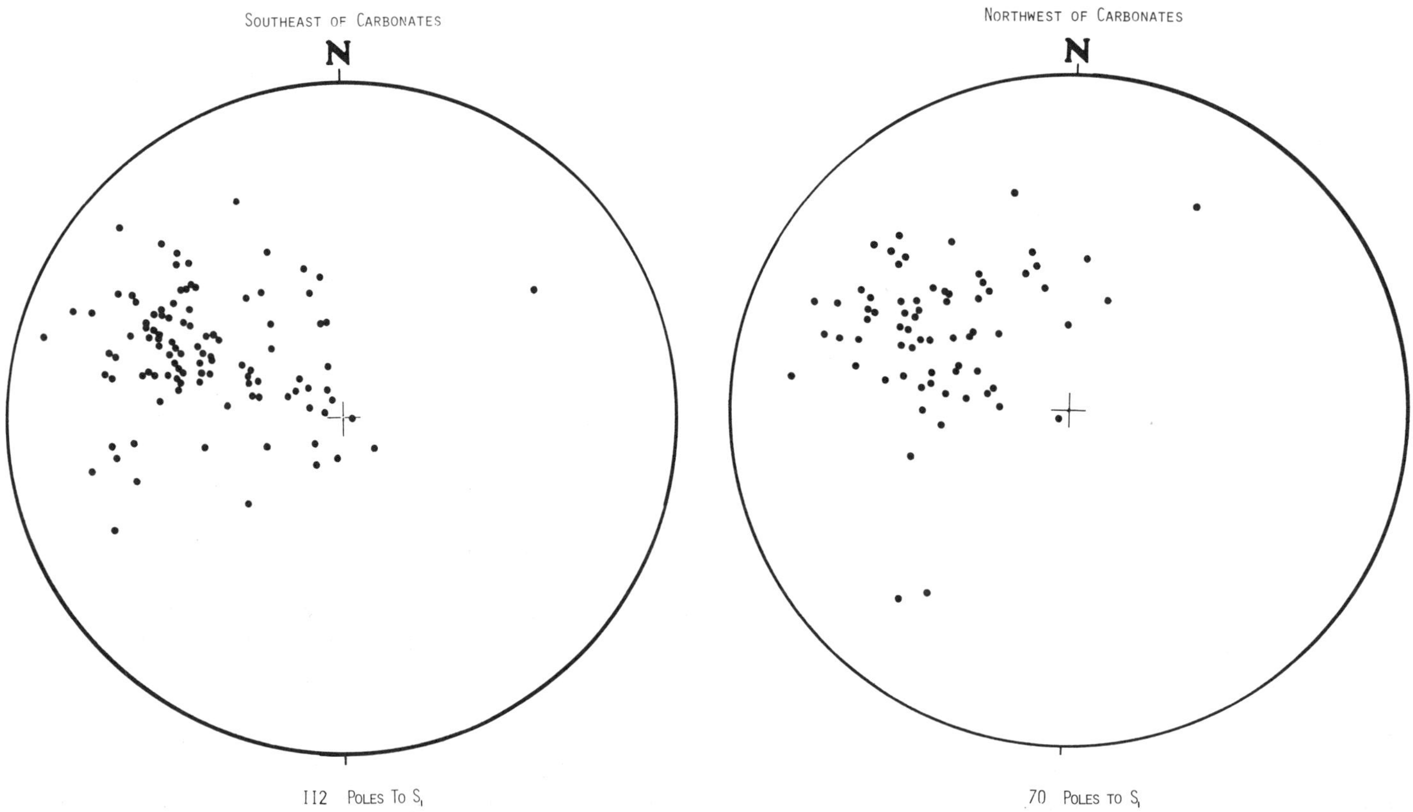

Figure 7. Structural data above and below the carbonate sequence.

or in both positions relative to the carbonate sequence. Carrington (1973) suggested that a tectonic contact occurs below his Mount Zion Formation.

SUMMARY

The carbonate sequence in the Talladega slate belt is overlain and underlain by major clastic sequences. The upper contact of the carbonates is a major regional unconformity known as the pre-Lay Dam Formation unconformity. The lower contact of the carbonate sequence is a gradational contact with the underlying Wash Creek Slate. Even though the pre-Lay Dam Formation unconformity bevels through the carbonate sequence in some areas, the carbonate sequence appears to have once been continuous along strike for at least 100 km (Fig. 1). The Sylacauga Marble Group continues southwestward from Sylacauga into the Marble Valley carbonates, which in turn continues into the Jumbo Dolomite. This implies that the Jumbo Dolomite represents a unit within the Sylacauga Marble Group. It is most probably at or near the base of the Sylacauga Marble Group.

Field relationships indicate no regional tectonic contact at the base or the top of the carbonate sequence in this area. There also appears to be no tectonic contact within at least the upper 75 m of the Wash Creek Slate in the study area.

The Middle to Early Devonian macrofossils found in the Jemison Chert and the Devonian potassium-argon whole rock ages from the slates in the Lay Dam Formation and the Wash Creek Slate suggest a pre-Devonian age for the carbonate sequence.

The Jumbo Dolomite contains rounded quartz grains and intraclasts, both of which indicate a shallow water depositional environment for this carbonate rock. The Lay Dam Formation appears to have been deposited as many separate subaqueous debris flows. The rock fragment suite found in the diamictites contain Grenville basement and Paleozoic cover sequence fragments (Telle and others, 1979). The source area for the rock fragments must have contained basement as well as cover rocks and must have been subjected to very high energy erosion, which would account for the large (1.5 m) boulders of basement and cover rock found within the diamictite.

If the Sylacauga-Marble Valley-Jumbo carbonate sequence is equivalent to part of the Cambro-Ordovician carbonate sequence found within the foreland fold and thrust belt, then the Talladega slate belt contains what appears to be a far-traveled eastern distal segment of the carbonate bank that developed along the trailing margin of the proto-North American continent. The clastic units below the carbonates in the Talladega slate belt would then be equivalent to the upper Precambrian to Lower Cambrian clastic rocks of the foreland fold and thrust belt and Blue Ridge belt, as originally suggested by Butts (1940) and Shaw (1970). Above the carbonate sequence, the pre-Lay Dam Formation unconformity, which bevels across the carbonates, may mark the end of the trailing margin tectonic environment and the beginning of a convergent margin tectonic environment.

ACKNOWLEDGMENTS

Partial support for this study was provided by the Geological Survey of Alabama and the Alabama Academy of Science. I appreciate the advice and assistance of J. F. Tull throughout this study. I also thank the other professors, graduate students, and staff of the Department of Geology at the University of Alabama in Tuscaloosa, Alabama. Assistance was also provided by D. N. Bearce at the University of Alabama in Birmingham.

REFERENCES CITED

Butts, Charles, 1926, The Paleozoic rocks, *in* Adams, G. I., and others, Geology of Alabama: Alabama Geological Survey Special Report 14, p. 41–230.

——1940, Description of the Montevallo and Columbiana quadrangles, Alabama: U.S. Geological Survey Atlas, Folio 226, 22 p.

Carrington, T. J., 1972, Meta-Paleozoic rocks, Chilton County, Alabama, *in* Tolson, J. S., ed., Guide to Alabama geology: Geological Society of America, Southeastern Section, 21st Annual Meeting, Guidebook, p. 1-1 to 1-25.

——1973, Metamorphosed Paleozoic sedimentary rocks in Chilton, Shelby, and Talladega Counties, Alabama, *in* Carrington, T. J., ed., Talladega metamorphic front: Alabama Geological Society, 11th Annual Field Trip, Guidebook, p. 22–38.

Kish, S. A., and Harper, C. T., 1973a, Potassium-argon geochronology of a portion of the southwestern Blue Ridge [abs.]: Geological Society of American, Abstracts with Programs, v. 5, p. 409.

Kish, S. A., and Harper, C. T., 1973b, unpublished open file report, *in* Copeland, C. W., and Beg, M. A., 1979, Geology of the Birmingham, Gadsden, and Montgomery 1° X 2° NTMS Quadrangles, Alabama: Alabama Geological Survey.

McCalley, Henry, 1897, Report on the valley regions of Alabama: Part 2, The Coosa Valley region: Alabama Geological Survey Special Report 9, 826 p.

Prouty, W. F., 1916, Preliminary report of the crystalline and other marbles of Alabama: Alabama Geological Survey Bulletin 18, 212 p.

Shaw, C. E., 1970, Age and stratigraphic relations of the Talladega slate: Evidence of pre-Middle Ordovician tectonism in central Alabama: Southeastern Geology, v. 11, no. 4, p. 253–267.

Telle, W. R., and others, 1979, Tectonic significance of the bouldery facies of the Lay Dam Formation, Talladega slate belt, Chilton County, Alabama: Geological Society of America, Abstracts with Programs, v. 11, p. 215.

Tull, J. F., 1978, Structural development of the Alabama Piedmont northwest of the Brevard zone: American Journal of Science, v. 278, p. 442–460.

——1979, Stratigraphic and structural relationships of the eastern Talladega slate belt of Alabama, *in* Tull, J. F., and Stow, S. H., eds., The Hillabee metavolcanic complex and associated rock sequences: Alabama Geological Society, 17th Annual Field Trip, Guidebook, p. 3–13.

——1982, Stratigraphic framework of the Talladega slate belt, Alabama Appalachians: Geological Society of America Special Paper 191 (this volume).

Tull, J. F., and Stow, S. H., 1980, Structural and stratigraphic setting of arc volcanism in the Talladega slate belt of Alabama, *in* Frey, R. W., ed., Excursions in southeastern geology: American Geological Institute, Geological Society of America Field Trip, v. 2, p. 543–569.

MANUSCRIPT ACCEPTED BY THE SOCIETY MAY 7, 1982

Printed in U.S.A.

Geological Society of America
Special Paper 191
1982

A reinvestigation of the paleobotanical evidence for the age of the Erin Shale

Robert A. Gastaldo
Robert B. Cook
Department of Geology
Auburn University, Alabama 36849

ABSTRACT

The age of the Erin Shale, Clay County, Alabama, has been variously interpreted based primarily upon the acceptance or rejection of the original Carboniferous age assignment. The failure to confirm the in-place occurrence of the reported Pennsylvanian megafossils, coupled with both regional and detailed mapping, has led to the recent placement of the Erin Shale in a stratigraphically equivalent position with the Lay Dam Formation of inferred Early Devonian age. An investigation of the previously illustrated Carboniferous fossils reportedly collected from the Erin Shale and suites of rock specimens collected during the present study has added new pieces to the Erin puzzle but has not resolved the age asignment problem. Petrographic and x-ray diffraction powder analyses of the adhering matrix on *Lepidostrobus hobbsii* D. White and phyllites collected from the Erin outcrop belt indicate that these rocks are similar if not indistinguishable. Identifiable coal fragments exist in the Erin as do poorly preserved, unassignable fossil plant fragments. Much additional work is needed to fully define the fossiliferous nature of the Erin.

INTRODUCTION

This paper is intended to review the known facts relative to one of the more historically interesting problems in Alabama geology and to present new data based on the reexamination of fossil material originally attributed to the Erin Shale and the collection of new rock sample suites. This work has not resolved the critical questions pertaining to the fossiliferous nature of the Erin but is presented in the hope that continued interest in the Erin problem will ultimately result in a thorough understanding of the nature of the Erin.

In the vicinity of its type location the Erin Shale is a fine-grained carbonaceous quartz phyllite generally occupying the valley of Talladega Creek along the eastern base of the Talladega Mountains, Clay County, Alabama. The unit is an enigma due to historically conflicting interpretations of its age and structural relationship to the Talladega slate belt. The problem is significantly more complex if lithologically similar and apparently stratigraphically equivalent units both to the northeast and southwest are correlative with the Erin.

The Erin problem is the result of the submission to the Alabama Geological Survey by Joshua Franklin of unusually well preserved, unquestionable Carboniferous lycopod fossils reported collected near the village of Erin (Smith, 1903). Subsequent workers for at least three decades were unable to collect additional fossil material equal to that of the original two samples but continued to regard all or part of the Erin sedimentary package as Carboniferous (Prouty, 1922, 1923; Buttes, *in* Adams and others, 1926; Brown, 1925; Jones, 1932; Park, 1935). However, recent mapping and interpretation within the Talladega belt as a whole suggest that the Erin is no younger than Early Devonian in age (Tull and Stow, 1980b).

Originally, the Erin Shale was considered as stratigraphically enclosed within the Talladega slate belt and was described by Smith (1903) as a lenticular mass equivalent to the Ocoee supergroup. Butts (*in* Adams and others, 1926) formally named the phyllitic rock sequence the Erin Shale and considered it an integral part of the upper Talladega Slate. Hays and White (*in* McCaskey, 1908) thought the Erin to be unconformable with the Talladega Slate. Jonas (1932) and Park (1935) proposed that these phyllites were exposed in a window through the Talladega. This latter interpretation provided a somewhat difficult-to-demonstrate explanation for the Carboniferous age assignment for the Erin within geologically older units. In addition, the use of a window model allowed the assumption that the Erin was originally deposited as a part of the Talladega assemblage to be discarded (Park, 1935).

Renewed interest in the Talladega slate belt since the late 1960s has resulted in various structural interpretations based, in part, upon regional tectonic models (Bearce, 1973; Carrington, 1973; Tull, 1978; Tull and Stow, 1980a, 1980b). The presence of Devonian (Emsian) fossils within the Jemison Chert of the Talladega slate belt (Butts, *in* Adams and others, 1926; Boucot, 1972, written commun.) and Mississippian and Pennsylvanian spores in low-rank metasediments within fault blocks northwest of the Talladega block (Carrington, 1973), and the results of regional mapping projects coupled with the failure to reestablish the presence of Pennsylvanian megafossils within the Erin Shale led to the rejection of the Carboniferous age assignment (Tull and Stow, 1980a, 1980b). In order to rectify the correlation problem, the Erin was placed in a stratigraphic position equivalent with the Lay Dam Formation of inferred Early Devonian age (Tull and Stow, 1980b).

The nagging question relative to the early published reports of Erin megafossils led to the reexamination of the best of the assumed original samples and to the collection of new suites of rock samples from railroad cuts (Fig. 1) and other exposures in the immediate Erin vicinity. These samples were subjected to paleobotanical, petrographic, and x-ray diffraction analyses. The results of this work indicate, but do not prove, that the original samples could have been collected at Erin and that currently collectable material contains identifiable coal fragments and poorly preserved stem and/or leaf fragments possibly attributable to *Cyperites*. Additionally, what may be poorly preserved allochthonous stem casts similar to material figured by Prouty (1923) and described by Smith (1903), Prouty (1922), and Butts (*in* Adams and others, 1926) as stems of Lepidodendrids, *Calamites* and *Artisia* can be collected from the railroad cut three-fourths of a mile southwest of Erin (Fig. 1).

THE ERIN SHALE

In the immediate Erin vicinity, the slate is poorly exposed except in railroad and locally restricted highway cuts. The unit is fine-grained. Typically, the only megascopically identifiable minerals are quartz, a white mica, and black carbonaceous matter generally thought to be graphite. In the best exposures the unit is a thinly foliated, variably carbonaceous phyllite with a dominant slaty cleavage. Siliceous boudins are locally abundant as are hollow, nodular silica masses containing pyrite, variscite, wavellite, and turquoise. A distinctive feature of the Erin is the bleaching and/or loss of carbonaceous matter upon weathering. Thus, beds of obviously high carbon content in deep exposures grade upward into a buff phyllite at outcrop surface or in saprolite.

Weakly deformed carbonaceous concretions occur locally within the phyllite exposed in the railroad cut south of Erin (Fig. 2). These are commonly up to 5 cm in diameter, phyllite encased, and appear as structures within the foliation. The interior of these concretions is usually composed of a solid mass of carbonaceous material of unknown origin. The carbonaceous masses are strongly anisotropic with reflectances varying between R_{max}4.6 and R_{min}2.8 (Ralph J. Gray, 1981, personal commun.), and contain substantial intergranular variscite. Smith (1903) noted the same general conditions for the concretions and attributed the highly coalified nuclei to badly preserved organic forms.

Examination of thin sections cut perpendicular to foliation of the phyllite matrix adhering to one of the unusually well preserved, assumed original samples (Fig. 9) indicates a metasiltite source for the original fossils (Figs. 13, 14). Comparison of these thin sections with those prepared from Erin Shale collected around the village of Erin shows that both the matrix and the Erin Shale are composed of approximatley 90 percent quarz, 9 percent muscovite, and accessory biotite(?), plagioclase, carbonaceous matter, and opaque minerals. Quartz occurs predominantly as mosaics of interlocking grains averaging 0.01 mm in diameter. Undulatory extinction and inclusions of muscovite and opaques are locally prevalent. Narrow zones of increased grain size have developed parallel to foliation by recrystallization. Other zones contain rounded, flattened, recrystallized quartz grains of apparently clastic origin. Maximum observed dimensions of these grains are 0.1 by 0.4 mm.

Phyllitic texture is well developed by parallel aggregates of muscovite 0.01 mm thick by as much as 1 mm long. These aggregates exhibit uniform optical orientation within individual thin sections, are stained by the oxidation products of extremely fine-grained pyrite, and are commonly warped or wrapped around the larger clastic quartz grains. Oriented muscovite aggregates at the longitudinal ends of larger clastic quartz grains are locally suggestive of pressure shadows. Fractured clastic quartz grains are locally veined by muscovite.

Comparison of x-ray diffractograms of the fossil matrix phyllite with known Erin Shale collected at the railroad cut and along the reported (Prouty, 1923) strike continuation of the fossil-bearing zone (NW¼ Sec. 22 and SW¼ Sec. 15, T19S, R7E) confirms the above primary mineralogy, and the diffractograms are essentially indistinguishable. Muscovite exhibits preferred orientation in all mounts with minor peaks suggesting a dominance of the $2M_1$ polymorph.

Polished sections prepared from the most carbonaceous portions of the Erin exposed in the railroad cut locality contain well-preserved, angular, medium- to low-volatile bituminous coal fragments exhibiting reflectances of $R_{max}1.63$ and $R_{min}1.03$ (Ralph Gray, 1981, personal commun.). Exinite bands containing possible spore-derived forms compose a part of this coal.

X-ray diffraction analysis of carbonaceous residue extracted from "graphite-rich" zones within the Erin indicates a very low degree of crystallinity with diffuse, offset (002) reflections at 3.44 (Å) and weak (001) reflections at 2.11 (Å). Comparison of these data with published parameters for carbon crystallinity from graphitic rocks of known metamorphic grade (Grew, 1974; Bluman and others, 1974) suggests that the Erin has undergone regional metamorphism at conditions that do not exceed the chlorite zone of Barrovian metamorphism.

PLANT FOSSILS

Lepidostrobus hobbsii D. White

The two specimens of lycopsid strobili reported by Smith (1903), figured by Prouty (1922, 1923), and named by David White (*in* Adams and others, 1926) as *Lepidostrobus hobbsii* are incomplete fructifications (Figs. 9, 10, 11). The strobili are preserved by extensive silicification of the axes (Fig. 12), sporophylls, and sporangial walls, and the sporangia contain aggregates of fine-grained muscovite with locally abundant carbonaceous material. The quartz is fine-grained with crystals attaining a maximum dimaeter of 1 mm, although smaller patches of unrecrystallized chalcedonic silica remain in sporangial walls. Both cones have been compressed in the transverse plane.

The longer and more compressed strobilar fragment originally believed to be a *Lepidodendron* stem (Prouty, 1922, Pl. 23, Fig. C; 1923, Pl. 1, Fig. C) is not as well preserved as the other specimen (Fig. 9). The distal and proximal parts of the cone, as well as the majority of laminar portions of the sporophylls, are not preserved. The strobilar axis is extensively silicified and there is no trace of internal detail. The pedicels and sporangial walls are discernible and spirally arranged around the axis.

The specimen is 8.7 cm in overall length, 5.2 cm in width as measured parallel to the plane of compression, and 3.0 cm in the plane perpendicular to the plane of compression. The axial diamerter is 6 mm in the plane perpendicular to compression and 8 mm in the plane of compression. The pedicel part of the sporophyll is attached at a right angle to the axis and is at least 1.6 cm in length. The lack of its exposure and possibly preservation precludes a description of the transitory zone between the pedicel and the lamina. In transverse section the distal part of the pedicel is triangular in shape, whereas the more proximal parts of the pedicel are slightly rhomboidal. The adaxial side of the pedicel is not as accentuated as the abaxial side. The pedicel may be slightly alate, but the wings do not appear to be upturned surrounding the sporangium. The distal pedicel measures at least 5 mm in width and 4 mm in height; the more proximal protion measures 5 mm in width and 3 mm in height. A single, terete vascular strand is positioned beneath the adaxial surface.

The laminar portions of the preserved sporophylls are fragmentary and do not overlap the supraadjacent sporophylls; therefore it is impossible to determine their orignal length. The lamina are at least 6 mm in width and posess a single vascular strand.

Sporangia are attached along the adaxial surface of the pedicel possibly by an elongate ridge which would be the adaxial side of the rhomboidal-shaped pedicel. The sporangia are large, elongate, and measure at least 12 mm in length, 5 mm in width,and 4 mm in height. In transverse section they appear cordate with the rounded arches of the heart-shaped sporangium adjacent to the pedicel. To date, spores have not been recovered from this specimen.

The smaller and least compressed specimen of *L. hobbsii* is the proximal part of a strobilus with attached peduncle (Figs. 10, 11). The specimen is well preserved and anatomical detail of the axis is discernible, as well as features of the sporophylls and sporangia. The strobilar fragment is 6.0 cm in length, 5.2 cm in width as measured radially and parallel to the plane of compression, and 4.4 cm in width as measured in a plane perpendicular to the plane of compression. The pedicel is attached at a right angle to the strobilar axis which measures 8 mm in diameter as measured in the plane of compression. A single vascular trace continues from the pedicel to the lamina, and the sporangium is adaxial to the pedicel. In transverse section the lamina is diamond-shaped with a centrally located terete vascular strand. The Sporangia in transverse section are elongate and measure at least 1.5 cm in length and 3 mm in width. The sporangia appear to preserve numerous, small megaspores less than 1 mm in diameter, the identity of which, to date, is unknown.

Duripartic Stem Casts

In addition to the specimens of *Lepidostrobus,* Smith (1903) described the collection of what he considered to be pieces of stems of *Lepidodendron, Calamites* and *Artisia.* These were illustrated by Prouty (1923, Pl. 2). Butts (*in* Adams and others, 1926) stated that most of the specimens are *Calamites* and *Lepidodendron* and referred to the strobili figured in Plate 70, but did not illustrate the stem fragments. A reexamination of the specimens curated at the

Figure 1. Erin Shale railroad cut (NE¼, NW¼, Sec. 28, T19S, R7E; Clairmont Springs 7½ minute quadrangle). Locations from which suites of magafloral remains have been collected are marked by the arrows.

Figure 2. *In situ* phyllitic concretion with coalified nucleus (boundary between coalification and phyllite at arrow). The concretions are lenticular in transverse section, oval to elliptical in polar view, and are lighter in specific gravity than the surrounding matrix.

Figure 3. *In situ* possible allochthonous stem casts in transverse section. The casts are typically lenticular to fusiform in outline and, for the most part, permineralized by silica. No internal structure is preserved.

Figure 4. Surface view of possible allochthonous flattened stem cast in the phyllitic matrix. The stem casts are commonly smooth and do not possess any of the foliation features characteristic of the surrounding phyllite.

Figure 5. Possible allochthonous flattened stem cast with adhering phyllitic matrix (scale in mm).

Figure 6. Liberated possible allochthonous stem cast showing some textural features. The textural features do not provide sufficient detail to warrant their assignment to any generic status (scale in mm).

1
2
4
5
3
6

Figure 7. Fragmentary coalified compression of unidentifiable plant material cross cut by crenulations of the phyllitic matrix (scale in mm).

Figure 8. Fragmentary coalified compression tentatively assignable to *Cyperites* Lindley and Hutton. Cellular detail is well preserved and appears as longitudinal striations on the specimen. The coalified compression is cross cut by the crenulations of the matrix.

Figure 9. *Lepidostrobus hobbsii* D. White. Permineralized lycopod cone originally reported from the Erin Shale by Smith (1903). The spirally arranged sporophylls appear whorled due to the tight helical arrangement. Pedicels and sporangia are preserved, but laminar units are not preserved. Adhering phyllite matrix (at arrow) is a metasiltite and is essentially indistinguishable from phyllites analyzed from the Erin Shale outcrop belt (scale in mm).

Figure 10. Transverse section of permineralized *Lepidostrobus hobbsii* D. White. The central strobilus axis is well preserved and cellular detail is recognizable. Spirally arranged pedicels and sporangia emerge from the strobilus axis.

Figure 11. Side view of *Lepidostrobus hobbsii* D. White (scale in mm).

Figure 12. Mosaic of crystalline quartz typical of silicified portion of strobilar axis of *Lepidostrobus hobbsii* (Fig. 9). Maximum grain diameter 0.9 mm. Crossed nicols.

Figure 13. Thin section micrograph of metasiltite matrix adhering to *Lepidostrobus hobbsii* (Fig. 9). Dark streaks are iron-stained aggregates of locally graphitic muscovite. Note lensoid quartz grains of probable clastic origin. Maximum grain diameter 0.1 mm. Plane polarized light.

Figure 14. Thin section micrograph of metasiltite matrix adhering to *Lepidostrobus hobbsii* (Fig. 9). Dark streaks are iron-stained aggregates of locally graphitic muscovite. Note lensoid quartz grains of probable clastic origin. Maximum grain diameter 0.1 mm. Crossed nicols.

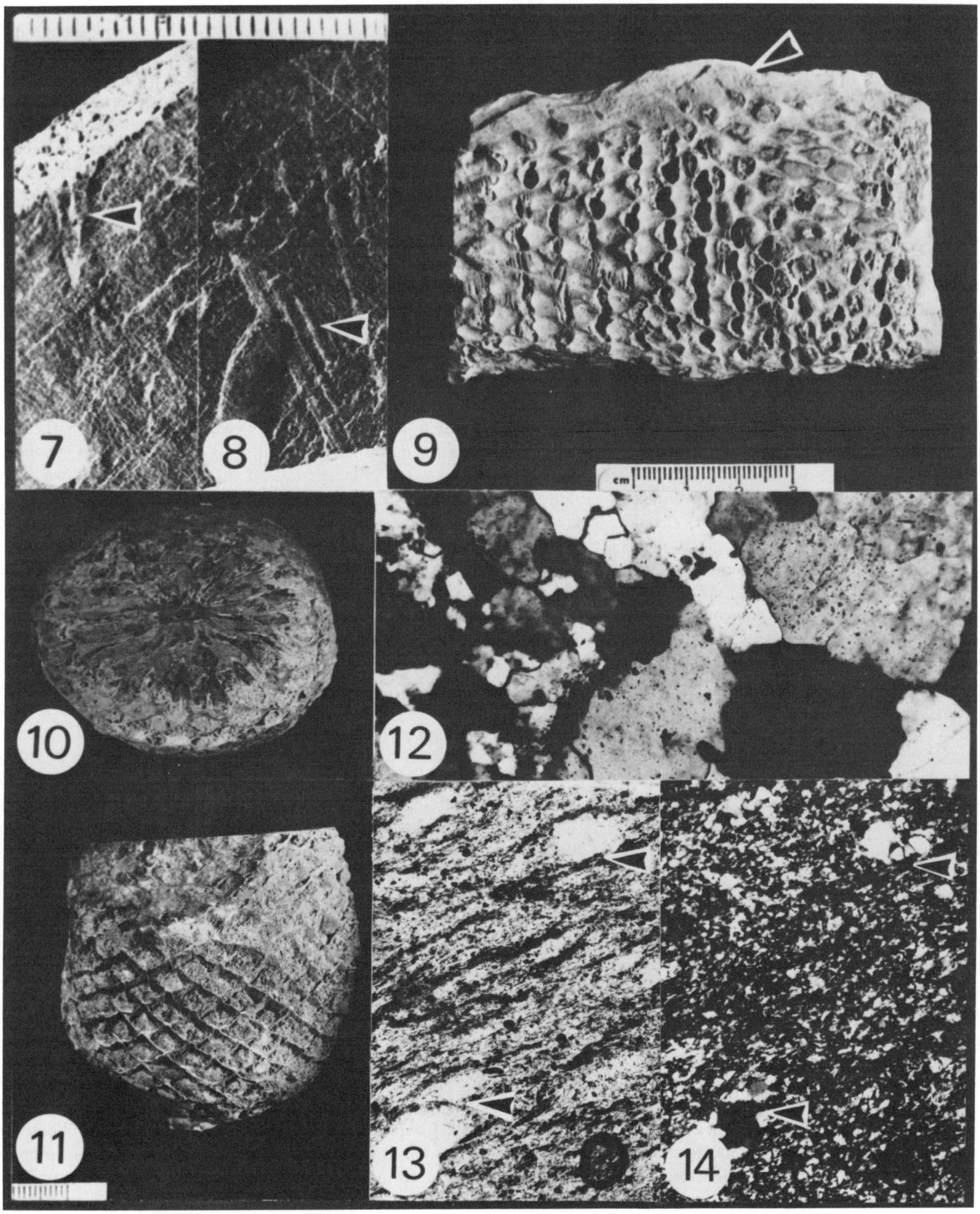
7
8
9
cm
10
12
11
13
14

Alabama Geological Survey and recent collections made from the outcrop suggest that at least some of the elongate, parallel-sided, elliptical in transverse section fragments may be stem casts of unidentifiable character (Figs. 3, 4, 5, 6). Most of the fragments examined have a carbonaceous layer surrounding the surfaces of the cast. These casts are preserved in various orientations within the matrix, a disoriented taphocenosis, and when collected the phyllite matrix normally breaks along the area of weakness associated with the cast. The casts are smooth or slightly textured and compressed into a fusiform cross section in a plane parallel to the compositional layering. The casts are primarily composed of silica, and the grain size of the original casting minerals is larger than the enclosing matrix. In addition, the casts do not possess any of the foliation features characteristic of the surrounding matrix.

Coalified Compression Debris

The phyllitic matrix locally contains coalified fragments of small stems (rachises?) and/or leaves (Fig. 7, 8). These compressed fragments are found in random orientation on foliation (bedding?) planes. The specimen in Figure 8 appears to be a fragment of a lycopsid vegetative leaf referable to *Cyperites* Lindley and Hutton. *Cyperites* is characterized as being a long, linear, univeined microphyll that may obtain a width of about 1 cm. Although the specimen is fragmentary, the apparent univeined character and preserved epidermal features of the coalified compression allow this tentative taxonomic assignment.

DISCUSSION

The recognition of the Erin Shale as a coal-bearing fossiliferous formation is significant in that this observation could allow for the acceptance or denial of the original Carboniferous age contention of Smith (1903). Unfortunately, the Erin Shale to date has been found to contain a poorly preserved allochthonous mixed taphocenosis which cannot be attributed to a specific age. Although the rocks are metamorphosed, the degree of metamorphism is slight and does not exceed the chlorite zone of Barrovian regional metamorphism as indicated by carbon crystallinity and coal reflectance. This low temperature deformation has allowed apparently for the partial preservation of the plant debris originally deposited with the sediments.

The elongated, parallel sided, elliptical in cross section structures believed to be stems by Smith (1903) and Prouty (1922, 1923) may be in part boundinage structures formed by the deformation of siliceous zones in a more plastic and argillaceous rock. We believe, at the present, that both boudinage structures and duripartic stem casts may exist in the Erin, since a portion of these structures are oriented in several planes and are scattered throughout the outcrop in various spatial configurations. These possible casts are not parallel to each other and are divergent, that is, they may be oriented in various alignments in the same area and in different planes. In addition, some casts possess a coalified outer surface that encompasses the entire circumference of the specimen. These specimens may be internal casts with the organic bark retained. The elliptical transverse shape of such casts can be explained by diagenesis and has been demonstrated to be a result of compression (Walton, 1936).

The homology in mineral composition of the adhering phyllite matrix on the *L. hobbsii* specimen and the phyllites of the Erin Shale suggests that the original samples described by Smith (1903) could have been collected in the Erin area. The acceptance of the specimens as originating from the Erin would, of course, allow for the acceptance of a carboniferous age assignment. Unfortunately, it now appears that the origin of these samples may never be resolved prior to the in-place discovery of essentially identical material.

Much paleobotanical study has ensued since the age assignment by D. White and, to date, the oldest lycopods with sporophylls aggregated into strobili are known from the Upper Mississippian. Devonian lycopods are herbaceous plants as small trees with fertile sporophylls interspersed among the vegetative leaves (Taylor, 1981). The development in the lycopods of compact strobili occurs sometime in the lower Carboniferous and is first reported from the Lower Mississippian (Phillips, 1979). *Lepidostrobus* has been reported from the Mississippian of Arkansas (Taylor and Eggert, 1968; White, 1937a, 1937b), Kentucky (Scott and Jeffrey, 1914; Mathews, 1940), Pennsylvania (Arnold, 1933), and Nova Scotia (Bell, 1944). To date, no known lycopsid strobili have been reported from the Devonian. This level of strobilar development is, therefore, uniquely Carboniferous, and the presence of *Lepidostrobus* in the Erin Shale would allow an assignment of at least Mississippian age to the formation.

It is evident that additional data are needed to satisfactorily resolve questions as to the fossiliferous nature of the Erin. Detailed palynological study of the "fossiliferous zones" within the Erin as determined by Prouty (1922) is in order as is the continued study of the material available in currently exposed portions of the Erin throughout its strike belt.

ACKNOWLEDGMENTS

We thank the Alabama Geological Survey and Charles Copeland for the loan of the specimens of *Lepidostrobus hobbsii* D. White and permission for preparation of the adhering phyllitic matrix. We also thank Ralph Gray, U.S. Steel Corporation, for his assistance with the reflectivity values for the carbonaceous nodules and coal fragments discovered in the Erin Shale. Acknowledgment is made to the donors of The Petroleum Research Fund, administered by the American Chemical Society, for partial support of this research.

REFERENCES CITED

Adams, G. I., Butts, C., Stephenson, L. W., and Cook, C. W., 1926, Geology of Alabama: Alabama Geological Survey Special Report 14, p. 208–230.

Arnold, C. A., 1933, A lycopodiaceous strobilus from the Pocono Sandstone of Pennsylvania: American Journal of Botany, v. 20, p. 114–117.

Bearce, D. N., 1973, Geology of the Talladega metamorphic belt in Cleburne and Calhoun counties, Alabama: American Journal of Science, v. 273, p. 742–752.

Bell, W. A., 1944, Carboniferous rocks and fossil floras of nothern Nova Scotia: Geological Survey of Canada Memoir 238, p. 1–276.

Bluman, B. A., Dyakonov, Yu. S. Krasavina, T. N., and Pavlov, M. G., 1974, Use of the thermal and x-ray diffraction characteristics of graphite for determining the facies and type of metamorphism: Zapiski Vsesoyugnogo Mineralogicheskogo Obshchestva, v. 103, p. 95–103.

Brown, J. A., 1925, Graphite deposits of Ashland, Alabama: Economic Geology, v. 20, p. 208–229.

Carrington, T. J., ed., 1973, Talladega metamorphic front: Alabama Geological Society, 11th Annual Field Trip, Guidebook, 122 p.

Grew, E. S., 1974, Carbonaceous material in some metamorphic rocks of New England and other areas: Journal of Geology, v. 82, p. 50–73.

Jonas, A. I., 1932, Structure of the metamorphic belt of the southern Appalachians: American Journal of Science, v. 244, p. 238–243.

Mathews, G. B., 1940, New lepidostrobi from central United States: Botanical Gazette, v. 102, p. 26–49.

McCaskey, H. D., 1908, Some gold deposits of Alabama: U.S. Geological Survey Bulletin, v. 340, p. 38.

Park, C. F., Jr., 1935, Notes on structure of the Erin Shale of Alabama: Journal of the Washington Academy of Sciences, v. 25, p. 276–279.

Phillips, T. L., 1979, Reproduction of heterosporous arborescent lycopods in the Mississippian-Pennsylvanian of Euramerica: Review of Palaeobotany and Palynology, v. 27, p. 239–289.

Prouty, W. E., 1922, Age of Talladega slates of Alabama: Pan American Geologist, v. 37 (5), p. 363–366.

——1923, Geology and mineral resources of Clay County: Alabama Geological Survey, County Report 1, 190 p.

Scott, D. H., and Jeffrey, E. C., 1914, On fossil plants, showing structure, from the base of the Waverly Shale of Kentucky: Royal Society of London Philosophical Transactions, B. 205, p. 315–373.

Smith, E. A., 1903, Carboniferous fossils in Ocoee slates in Alabama: Science, v. 18, p. 244–246.

Taylor, T. N., 1981, Paleobotany. An introduction to fossil plant biology: New York, McGraw-Hill, 588 p.

Taylor, T. N. and Eggert, D. A., 1968, Petrified plants from the Upper Mississippian of North America. II. *Lepidostrobus fayettevillense* sp. n.: American Journal of Botany, v. 55 (3), p. 306–313.

Tull, J. F., 1978, Structural development of the Alabama Piedmont northwest of the Brevard zone: American Journal of Science, v. 278, p. 442–460.

Tull, J. F., and Stow, S. H., 1980a, The Hillabee Greenstone: A mafic volcanic complex in the Appalachian Piedmont of Alabama: Geological Society of America Bulletin, v. 91, p. 27–36.

——1980b, Structural and stratigraphic setting of arc volcanism in the Talladega slate belt of Alabama, *in* Frey, R. W., ed., Excursions in southern geology: American Geological Institute, Guidebook for Geological Society of America 1980 Annual Meetings, v. 2, p. 545–581.

Walton, J., 1936, On the factors that influence the external form of fossil plants, with descriptions of the foliage of some species of the Paleozoic equisetalean genus *Annularia* Sternberg: Royal Society of London Philosophical Transactions, B. 226.

White, D., 1937a, Fossil flora of the Wedington Sandstone Member of the Fayetteville Shale: U.S. Geological Survey Professional Paper 186-B.

——1937b, Fossil plants from the Stanley Shale and Jackfork Sandstone in southeastern Oklahoma and western Arkansas: U.S. Geological Survey Professional Paper 186-C.

MANUSCRIPT ACCEPTED BY THE SOCIETY MAY 7, 1982

Printed in U.S.A.

Geological Society of America
Special Paper 191
1982

Igneous petrology of the Hillabee Greenstone, Northern Alabama Piedmont

Stephen H. Stow
*P.O. Box X, Oak Ridge National Laboratory**
Oak Ridge, Tennessee 37830

ABSTRACT

The Hillabee Greenstone is a mafic metavolcanic sequence at the stratigraphic top of the Talladega Group in the Northern Alabama Piedmont. The Hollins Line fault forms the upper contact of the Hillabee, and an undetermined amount of the Greenstone has been removed by faulting. The igneous protolith consisted of tholeiitic rocks (ash and lava) with minor calc-alkaline dacites. Volcanism was arc-related, based on geochemical evidence and the associated lithologies. Greenschist facies metamorphism and alteration have resulted in changes in the chemistry and mineralogy of the rocks. However, a variety of remnant igneous textures can be observed, and through statistical means, samples that have experienced minimal alteration can be identified, thereby allowing some of the igneous evolution of the Hillabee to be deduced. The chemistry and mineralogy (norm and mode) indicate tholeiitic fractionation representative of relatively shallow crustal depths with typical enrichment in Fe, V, Ti, and quartz and depletion of Mg, Cr, Ni, Co, and olivine. The dacites are associated with more highly fractionated basalts but do not appear to be a direct product of fractionation from the basalts. Some mafic rocks are typically gabbroic in texture and are interpreted to represent lava flows. Geochemical and field evidence indicates that the exposed Hillabee Greenstone represents an immature arc; plagioclase and olivine were highly fractionated prior to extrusion, and early stages of fractionation are not represented in surface exposures. Temporally earlier volcanics may occur downdip to the southeast toward the volcanic source; the exposed Hillabee Greenstone represents volcanics deposited somewhat distally from the volcanic source on the edge of the basin.

INTRODUCTION

The Talladega slate belt, a sequence of lower greenschist grade metamorphic rocks, occurs in the Northern Alabama Piedmont. At the top of this belt is a metavolcanic unit, the Hillabee Greenstone, which outcrops discontinuously for 170 km along strike (Fig. 1). Tull and Stow (1980a) have previously discussed the geologic setting and general geochemistry of the Hillabee Greenstone.

The Hillabee Greenstone represents a series of tholeiitic basalts and minor, but significant, quartz dacites that were extruded along an actively developing continental margin (Andean-type arc system). The volcanism, which was interpreted by Tull (1979) and Tull and Stow (1980a, 1980b) to be Devonian in age, spread from the southeast over marine sediments, which now compose the bulk of the Talladega Group. Thus, the lower Hillabee contact is stratigraphic and comformable; the sequence is considered to be upright based on field observations. Late faulting resulted in tectonic removal of an unknown amount from the top of the Greenstone and, thus, the present discontinuous and highly variable outcrop pattern (Fig. 1). Medium to high

*Operated by Union Carbide Corporation under contract W-7405-eng-26 with the U.S. Department of Energy.

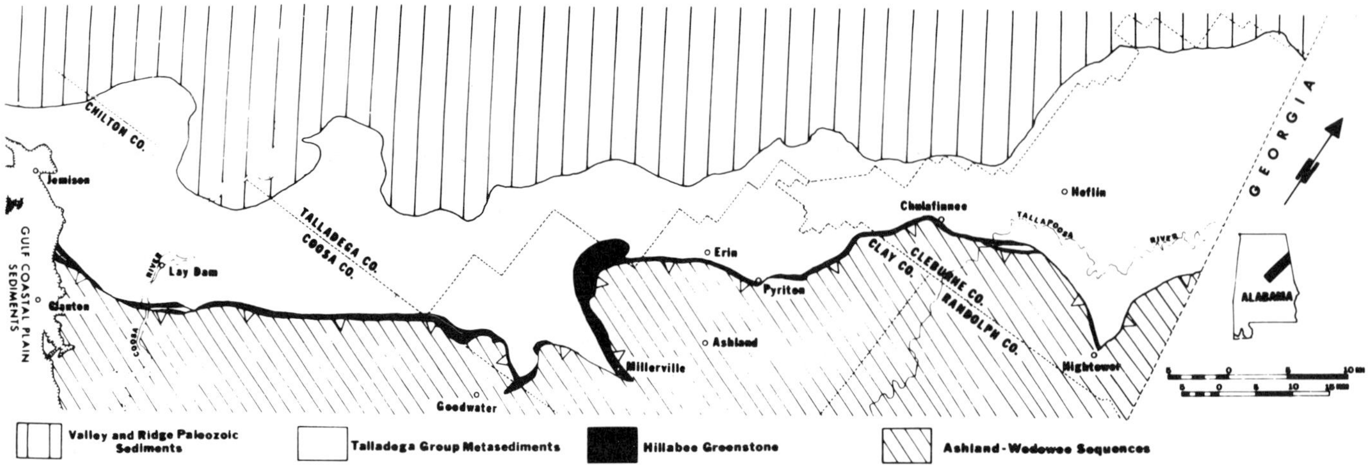

Figure 1. Location map of the Hillabee Greenstone.

metamorphic-grade rocks—predominantly metasedimentary Ashland-Wedowee sequences—occur above the Hollins Line fault at the top of the Greenstone. A more complete discussion of the tectonic, structural, and stratigraphic framework of the Hillabee Greenstone and the associated metasediments of the Talladega slate belt can be found in Tull (1979) and Tull and Stow (1980a, 1980b). A discussion of the sulfide deposits associated with the greenstone can be found in Stow and Tull (1979, 1982) and Tull and Stow (1980b, 1982).

This paper presents a detailed discussion of the igneous petrogenesis of the Hillabee Greenstone. As is typical of mafic igneous rocks that have undergone greenschist facies metamorphism, the Hillabee protolith suffered hydration and is altered somewhat chemically. Therefore, the relatively unaltered samples must be identified to discern the igneous petrogenesis of the volcanics. General geochemistry of the Hillabee Greenstone, previously described in part by Tull and Stow (1980a), will be reviewed prior to the discussion of the igneous petrogenesis. The methods of chemical analysis used are detailed in Tull and Stow (1980a).

GENERAL GEOCHEMICAL CHARACTERISTICS.

Eighty-eight samples of the Hillabee Greenstone have been analyzed for major, minor, and trace elements. The complete 170-km outcrop extent of the volcanics is represented in this sampling; however, the central area in Clay and Talladega Counties, where the formation has a wide outcrop (Fig. 1), is represented by approximately half the samples. Composition of the Greenstone is highly variable; for example, SiO_2 ranges from 35% to 71% with a bimodal distribution; modes occur at 48% to 50% and at 66%. Other components show similar extreme ranges and bimodal distributions.

The mafic rocks, which represent the SiO_2 mode of 48% to 50%, occur as phyllites and greenstones and are compositionally similar to low-K tholeiites (Table 1). Average compositions of both lithologies are similar and overlap within one standard deviation for all components analyzed. Some samples of phyllite and greenstone have low SiO_2 values (<45%). These values represent altered greenstone and mafic phyllite, many of which are highly epidotized and have high CaO contents (16% to 20%); the values do not represent ultramafic rocks.

The third major lithology, metaquartz dacite, has an average SiO_2 content of 67.4% (Table 1) and a range of 65% to 71% (Long, 1981). Trace element contents are distinctly different from those of basaltic rocks and generally reflect the felsic nature of the lithology. None of the samples identified represents compositions between those typical of basaltic and felsic lithologies. This compositional gap is not interpreted to result from sampling bias. Thus, the well-recognized bimodal distribution of volcanic rock types (Chayes, 1963) occurs in the Hillabee Greenstone. In some cases, mafic and felsic lithologies are interlayered through a distance of a few meters, but contacts commonly are knife sharp, and neither field nor chemical evidence exists for gradation between rock types.

Rocks of spilitic composition occur in Coosa and Chilton counties, near the southern extremity of the Hillabee outcrop. These rocks have high Na_2O content (average 5.61%, Table 1) and low CaO and MgO contents (average 4.28% and 4.88%, respectively) and are similar to the average spilite discussed by Turner and Verhoogen (1960).

THE HILLABEE PROTOLITH

Identification of Relatively Unaltered Samples

Samples representative of relatively unaltered igneous protolith of the Hillabee Greenstone were identified on the basis of statistical treatment of geochemical data. Any

TABLE 1. AVERAGE COMPOSITIONS OF VARIOUS LITHOLOGIES OF THE HILLABEE GREENSTONE

	Greenstones		Mafic Phyllites		Metaquartz Dacites		High Na_2O Spilites	
SiO_2	49.4	(2.39)	48.9	(2.19)	67.4	(2.19)	50.2	(1.18)
Al_2O_3	15.1	(1.37)	15.8	(1.12)	14.9	(0.70)	16.7	(1.25)
TiO_2	1.24	(0.37)	1.22	(0.45)	0.32	(0.05)	1.41	(0.20)
Fe_2O_3	4.32	(1.42)	4.24	(1.33)	2.07	(0.28)	5.72	(2.69)
FeO	7.15	(1.79)	6.26	(1.46)	2.16	(0.20)	7.52	(1.26)
MgO	6.26	(1.00)	6.38	(1.08)	1.76	(0.68)	4.88	(1.22)
CaO	10.8	(3.13)	9.91	(2.00)	3.63	(1.11)	4.28	(1.35)
Na_2O	2.92	(0.98)	2.81	(0.73)	3.08	(0.08)	5.61	(0.82)
K_2O	0.30	(0.37)	0.26	(0.21)	3.48	(1.19)	0.28	(0.16)
MnO	0.177	(0.040)	0.159	(0.025)	0.04	(0.02)	0.252	(0.096)
P_2O_5	0.17	(0.05)	0.16	(0.06)	0.12	(0.07)	-	-
LOI	2.25	(0.98)	3.03	(1.45)	1.32	(0.15)	2.72	(0.75)
Total	100.15		100.03		100.38		99.57	
FeO*	11.1	(1.66)	10.1	(1.27)	3.80	(0.72)	13.4	(1.51)
Ag	0.05	(0.02)	0.07	(0.03)	-	-	0.04	(0.01)
Ba	71	(39)	73	(50)	531	(36)	133	(57)
Cu	234	(287)	132	(79)	43	(42)	67	(35)
Cd	0.08	(0.01)	0.08	(0.01)	-	-	0.09	(0.02)
Co	38	(7)	36	(5)	19	(5)	41	(2)
Cr	158	(98)	206	(151)	43	(12)	86	(116)
Li	11	(8)	17	(7)	-	-	9	(4)
Nb	4.8	(3.5)	3	(28)	-	-	3.5	-
Ni	48	(27)	60	(39)	13	(3)	19	(5)
Rb	13	(36)	2	(1)	-	-	-	-
Pb	20	(5)	20	(7)	17	(4)	-	-
Sr	198	(89)	199	(63)	178	(33)	105	(40)
V	294	(95)	253	(63)	64	(20)	356	(70)
Zn	81	(47)	70	(26)	26	(8)	203	(176)
Zr	72	(17)	66	(27)	-	-	-	-
Y	25	(9)	26	(8)	-	-	-	-
No. of samples averaged†	47		26		4		5	

Note: Data for major and minor elements given in percent; trace element data given in ppm. standard deviations shown in parentheses, for sample populations greater than 3.

†Number may be less than indicated for P_2O_5, Ag, Rb, Pb, V, Zr, and Y.

procedure of this type that discriminates relatively unaltered from more highly altered samples may be hindered by lack of knowledge of the exact composition of the rocks prior to alteration, as well as by different processes of alteration that may have been operative to various degrees in different samples. For such a procedure to be successful, chemical criteria used to identify the more altered rocks must be reflective of alteration processes that can be identified within samples.

Identification of more altered mafic samples can be achieved by study of concentrations of components known to be especially mobile during alteration. Elimination of samples yielding evidence of mobility of these components results in a set of "residual" samples, which are representative of relatively unaltered mafic rock. Examination of geochemical data showed that some samples have compositions not representative of igneous rocks and that most samples have undergone hydration. Assuming the degree of hydration to be an indicator of progressive low-grade alteration (Floyd, 1976; Humphris and Thompson, 1978), all samples having loss of ignition (LOI) of 2% or greater were eliminated. This step identified 39 samples with LOI (water) less than 2%; the range of SiO_2 in these samples was 46.7% to 53.9%. Most samples having extreme compositions (low SiO_2, epidote-rich and spilitic samples) were eliminated.

A second screening step involved statistical treatment of CaO and Na_2O concentrations, two of the most mobile elements during basalt alteration. Epidotization and spilitization, evident in the Hillabee Greenstone, operate in opposite directions with respect to Na_2O and CaO. Spilitization involves Na_2O enrichment and depletion of CaO (and MgO; Turner and Verhoogen, 1960); epidotization causes the opposite changes. The 39 samples having low LOI were divided into basalts ($SiO_2 < 52\%$) and basaltic andesites ($SiO_2 > 52\%$); samples having Na_2O and/or CaO contents outside two standard deviations of the mean for Na_2O or CaO in each group were eliminated. The 32 remaining "select" samples are considered to represent the relatively unaltered basaltic protolith. The majority of these samples (28) are from within the wide outcrop area of the Hillabee in Clay and Talladega Counties.

To test further the relic igneous chemical character of the suite of 32 samples, data were plotted on the "igneous spectrum" of Hughes (1973). The vast majority (30) fell within the boundaries of igneous rocks (Fig. 2). Samples eliminated on the basis of high Na_2O content fell in the realm of spilites, and two samples eliminated on the basis of high CaO and low Na_2O concentration were confirmed petrographically to be highly enriched in epidote. Four dacitic rocks,

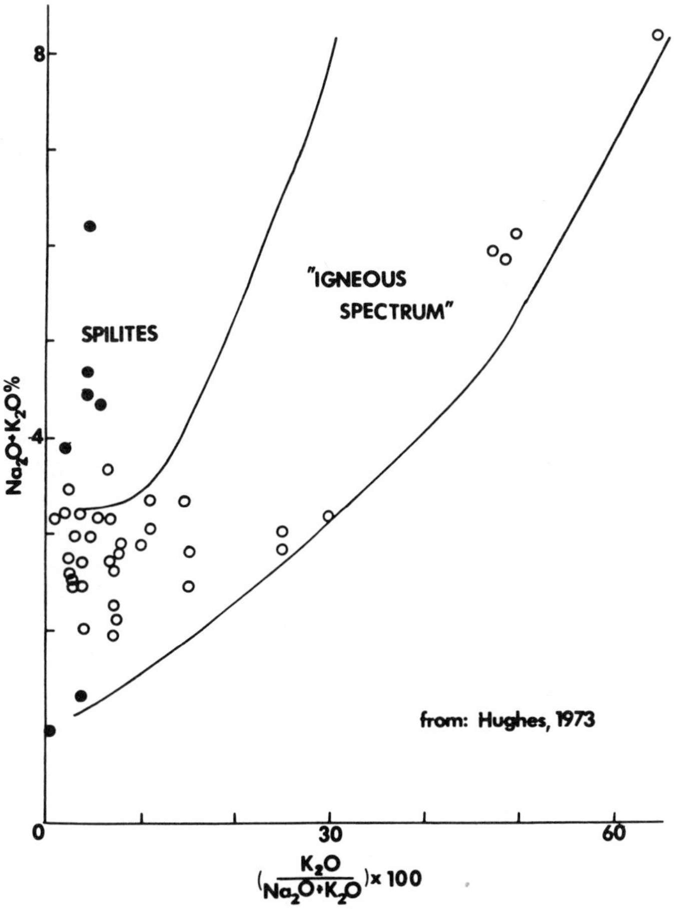

Figure 2. Igneous spectrum of Hughes (1973) with Hillabee samples containing less than 2% water (LOI) plotted; filled circles represent samples eliminated from select group on basis of extreme Na_2O and/or CaO contents.

which were not included in the discrimination procedure but which have LOI less than 2%, also occur in the igneous field. Distributions of relatively unaltered samples in relation to more altered samples were examined on variation diagrams in further effort to discern the effectiveness of this procedure for selecting relatively unaltered samples. The results, described subsequently, indicate that the selection procedure was effective.

Petrography of Select Samples

A detailed description of metamorphic minerals present in the Hillabee was given by Tull and others (1978). The 32 mafic samples petrographically display metamorphic textures typified by various degrees of recrystallization. In some samples, larger grains are highly sheared. Original igneous plagioclase and pyroxene are extensively to completely altered. In spite of metamorphic overprinting, original igneous textures (gabbroic, diabasic, and porphyritic) may be observed in many samples. Deformation and alteration have affected some samples so that identification of original textures is uncertain. Stratigraphic and field relationships indicate that the base of the Hillabee is dominated by a zone of pyroclastics (Tull and Stow, 1980b), but pyroclastic textures apparently are not preserved.

Gabbroic rocks generally are represented as massive greenstones in hand specimen; however, highly deformed gabbroic rocks are also represented by mafic phyllites. Relic shapes of igneous pyroxene and plagioclase can be identified in unsheared gabbroic samples; a hypidiomorphic granular texture dominates. Because of alteration, igneous grain characteristics such as zoning and diallage, if existent originally, cannot be observed. Plagioclase, completely recrystallized to albite and/or epidote and zoisite, composes 40% to 70% of the rock; plagioclase occurred originally as laths up to 3 mm long. Undeformed gabbroic rocks contain small (<0.3 mm) laths of epidotized plagioclase in altered pyroxenes. Pyroxene—altered to tremolite and, more rarely, to actinolite—composes up to 60% of gabbroic rocks as grains up to 3 mm across. Most gabbroic samples display poikilitic texture. In rare cases, pyroxenes that have not been amphibolitized completely contain a small core of clino(?)-pyroxene. Opaques are rare, and hematite, chlorite, and actinolite are present in trace quantities.

Diabasic rocks are represented only by greenstones that exhibit diabasic textures consisting of altered plagioclase laths up to 0.5 mm long, which form a mosaic pattern in which actinolitized pyroxenes fill interstices among plagioclases, with the latter altered to epidote and zoisite. Some samples display a deformed diabasic texture with subparallel alignment of some plagioclase laths; however, the original mosaic pattern is recognizable in parts of these samples. In both diabasic and gabbroic rocks, grain-boundary relationships suggest that plagioclase and pyroxene crystallized together but that plagioclase crystallization began prior to that of pyroxene. Opaques may occur as vuggy concentrations in diabasic rocks along postmetamorphic fractures.

Amphibolitized pyroxene is found as relic anhedral to subhedral phenocrysts up to 2 mm across in prophyritic samples. These amphiboles may compose up to 50% of the sample but more commonly constitute only 20%. The groundmass is very fine-grained, altered plagioclase and amphibole.

Samples (some mafic phyllites) interpreted to represent pyroclastic rocks, on the basis of stratigraphic and field criteria, exhibit strong foliation and, in many places, delicate compositional layering defined by felsic (plagioclase and quartz) zones up to 1 mm thick, interlayered with mafic minerals.

Determination of Basalt Type

A variety of evidence indicates that the protolith of the mafic phyllite and greenstone was of tholeiitic composition. The average composition of the 32 samples (Table 2) is similar to island arc tholeiites of Jakes and White (1973). Previous demonstrations (Tull and Stow, 1980a) showed that the mafic rocks of the Hillabee are of tholeiitic character based on alkali-SiO_2 and Y/Nb discrimination relations.

Based on CIPW normative mineralogy, most Hillabee basaltic rocks range from olivine tholeiites to quartz tholeiites. Normative mineralogy of the average Hillabee Greenstone (Table 3) is that of an olivine tholeiite having high hypersthene content. Individual samples with high normative olivine (generally >12%) have small amounts of normative nepheline (<2%) and no normative hypersthene; these could be termed "alkaline olivine basalts." No samples exist with more than 2% normative Ne. Ne-normative samples may be the result of a minor Na-metasomatic addition, as discussed subsequently. Forsterite generally dominates over fayalite in rocks containing normative olivine; the average molecular ratio of Fo/Fa is 1.26. Samples from the vicinity around and northwest of Millerville in Clay County (Fig. 1) have normative quartz (2.22%) and greater amounts of hypersthene (23.2%) and less diopside (14.4%) than do samples from other outcrop areas of the Hillabee Greenstone (5.45% and 26.0%, respectively). This suite, referred to as the Millerville samples, represents quartz tholeiites, whereas samples from other areas are predominantly olivine tholeiites. However, one of the thickest sequences (1.4 km) of the Hillabee occurs in the Millerville area. Thus, to a large degree, Millerville samples represent levels stratigraphically different from those found in other outcrop areas.

On the basis of the AFM relationship, mafic rocks of the Hillabee Greenstone lie on a tholeiitic trend of moderate iron enrichment (Fig. 3). Millerville samples,which are quartz tholeiites, are slightly more fractionated than those from other outcrop areas. Dacitic rocks correspond closely

TABLE 2. CHEMICAL ANALYSES OF RELATIVELY UNALTERED "SELECT" SAMPLES OF HILLABEE LITHOLOGIES AND COGNATE XENOLITHS

	Relatively Unaltered Select Mafic							Jakes and White (1972)		Cognate Xenoliths	
	Mafic Phyllite		Greenstone		All	Miller-ville Area	Other Areas	Island Arc Tholeiite	Calc-alkaline Dacite	S197A	S197B
SiO_2	50.8	(1.69)	49.5	(1.28)	50.0	51.4	48.9	51.6	66.8	43.5	48.9
Al_2O_3	15.7	(0.68)	15.1	(0.81)	15.3	15.5	15.1	15.9	18.2	17.7	10.0
TiO_2	1.17	(0.25)	1.23	(0.30)	1.21	1.28	1.16	0.80	0.23	2.45	1.26
Fe_2O_3	3.73	(1.14)	3.80	(1.06)	3.78	3.53	3.97	2.74	1.25	-	-
FeO	6.01	(2.35)	7.59	(0.87)	7.32	7.35	7.30	7.04	1.02	10.8*	12.1*
MgO	6.80	(0.77)	6.73	(0.77)	6.75	6.21	7.17	6.73	1.50	3.87	9.66
CaO	10.6	(1.19)	11.5	(1.67)	11.0	9.62	12.2	11.7	3.17	17.7	14.4
Na_2O	2.54	(0.31)	2.61	(0.46)	2.59	2.66	2.61	2.41	4.97	1.24	1.23
K_2O	0.31	(0.30)	0.27	(0.34)	0.25	0.36	0.18	0.44	1.92	0.03	0.06
P_2O_5	0.17	(0.05)	0.17	(0.05)	0.17	0.17	-	0.11	0.09	-	-
MnO	0.180	(0.017)	0.187	(0.020)	0.185	0.194	0.178	0.17	0.06	0.166	0.234
LOI	1.74	(0.19)	1.50	(0.27)	1.58	1.62	1.55	0.45	0.26	2.30	2.54
Total	99.75		100.19		100.14	99.88	100.32			99.8	100.08
FeO*	10.1	(0.86)	11.0	(0.90)	10.7	10.5	10.9			10.8	12.1
Ag ppm	0.007	(0.03)	0.05	(0.20)	0.055	-	0.055	-	-	-	-
Ba	81	(54)	87	(144)†	83†	84	83†	75	520	2	<1
Cd	0.07	(0.01)	0.08	(0.01)	0.073	-	0.073	-	-	0.031	0.030
Co	33	(7)	37	(4)	36	32	38	-	-	33	38
Cr	238	(1945)	214	(105)	213	166	249	50	13	108	280
Cu	174	(117)	195	(186)	185	210	165	-	-	93	81
Li	15	(6)	15	(17)	15	15	15	-	-	0.4	0.7
Nb	1	-	5	(3)	4	-	4	-	-	-	-
Ni	68	(49)	79	(76)	69	65	72	30	5	58	80
Pb	21	(6)	20	(6)	20	20	21	-	-	24	16
Rb	3	-	4	(1)	4	-	4	5	45	-	-
Sr	179	(36)	194	(41)	188	179	197	200	460	737	199
V	267	(37)	269	(35)	268	281	257	270	68	507	263
Zn	75	(16)	84	(16)	81	78	84	-	-	68	39
Zr	61	-	74	(9)	70	-	70	70	100	-	-
Y	22	-	46	(30)	24	-	24	-	20	-	-
No. of samples averaged	10		22		32	14	18	-	-	1	1

Note: Standard deviations shown in parentheses for sample populations greater than 3; FeO* refers to total Fe as FeO.

†With one high Ba value (710 ppm) omitted, average Ba is 56(14) for greenstones, 62 for all samples, and 46 for samples from other areas.

*Refers to total Fe as FeO.

TABLE 3. NORMATIVE ANALYSIS OF AVERAGE HILLABEE LITHOLOGIES (DATA IN PERCENT)

	Relatively Unaltered Mafic Rocks	Quartz Dacites	Millerville Area	Other Areas	Cognate Xenoliths	
					S197A	S197B
Q	-	25.0	2.22	-	-	-
Or	1.48	20.5	2.14	1.06	0.18	0.37
Ab	21.9	25.9	22.6	22.1	5.53	10.8
An	29.4	16.4	29.4	29.0	44.2	22.3
Ne	-	-	-	-	2.90	-
Di	19.9	0.68	14.4	26.0	39.0	42.1
Wo	10.1	0.35	7.26	13.2	19.6	21.5
En	15.2	0.21	3.64	6.88	8.89	12.5
Fs	4.64	0.12	3.46	5.92	10.5	7.99
Hy	18.9	6.49	23.2	5.45	-	20.0
En	9.99	4.16	11.9	2.93	-	12.2
Fs	8.96	2.33	11.3	2.52	-	7.79
Fo	1.18	-	-	5.64	0.76	0.11
Fa	1.16	-	-	5.35	1.01	0.07
Mt	1.73	2.71	1.70	1.75	1.62	1.82
Il	2.30	0.60	2.44	2.20	4.82	2.48
Ap	0.40	0.27	0.40	-	-	-
Number of samples represented	32	4	14	18	1	1

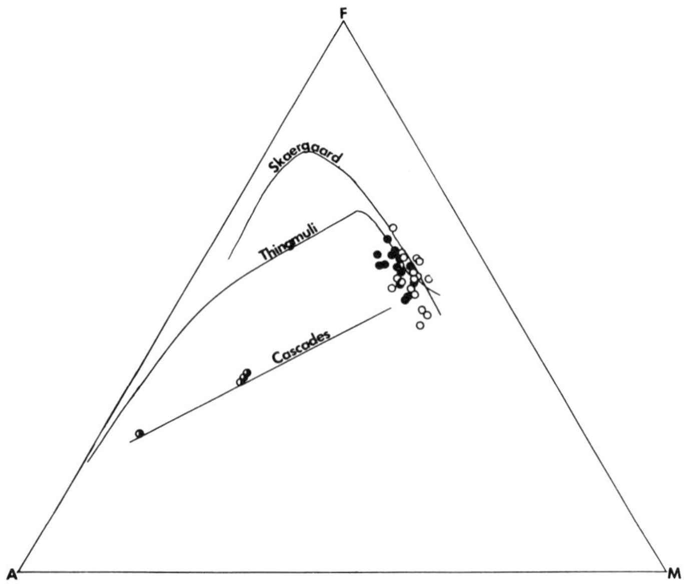

Figure 3. AFM diagram for select, relatively unaltered Hillabee samples. Filled circles represent Millerville mafic samples; half-filled circles represent dacitic rocks; unfilled circles represent mafic rocks from areas other than Millerville.

to a calc-alkaline trend (Long, 1981) and show no evidence of iron enrichment. Thus, tholeiitic and dacitic rocks do not seem directly related through fractionation. Hillabee dacites also compare more closely with dacites of the calc-alkaline trend than those of the island arc tholeiitic trend (Tables 1 and 2).

FRACTIONATION TRENDS

By examining data for basaltic rocks, in the form of variation diagrams, evidence of fractionation can be seen. The tholeiitic trend of moderate iron enrichment toward ferrobasaltic composition (Fig. 3) corresponds closely to well-established iron enrichments for Thingmuli (Carmichael, 1964) and the Skaergaard, but does not show the extent of fractionation seen for those two examples. Hillabee Greenstone magmatism may simply not have undergone extensive fractionation toward more extreme iron and alkali enrichment. However, more highly fractionated rocks may have been removed by or are concealed beneath the tectonic upper contact of the Hillabee.

Evidence for fractionation is seen in the MgO/FeO + Fe_2O_3 graph (Fig. 4). Mafic rocks define two separate tholeiitic trends, one dominated by Millerville samples, which have relatively low MgO values for equivalent iron values when compared with other basaltic rocks; both trends are subparallel to the trend of the Skaergaard and are divergent from the trend of the dacites, which correspond to well-established calc-alkaline lines of descent. No evidence for magmas less fractionated than these (Fig. 4) has been found in the Hillabee. One might presume that the quartz dacites represent an extreme fractionation product of tholeiitic magmas and should be on a trend somewhat similar to, but displaced toward higher MgO values than, the Thingmuli trend. No compelling evidence supports this contention; in fact, the distinct bimodal SiO_2 content of Hillabee rocks and field relationships, noted previously, indicate that the dacites are not a product of the magma series that produced the tholeiites.

Distribution of Hillabee samples in the normative Ne-An-Ol-Hy-Q system confirms fractionation and defines a distinct trend from Ol- (and rare Ne-) normative tholeiites, across the plane of SiO_2-saturation, into the field of Q-normative tholeiites (Fig. 5). A trend away from the An apex indicates a general decrease in Al_2O_3 content. The FeO*/MgO ratios mirror the differentiation and increase

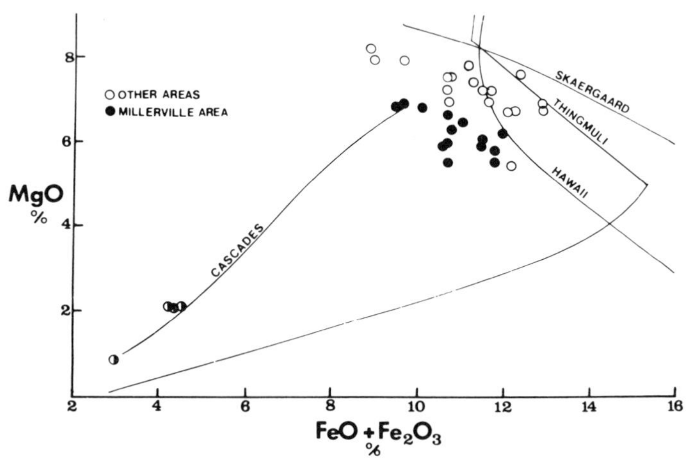

Figure 4. Distribution of mafic samples of the Hillabee on MgO-/FeO + Fe_2O_3 diagram. Filled circles represent Millerville samples; half-filled circles represent dacitic rocks.

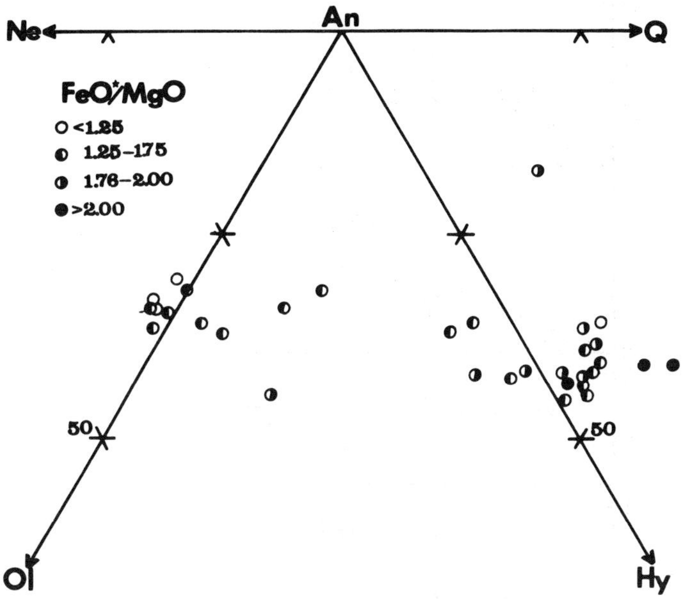

Figure 5. Distribution of select, relatively unaltered mafic Hillabee samples on normative Ne-An-Ol-Hy-Q diagram.

across the An-Ol-Hy field into the Q-normative field. Basalts that are highly fractionated with regard to FeO*/MgO (>1.75) generally are Q-normative. Primitive samples (FeO*/MgO < 1.25) lie near the critical plane of undersaturation, but intermediately fractionated basalts (FeO*/MgO = 1.25 to 1.75) are spread from the Ne- to Q-normative fields.

Chemical effects of fractionation in basaltic rocks can be depicted more clearly on MgO-variation diagrams (Fig. 6A). General trends of fractionation are similar to those reported for other tholeiitic systems (Carmichael, 1964; Batiza and others, 1977); well-defined trends are seen for most major and minor elements. Trends for iron (FeO* and FeO*/MgO) show distinct separation of Millerville samples from other samples at approximately 7% MgO. K_2O and Na_2O suggest only a general enrichment with decreasing MgO; Al_2O_3 shows no obvious relationship to MgO for the bulk of the samples. When Millerville samples are considered separately, however, two broad trends of decreasing Al_2O_3 may be observed.

Some trace element distributions are controlled by fractionation of the Hillabee Greenstone. Elements that most readily enter structures of early crystallizing minerals and generally follow Fe and Mg in basaltic systems (i.e., Cr, Ni, V, Co) show definite trends. Greatest changes are for Cr and Ni, which decrease by a factor of about 3 from the most primitive to the most fractionated rocks (Fig. 6B). Co and V show fairly well-defined changes, although of relatively small magnitude, during fractionation. These changes are the same as those discussed for tholeiitic fractionation by Miyashiro and Shido (1975).

Large ion lithophile (LIL) trace elements do not exhibit fractionation changes as consistent as those seen for other trace elements. Ba shows only little change with MgO, but does seem to be somewhat related to K_2O. High Ba-containing samples (128 to 710 ppm) are enriched in K_2O (0.21% to 0.89%), but not all high-K_2O samples are necessarily enriched in Ba. Distribution of Sr values shows a slight, but not well-defined, decrease with MgO. This trend is analagous to CaO, which controls Sr in differentiating basalt. Sr does seem to be controlled by CaO in plagioclase in the Hillabee Greenstone. A cognate xenolith (sample S197A, Tables 2 and 3), presumed to represent early crystallization of plagioclase and pyroxene (see subsequent discussion) and containing a very high CaO content (17.7%) and abundant normative anorthite (44.2%), contains 737 ppm Sr. This Sr concentration is far in excess of that found in any of the samples depicted in Figure 6B. General lack of relationship between degree of fractionation and abundance of LIL elements may be the result of a limited MgO (and SiO_2) range of concentrations, and the fact that LIL elements are relatively mobile during weathering and metamorphism (Hart, 1970; Pearce, 1975). Any primary fractionation trends that occurred possibly have been changed by alteration.

Abundances of Zn and Li are not well related to fractionation. The range of Zn abundance does not vary significantly, and Li is somewhat enriched in less primitive rocks. Zr, for which few data are available, increases with decreasing MgO, as is typical in fractionating tholeiite.

The relationship between rock composition and geographic distribution has been noted with regard to chemical distinction of Millerville samples. Relatively highly fractionated Millerville samples (filled circles) are depicted further in MgO-variation diagrams (Fig.6) with regard to constituents such as SiO_2, CaO, Cr, and Ni. A geographical distribution of the data (Fig. 7) indicates a rather consistent increase in MgO values from Millerville (subarea I) north along the outcrop of the Hillabee; the most mafic sample (i.e., highest MgO%, 8.20) occurs at Pyriton (subarea VI). Constituents other than MgO generally vary to the northeast, in accord with their fractionation behaviors relative to MgO. Chemical data and outcrops are insufficient to allow extension of this trend into Chilton and Coosa counties to the south and Cleburne County to the north. Except for the single, very mafic sample from Pyriton (Fig. 7), the most mafic region occurs at the Clay-Talladega county line, which is north of the extent of quartz dacites. The most fractionated (lowest MgO%) mafic rocks are associated with rather extensive outcrop areas of dacitic rocks near Millerville.

Relic igneous textures seen in some of the select samples also correlate closely with spatial distribution of chemical compositions of the Hillabee Greenstone (Fig. 7). Relatively mafic rocks display relic gabbroic textures. These rocks may

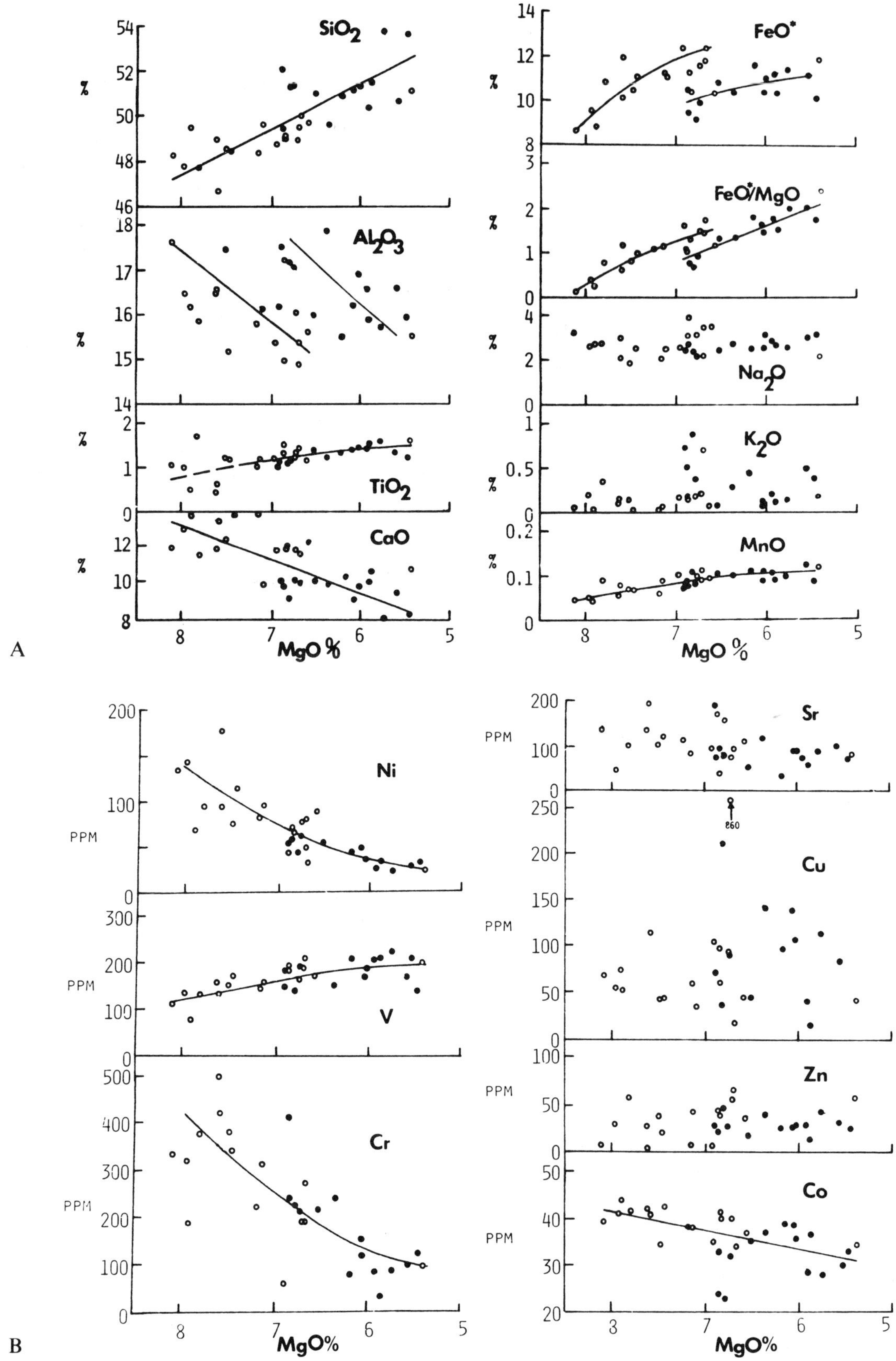

Figure 6. MgO-variation diagrams for the select, relatively unaltered mafic Hillabee samples. Filled circles represent Millerville samples.

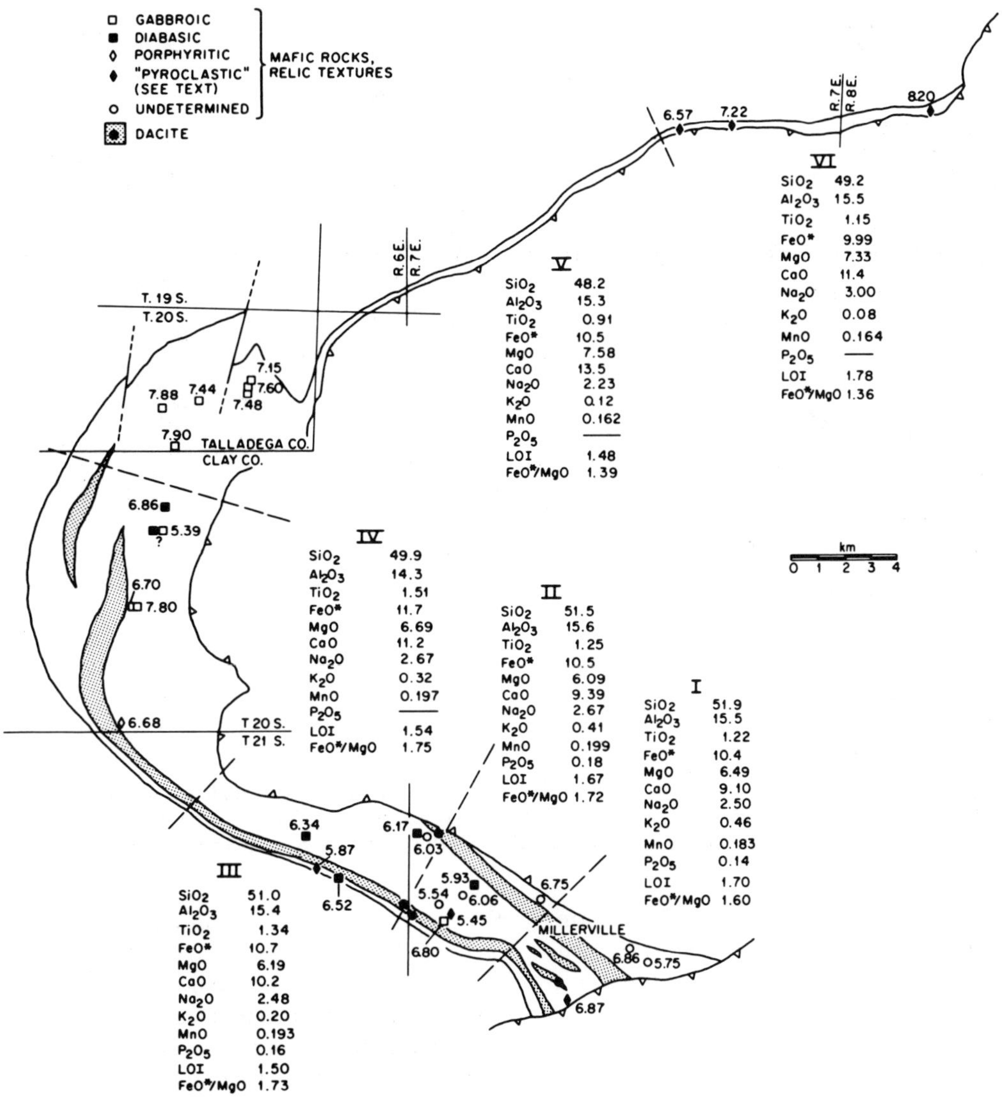

Figure 7. Map of Hillabee outcrop in Clay and Talladega counties, northwest of Millerville, showing average composition of mafic select samples in six subareas and the percentage of MgO for each sample. Relic igneous textures, where recognizable, are also shown, as is the distribution of the meta-quartz dacites. Apparent pyroclastic rocks include samples whose pyroclastic nature is based on field and stratigraphic data.

represent thick flows. As one proceeds south toward Millerville, mafic rocks become more fractionated, quartz dacites become abundant, gabbroic textures become rare, and diabasic textures become more common. Although sufficient textural and chemical data are not available to allow identification and tracing of individual strata within mafic lithologies of the Hillabee, pyroclastic rocks seem to be dominant at the base. The occurrence of pyroclastics is well documented on the basis of field and stratigraphic data at various places along the outcrop length of Hillabee volcanics.

The distribution of all mafic samples, including those eliminated previously, on MgO-variation diagrams reveals whether fractionation trends can be observed for the entire suite of samples (Fig. 8). CaO, SiO_2, and MnO contents and the FeO*/MgO ratio are representative of the entire spectrum of element-MgO relationships. Outlined in each graph is the field of fractionation trend defined by the 32 select samples (Fig. 6). When all samples are considered, fractionation trends of the Hillabee magma are either obscured or totally hidden. Greatest dispersion of data points occurs at low MgO values. In addition, the 7 samples eliminated on the basis of extreme CaO and/or Na_2O contents, identified by barred circles, are of reasonably low MgO values, suggesting that alteration may have resulted in some loss of MgO in addition to hydration and mobility of CaO and Na_2O. Many samples eliminated on the basis of high water (LOI) content consistently lie within the confines of the relatively unaltered suite, indicating that they may have

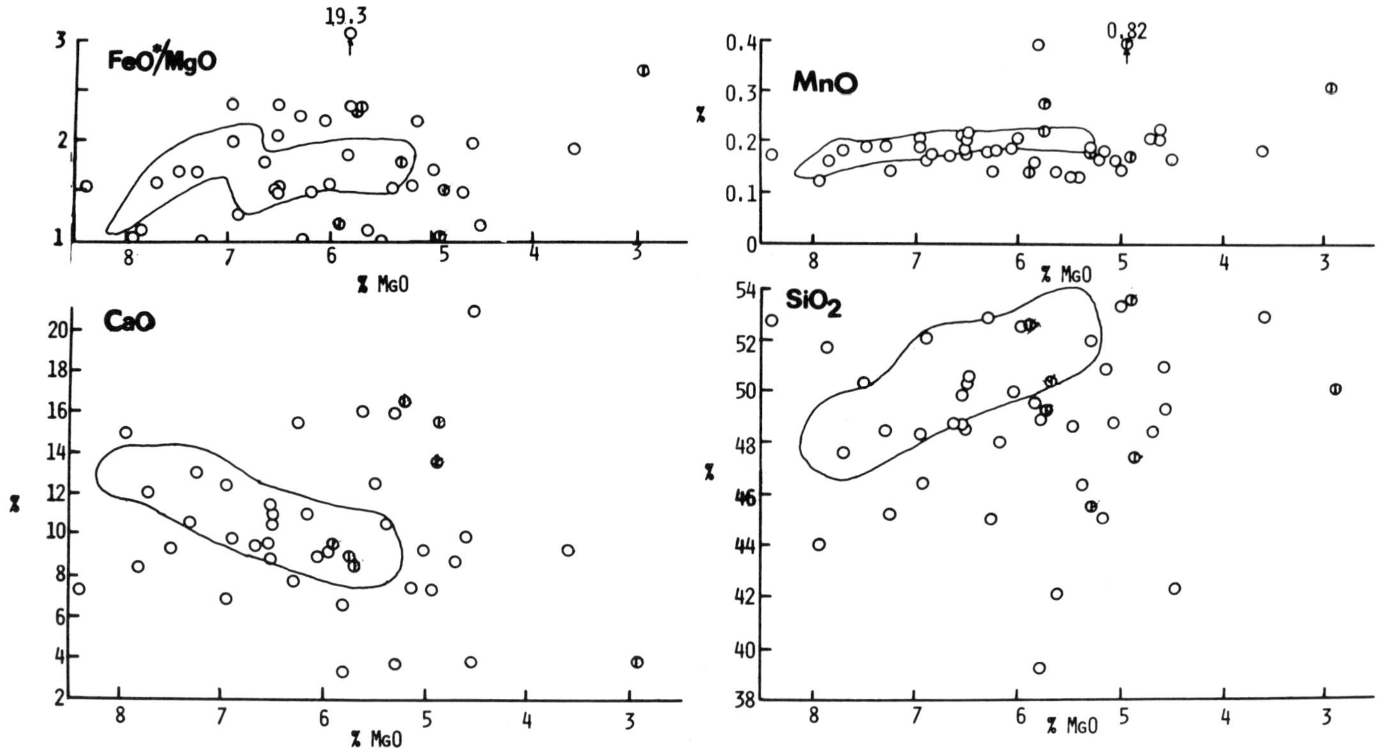

Figure 8. MgO-variation diagrams showing distribution of Hillabee samples identified as being relatively altered on the basis of high water (LOI) and/or extreme CaO and/or Na_2O contents. Area within the solid line represents the distributions of the select samples, as shown in Figure 6.

experienced only minimal alteration other than hydration. Thus, for the Hillabee Greenstone, simple statistical methods of discriminating between relatively altered and unaltered samples have been reasonably effective.

Fractionation trends seen in the Hillabee represent a rather limited range of SiO_2 values, from 47% to 54% (8.20 to 5.2% MgO). This narrow range is in contrast to that of numerous other studies (e.g., Carmichael, 1964; McCall, 1973; Floyd, 1976; Byerly and others, 1976), some of which were conducted on greenstone or meta-igneous complexes where rock types available for study generally ranged from ultramafic to at least intermediate and often to felsic compositions. The present study shows that one can "look through" lower greenschist facies metamorphism to observe subtle geochemical trends such as the limited fractionation that characterizes the Hillabee Greenstone, despite chemical effects of alteration on protolithologic chemistry.

MINERALOGIC CONTROL OF FRACTIONATION

To assess the control of crystallizing minerals on the fractionation trends described previously, a mixing model has been applied to determine mineralogic changes required for fractionation from a primitive to a more fractionated Hillabee basalt. Four primitive samples having highest MgO contents and four relatively highly differentiated samples having lowest MgO contents were selected; analyses for each group were averaged to represent the two end members of the fractionation sequence (columns 1 and 2, Table 4). The fractionated product can be derived from the more primitive tholeiite by removing a crystal extract (column 3, Table 4). To produce the fractionated product, up to 90% of the primitive tholeiite must crystallize fractionally. Mixing ratios for plagioclase (Pc), clinopyroxene (CPx), olivine (Ol), and ilmenite (Il) required to satisfy the bulk composition of the crystal extract are: An_{82}, 44.3%; $En_{41}Fs_{28}Wo_{31}$, 44%; Fo_{59}, 10%; and Il, 1.7%. The Ol/CPx/Pc ratio is 1/4.4/4.4. Similar calculations for fractionated tholeiite by other investigators have shown highly variable Ol/CPx/Pc ratios, but many have a fairly equal ratio of CPx/Pc, both of which generally are more abundant as crystallizing phases than is Ol. Other ratios are: 1/1/2 (Carmichael, 1964); 1/0/4.4 (Kay and others, 1970); 1/7.7/9.3 (Clague and Bunch, 1976); 1/3.3/5 (Byerly and others, 1976); 1/6.7/6.6 (Batiza and others, 1977).

Mixing models may be based on analysis of minerals obtained from fractionated rocks. For the Hillabee Greenstone, however, unaltered igneous minerals could not be obtained for analysis. Accordingly, the plagioclase composition used in the model (column 4, Table 4) represents a plagioclase from the Thingmuli tholeiitic system (Carmichael, 1964), and the olivine and ilmenite compositions are purely stoichiometric. Composition for the clinopyroxene (column 5, Table 4) represents the average composition of two mineral separates from lithic fragments (Tull and Stow, 1980b; described subsequently) which are assumed to represent crystallization of Hillabee magma. The mineral separates represent clinopyroxenes that have undergone tremolitization (based on x-ray diffraction and optical data), with a slight molar increase in SiO_2 over typical pyroxenes (54.4% SiO_2 on a volatile-free basis). Although the transformation from clinopyroxene to tremolite certainly has not been isochemical, use of these data for the mixing model is a better alternative than using the composition of a stoichiometric pyroxene for the calculations—a step that cannot account for significant amounts of Al_2O_3, TiO_2, and MnO contained in pyroxene. Use of a lower SiO_2 content for the "pyroxene" would result in a decrease in the amount of olivine needed for fractionation and would lower the Ol/CPx + Pc ratio slightly for the Hillabee fractionation.

A measure of the accuracy of the mixing model calculation is the magnitude of the difference between bulk composition of the crystal extract (column 3, Table 4) and the calculated composition resulting from mixing crystallizing phases (column 6). These differences (column 7) are either nil or quite small for all constituents except Na_2O. Excluding Na_2O, the sum of squares of differences between columns 3 and 6 is much less than 0.1, a figure suggested by Wright and Doherty (1970) to be an upper limit of error for a reliable mixing model solution. The large difference for Na_2O results from an unusually high Na_2O content for the primitive rocks compared with the differentiate (2.47% versus 2.68%, respectively). To account for the difference in the mixing model, the Pc composition would have to be An_{50}, an unusually Na-rich plagioclase. This composition neither works in the model nor is typical of tholeiitic basalts. Thus, the Na_2O content of the rocks may reflect either some metasomatic addition or redistribution of Na throughout the Hillabee, at a level undetected by the method used for identifying relatively unaltered samples. Indeed, rather extreme Na addition was noted for spilitized rocks. Existence of 7 samples having low normative nepheline contents (<2%) leads one to suspect that small amounts of Na may have been added to the Hillabee Greenstone. Reduction of Na_2O contents by only 0.1% to 0.4% eliminates normative Ne.

TABLE 4. DATA USED IN MIXING MODEL FOR HILLABEE FRACTIONATION

	1	2	3	4	5	6	7
SiO_2	48.9	53.7	48.6	47.6	54.4	48.7	0.1
Al_2O_3	15.8	15.3	15.9	32.7	3.63	16.0	0.1
TiO_2	1.10	1.41	1.17	-	0.59	1.15	0.02
FeO*	10.6	11.5	10.7	0.74	13.7	10.7	0.0
MgO	8.06	5.66	8.22	-	12.4	8.29	0.07
CaO	12.8	9.36	13.0	16.5	12.8	13.0	0.0
Na_2O	2.47	2.68	2.45	2.14	0.62	1.22	1.23
K_2O	0.09	0.30	-	-	-	-	-
MnO	0.173	0.204	0.160	-	0.288	0.127	0.033

Note: Column 1 is average composition of primitive suite on water-free basis (M175, M255, M265A, and T256). Column 2. is average composition of differentiated suite on water-free basis (G7, G8, G42, and M271). Column 3 is composition of crystal extract on water-free basis. Column 4 is composition of plagioclase, An_{82} (Carmichael, 1964). Column 5 is composition of tremolitized pyroxene, $En_{41}Fs_{28}Wo_{31}$, from lithic fragments. Column 6 is calculated composition of 44% $En_{41}Fs_{28}Wo_{31}$, 44.3% An_{82}, 10% Fo_{59}, and 1.7% Il. Column 7 is difference between bulk composition of material required to produce fractionation (column 3) and calculated composition of crystallizing minerals (column 6).

Amphibolitized pyroxenes discussed previously were obtained from lithic fragments that occur at the base of the Hillabee Greenstone in Cleburne County. These samples were not included in the select group. The lithic fragments are of coarse gabbroic texture (Tull and Stow, 1980b), vary in diameter up to 30 cm, and are wrapped by foliation of the enclosing pyroclastic strata. These xenolithic fragments might represent early crystalline phases associated with Hillabee fractionation. Two of the fragments were analyzed chemically (samples S197A, S197B, Table 2); each has a unique composition. Sample S197A represents the highest Al_2O_3- and CaO-bearing rock and lowest MgO-bearing rock of any Hillabee sample; both samples are low in Na_2O. Normative analyses of these two xenoliths (Table 3) show an unusually high Di content (39% to 42%) and, for S197A, a high An content (44% An_{88}). Normative Ol is low in both samples (0.2% to 1.8% Fo_{48-66}). The high normative Il content of sample S197A (4.8%) reflects the high TiO_2 content (2.45%), the highest of any Hillabee sample.

Gabbroic lithic fragments contain large (up to 5 mm) relic anhedral to subhedral altered pyroxenes, which compose approximately 40% to 60% of the rock; they generally are colorless, although some are light green. Along fractures, formation of fibrous amphibole is evident. Some pyroxenes have been highly altered to very fine-grained, low-birefringent material; others show occasional twinning. No evidence exists for zoning in these grains. Plagioclase ranges from fine-grained (0.1 mm) to large (2 mm) anhedral grains. The plagioclases have been highly epidotized; many large grains exhibit Carlsbad and albite twinning, and extinction angles less than 10°. No sphene was noted. Opaques are rare and constitute less than 1%, thus suggesting that a large part of the TiO_2 in S197A was incorporated in the pyroxene structure. No evidence for the existence of olivine was noted. Effects of shearing are evident. Altered pyroxenes have been rotated and deformed intensely into elongate grains having granulated and sheared edges.

Although composition of the two lithic fragments (Table 2) bears little relationship to the composition of the crystal extract required for fractionation (Table 4), additional information regarding the sequence of mineral formation can be acquired from the fragments. The CPx and Pc stages of crystallization are well displayed; the Ol stage is not represented by modal analysis and is minor in the normative analysis (Table 3). Throughout the fractionation series, the main phase of plagioclase crystallization was confined to relatively primitive rocks, such as gabbroic. However, continued crystallization of CPx—now amphibolitized—is evident throughout the fractionation series, as noted previously. Also, poikilitic inclusion of Pc grains within CPx grains in gabbroic rocks may indicate relatively early crystallization of Pc relative to CPx. Based on chemical data for the Greenstone, Ol crystallization perhaps preceded the Pc and CPx stages. Thus, the apparent sequence of mineral formation leading to fractionation of the Hillabee is Ol, Pc and CPx, with an undetermined amount of temporal overlap between phases. Trends of decreasing Cr and Ni with increasing fractionation (Fig. 6B) correspond with trends for oceanic tholeiites in which the crystallization sequence is Ol, Pc, CPx (Thompson and others, 1971); these Cr and Ni trends do not occur when the sequence is Pc, Ol, and CPx.

DEPTH OF FRACTIONATION

The Hillabee Greenstone represents a fractionated tholeiitic basalt, although the parental magma apparently is not represented by available samples. Experimental data of Green and Ringwood (1967) indicate that FeO*/FeO* + MgO ratios of 0.4 are characteristic of equilibrium partial melts of mantle periodotite generated at relatively shallow depths. Ratios in the Hillabee are consistently higher than 0.4, from 0.51 to 0.74, and thus suggest differentiation from a magma more primitive than that represented by the exposed Hillabee. In addition, the relatively high MgO content (8.20%) of the most primitive select sample of the Hillabee is considerably lower than that of mantle-derived primitive tholeiites of the Pacific (about 9.50% to 10.0% MgO; Batiza and others, 1977).

The experimental data of Green and Ringwood (1967) show that fractionation of Ol-tholeiite at pressures less than 5 kb produces an Fe-enriched (Hy-enriched) Q-tholeiite having at least 50% SiO_2 and 12% to 14% Al_2O_3. The sequence of crystallization observed is Ol, Pc, and CPx. Very early CaO, Al_2O_3, and SiO_2 enirchment of the magma is due to Ol formation prior to early development of Ca-Pc, at which time the more characteristic trends of decreasing CaO and Al_2O_3 contents ensue. Fractionation at higher pressure (9 kb) produces differentiates having relatively constant SiO_2 contents (47% to 48%), Ol and CPx coprecipitating prior to Pc precipitation, and evolution of high-Al_2O_3 Ol-tholeiites having high CaO and Na_2O contents. At still higher pressures (13.5 to 18.0 kb), fractionation produces Ne-normative Ol-rich magmas with decreasing SiO_2 and increasing alkali contents.

Application of data of Green and Ringwood (1967) indicates Hillabee Greenstone fractionation occurred at shallow, crustal levels at low pressure. Fractionation proceeded from Ol-tholeiites, across the plane of SiO_2-saturation, into the field of Q-tholeiites (Fig. 5). The FeO* and SiO_2 enrichment occurred (Fig. 6A), with increasing normative Hy and no evidence of Al_2O_3 enrichment, during differentation; in fact, Al_2O_3 content decreased (Figs. 5, 6A) during the evolutionary process. This trend, coupled with decreasing CaO (Fig. 6A), indicates that the major phase of Pc precipitation was prior to the time represented by our Hillabee samples. This interpretation of the sequence of Hillabee crystallization (Ol, Pc, CPx) parallels that for low-pressure fractionation. No evidence exists for late Pc formation or of significant coprecipitation of Ol and CPx, as occurs at higher pressures.

DISCUSSION

Geochemical observations regarding rock types and fractionation allow some constraints to be placed on igneous evolution of the Hillabee Greenstone. More complete interpretation is hampered chiefly by incomplete exposure of Hillabee volcanics. Variation in rock type and composition downdip, toward the southeast, are unknown. Evidence discussed by Tull and Stow (1980a) indicates that the Hillabee Greenstone represents arc volcanic rocks that spread from the southeast during Early to Middle Devonian time.

Early stages of arc development are characterized by tholeiitic volcanism; as subduction proceeds, calc-alkaline volcanism increases and eventually dominates (Mitchell and Reading, 1971; Miyashiro, 1974). For the exposed Greenstone, tholeiites represent the dominant rock type; quartz dacites represent only 15% of the maximum exposed thickness (Tull and Stow, 1980a), and these are calc-alkaline. Thus, the relative abundance of rock types suggests a fairly immature arc system for the Hillabee Greenstone, which formed over either very thin continental crust (Miyashiro, 1974) or oceanic crust. Assuming that effects of K_2O-metasomatism have been minimal, the low K_2O content of Hillabee tholeiites compared with that of typical island arc tholeiites (0.25% versus 0.44%; Jakes and White, 1973) can be interpreted to indicate minimal, continental crustal influence during volcanism.

The MgO content of the Hillabee Greenstone is relatively low, and the very early stages of fractionation are probably not represented by the exposed rocks. No evidence of olivine is seen. Indeed, olivine may have been altered severely; however, no alteration products have been observed. Also, decreases in CaO and Al_2O_3 contents (Fig. 6A; represented by decreasing An content, Fig. 5) during fractionation and during the interpreted sequence of crystallizing minerals (Ol, Pc, and CPx) indicate that the exposed Hillabee represents a magma from which Pc (and Ol) had been highly fractionated.

Occurrence of calc-alkaline dacites near the stratigraphic base of the Hillabee north of Millerville (Fig. 7) seems anomalous because calc-alkaline rocks normally do not appear during early arc volcanism. As noted previously, tholeiites of the Hillabee are fractionated from a more primitive magma, which is not represented, and basalts associated with the dacites are relatively highly fractionated. Dacites occur near the base of the Hillabee because the exposed part of the Greenstone represents material deposited somewhat distally from the volcanic center, as the depositional basin gradually filled. Earlier volcanics, deposited closer to the volcanic center (and perhaps more primitive ones having less calc-alkaline rocks), would be found downdip in the Hillabee at stratigraphically lower levels.

The reason for geochemical and textural differences between Millerville samples and those from other parts of the Hillabee Greenstone is not entirely evident at present. Certainly, the rather highly fractionated Millerville samples and their association with dacites are significant. With regard to the majority of constituents, however, Hillabee tholeiites represent continuous fractionation typical of tholeiitic systems; Millerville samples deviate from fractionation trends for only a small number of constituents. As noted previously, however, samples from the Millerville area represent different stratigraphic levels than those from other areas. Apparently, fractionation of tholeiitic magmas in general has been controlled by predictable physiochemical factors such as temperature and pressure. Different parts of the Hillabee Greenstone could represent magmas from different magma chambers, possibly at somewhat different levels of fractionation, as has been seen among Hawaiian volcanics (Wright and Fiske, 1971), thus resulting in chemical variations related to spatial distribution of Greenstone. Deviation of Millerville samples from fractionation trends of other samples possibly results from minor mixing of tholeiitic and dacitic magmas or assimilation of foreign material by the tholeiitic magma. Certainly, more work is needed to resolve this question.

A further unanswered question concerns the genetic relationship between tholeiitic and calc-alkaline rocks of the Hillabee Greenstone. Field and geochemical data do not suggest that the calc-alkaline dacites represent a direct fractionation product of tholeiitic magma. Association of the two rock types is typical of subduction zone volcanism, as discussed by Ringwood (1974) and Miyashiro (1974). Solution of this problem must await further work and must involve additional trace element studies of rock types and examination of data relative to depth of formation of tholeiitic and calc-alkaline magmas.

ACKNOWLEDGMENTS

The majority of this research was supported by the Mineral Resources Institute of The University of Alabama. The author is endebted to J. F. Tull, Mike Naney, C. S. Haase, and D. R. Wones who have reviewed this manuscript and who have provided many helpful suggestions. The aid of R. A. Lightsey is sincerely appreciated. Publication No. 1981, Environmental Sciences Division, Oak Ridge National Laboratory, Oak Ridge, Tennessee.

REFERENCES CITED

Batiza, R., Rosendahl, B. R., and Fisher, R. L., 1977, Evolution of oceanic crust 3. Petrology and chemistry of basalts from the east Pacific rise and the Siqueiros transform fault: Journal of Geophysical Research, v. 82, p. 265–276.

Byerly, G. R., Melson, W. G., and Vogt, P. R., 1976, Rhyo-dacites, andesites, ferrobasalts and ocean tholeiites from the Galapagos spreading center: Earth and Planetary Science Letters, v. 30, p. 215–221.

Carmichael, I.S.E., 1964, The petrology of Thingmuli, a Tertiary volcano in eastern Iceland: Journal of Petrology, v. 5, p. 435–460.

Chayes, F., 1963, Relative abundance of intermediate members of the oceanic basalt-trachyte association: Journal of Geophysical Research, v. 68, p. 4247–4256.

Clague, D. A., and Bunch, T. E., 1976, Formation of ferrobasalt at East Pacific mid-ocean spreading centers: Journal of Geophysical Research, v. 81, p. 4247–4256.

Floyd, P. A., 1976, Geochemical variation in the greenstones of S.W. England: Journal of Petrology, v. 17, p. 522–545.

Green, D. H., and Ringwood, A. E., 1967, The genesis of basaltic magmas: Contributions to Mineralogy and Petrology, v. 15, p. 103–190.

Hart, R., 1970, Chemical exchange between sea water and deep ocean basalts: Earth Planetary Science Letters, v. 9, p. 269–279.

Hughes, C. J., 1973, Spilites, keratophyres and the igneous spectrum: Geological Magazine, v. 109, p. 513–527.

Humphris, S. E., and Thompson, G., 1978, Hydrothermal alteration of oceanic basalts by seawater: Geochimica et Cosmochimica Acta, v. 42, p. 107–125.

Jakes, P., and White, A.F.R., 1973, Major and trace element abundance in volcanic rocks of orogenic areas: Geological Society of America Bulletin, v. 83, p. 29–40.

Kay, R., Hubbard, N. J., and Gast, P. W., 1970, Chemical characteristics and origin of oceanic ridge volcanic rocks: Journal of Geophysical Research, v. 75, p. 1585–1611.

Long, L., 1981, The relationship between the structure and geochemistry of the copper deposits of the Hillabee Greenstone in the Millerville region, Clay County, Alabama [M.S. thesis]: Tuscaloosa, University of Alabama, 233 p.

McCall, G.J.H., 1973, Geochemical characteristics of some Archean greenstone suites of the Yilgarn structural province, Australia: Chemical Geology, v. 11, p. 243–269.

Mitchell, A.H.G., and Reading, H. G., 1971, Evolution of island arcs: Journal of Geology, v. 79, p. 253–284.

Miyashiro, A., 1974, Volcanic rock series in island arcs and active continental margins: American Journal of Science, v. 274, p. 21–355.

Miyashiro, A., and Shido, F., 1975, Tholeiitic and calc-alkaline series in relation to the behaviors of titanium, vanadium, chromium, and nickel: American Journal of Science, v. 275, p. 265–277.

Pearce J. A., 1975, Basalt geochemistry used to investigate past tectonic environments on Cyprus: Tectonophysics, v. 25, p. 41–67.

Ringwood, A. E., 1974, The petrologic evolution of island arc systems: Journal of the Geological Society of London, v. 130, p. 183–204.

Stow, S. H., and Tull, J. F., 1979, Sulfide mineralization, *in* Tull, J. F., and Stow, S. H., eds., The Hillabee metavolcanic complex and associated rock sequences: Alabama Geological Society, 17th Annual Field Trip, Guidebook, p. 33–36.

Stow, S. H., and Tull, J. F., 1982, Geology and geochemistry of the strata-bound sulfide deposits of the Pyriton District, Alabama: Econ. Geology, v. 77, p. 322–334.

Thompson, G.,, Shido, F., and Miyashiro, A., 1971, Trace-element distributions in fractionated oceanic basalts: Chemical Geology, v. 9, p. 89–97.

Tull, J. F., 1979, Stratigraphic and structural relationships of the eastern Talladega slate belt of Alabama, *in* Tull, J. F., and Stow, S. H., eds., The Hillabee metavolcanic complex and associated rock sequences: Alabama Geological Society, 17th Annual Field Trip, Guidebook, p. 3–13.

Tull, J. F., and Stow, S. H., 1980a, The Hillabee Greenstone: A mafic volcanic complex in the Appalachian Piedmont of Alabama: Geological Society of America Bulletin, v. 91, Pt. 1, p. 27–36.

——1980b, Structural and stratigraphic setting of arc volcanism in the Talladega slate belt of Alabama, *in* Frey, R. W., ed., Excursions in southeastern geology, Volume II: American Geological Institute, Guidebook, p. 545–581.

——1982, Geologic setting of the Hillabee metavolcanic complex and associated strata-bound sulfide deposits in the Appalachian Piedmont of Alabama: Econ. Geology, v. 77, p. 312–321.

Tull, J. F., Stow, S. H., Long, L., and Hayes-Davis, B., 1978, The Hillabee Greenstone: Geochemistry, structure, mineralization and theories of origin: University of Alabama, Mineral Resources Institute, Research Report 1, 100 p.

Turner, F. J., and Verhoogen, J., 1960, Igneous and metamorphic petrology: New York, McGraw-Hill, 694 p.

Wright, T. L., and Fiske, R. S. 1971, Origin of the differentiated and hybrid lavas of Kilauea volcano, Hawaii: Journal of Petrology, v. 12, p. 1–65.

Wright, T. L., and Doherty, P. C., 1970, A linear programming and least squares computer method for solving petrologic mixing problems: Geological Society of America Bulletin, v. 81, p. 1995–2008.

MANUSCRIPT ACCEPTED BY THE SOCIETY MAY 7, 1982

Printed in U.S.A.

Geological Society of America
Special Paper 191
1982

Part 2

The Carolina slate belt: Origin and evolution of an ancient volcanic arc

Introduction

Stephen A. Kish
Department of Geology
University of North Carolina
Chapel Hill, North Carolina 27514

William W. Black
Department of Geology
University of Akron
Akron, Ohio 44325

A symposium, "Carolina Slate Belt: Origin and Evolution of an Ancient Volcanic Age" was held at the 28th Annual Meeting of the Southeastern Section of the Geological Society of America at Blacksburg, Virginia, on April 26, 1979. The goal of the symposium was to look at the slightly metamorphosed portion of the Appalachian eugeosyncline and evaluate its premetamorphic geologic setting both in terms of chronology of events and original environment. As in any symposium, the topics discussed were those of the current research interest among the participants, and a variety of subjects were covered, but emphasis was on the timing and nature of extrusive igneous activity preserved in this region. The collection of papers in this volume is intended to provide a broad sampling of current research within the slate belt and adjacent portions of the central and southern Appalachian region.

The southern Appalachian Piedmont has been subdivided into several belts, each of which has a characteristic association of lithologies and metamorphic grade. King (1955) divided the Piedmont province into five belts that trend parallel to the Blue Ridge and are named Brevard zone (westernmost), Inner Piedmont, Kings Mountain, Charlotte, and Carolina slate belt (Fig. 1). The Brevard zone is now recognized as a major fault zone that separates two distinctly different geologic terranes (Odom and Fullagar, 1973; Rankin, 1975). These belts also contain numerous plutonic bodies. Of the belts, only the Carolina slate belt is relatively unmetamorphosed and in many places relatively undeformed. King recognized that the volcanic-rich materials of the slate belt may be contemporaneous with, and stratigraphically equivalent, to the gneisses and schists typical of the belts in the western portion of the Piedmont. An eastern slate belt is present along the Coastal Plain of the Carolinas and Georgia and is separated from the main slate belt by a relatively narrow but highly deformed terrane of plutons and high-grade metamorphic rocks of the Raleigh belt in North Carolina and the Kiokee belt in South Carolina and Georgia. A detached segment of the eastern slate belt in South Carolina and Georgia is referred to as the Belair belt. Many workers consider both portions of the slate belt to be synclinorial structures flanked by anticlinoria, which contain higher metamorphic grade sequences. The higher grade rocks may be equivalent to the stratigraphic sequence in the slate belt but from deeper structural levels, exposed by extensive vertical uplift.

Geologic investigations within the Carolina slate belt prior to 1970 have been summarized by Sundelius (1970). Toward the end of the past century, it was recognized that much of the slate belt was characterized by semischistose volcanic rocks rather than metasedimentary slates. During the mid-1950s, the first detailed stratigraphic studies were completed in the central portion of the slate belt near Albemarle, North Carolina (Fig. 1). In the 1970s, several geochronological (Table 1) as well as paleontological studies (St. Jean, 1973; Maher and others, 1981) established a late Precambrian (northern portion) to Cambrian (central and southern portion) age for rocks of the Carolina slate belt. Geochemical studies by Black (1977), Seiders (1978), and

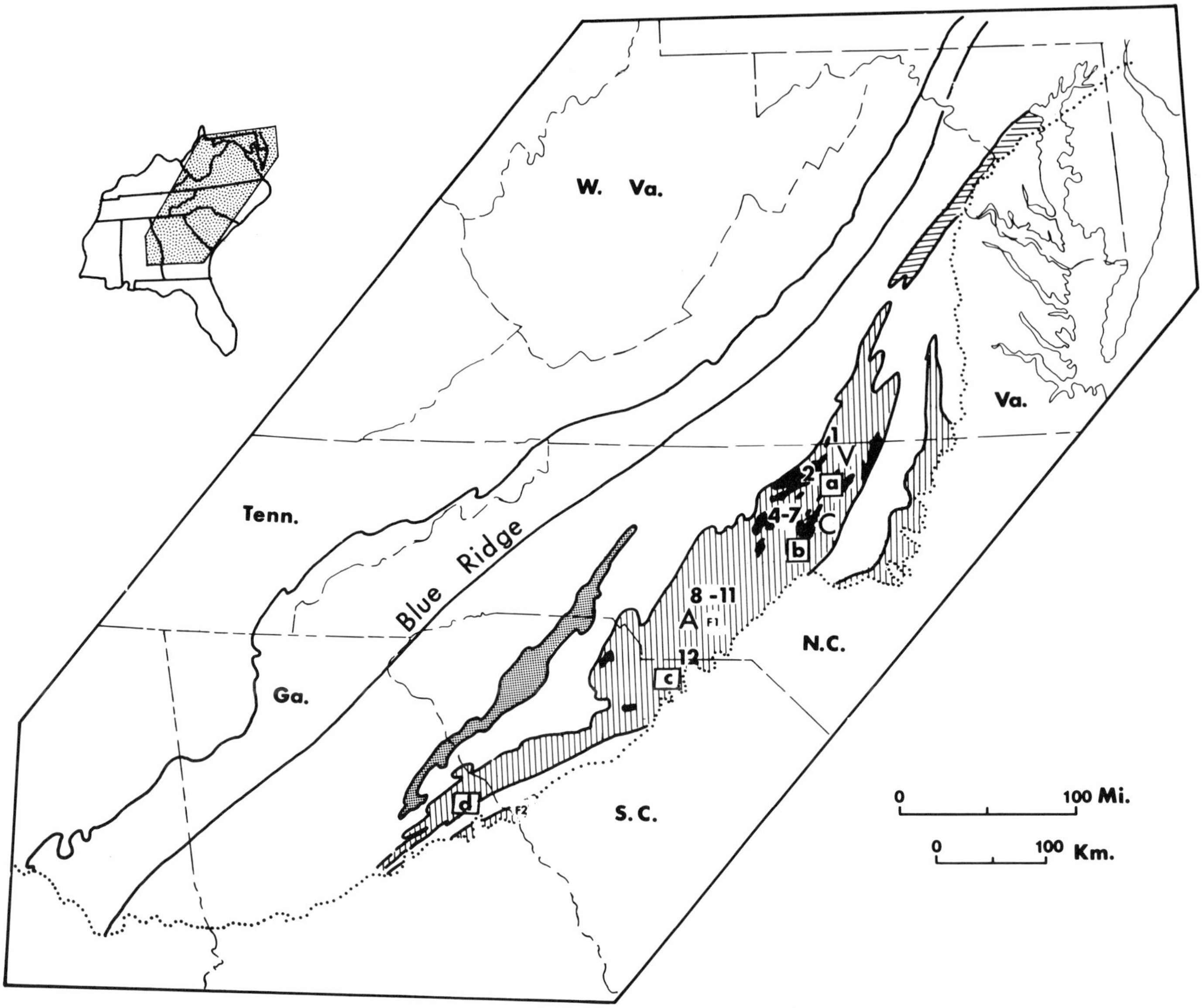

Figure 1. Generalized index map of the southern Appalachian Piedmont showing belts of low metamorphic grade that contain metavolcanic rocks. Vertical lines—Carolina slate belt; horizontal lines—Chopawamsic Formation in Virginia; dotted pattern—metavolcanic and metasedimentary rocks of the Kings Mountain belt. Dotted line indicates the extent of Coastal Plain onlap. Capital letters: V—Virgilina area: C—Chapel Hill area; A—Albemarle area. Small letters are areas of investigation in this volume: a—McConnell and Glover; b—Green and others; c—Bourland and Rigby; d—Carpenter and others. Papers by Rodgers and Feiss cover the entire region. The numbered locations are keyed to the radiometric dates given in Table 1. F1 is the *Paradoxides* locality of St. Jean (1973); F2 is the *Paradoxides* locality of Maher and others (1981). Black areas are plutons believed to be contemporaneous with volcanic activity.

Black (1980) recognized the bimodal nature of volcanic rocks preserved in the slate belt. Glover and Sinha (1973) noted the possibility of large-scale, late Precambrian folding and faulting in the northern (older) portion of the slate belt. This period of tectonic activity has been named the Virgilina deformational event. It is roughly equivalent in timing with the Avalonian event of the northern Appalachians.

Recent tectonic models for the southern Appalachians (Hatcher, 1972; Odom and Fullagar, 1973; Rankin, 1975) have related the development of the slate belt stratigraphic sequence to a volcanic arc environment associated with subduction. Based upon both geophysical and geochemical evidence, a variety of arc models have been proposed, including both eastward- and westward-dipping subduction

TABLE 1. SUMMARY OF RADIOMETRIC DATING FOR IGNEOUS ROCKS OF THE CAROLINA SLATE BELT

Map location	Geologic unit	Age (m.y.)	Method	Reference
Northern Carolina slate belt				
1	Tuffaceous volcanics, Virgilina area	620±20	U-Pb	Glover and Sinha (1973)
2	Roxboro granodiorite	575±20	U-Pb	Glover and Sinha (1973)
3(a)	Flat River complex	650±30	U-Pb	McConnell and Glover (1982)
4	Meadow Flats pluton	625±36	Rb-Sr	Black (1977)
5	Chapel Hill pluton	690±15 600±23	Rb-Sr Rb-Sr	Black (1977)
6	Dacite metavolcanics*	600±8	Rb-Sr	Black (1977)
7	Farrington pluton East pluton West pluton	 519±38 568±29	 Rb-Sr Rb-Sr	 Fullagar (1971, and pers. comm., 1982)
Central Carolina slate belt				
8	Uwharrie Formation	586±10	U-Pb	Wright and Seiders (1980)
9	Uwharrie Formation	553±49	Rb-Sr	Hills and Butler (1969)
10	Yadkin Member Millingport Formation	540±7	Rb-Sr	Black (1978)
11	Morrow Mountain rhyolite*†	514±15	Rb-Sr	Hills and Butler (1969)
12	Metarhyolite*	511±23	Rb-Sr	Butler and Fullagar (1975)
Southern Carolina slate belt				
13(d)	Lincolnton metadacite	566±15 554±20	U-Pb Rb-Sr	Carpenter and others (1982)

Note: U-Pb, uranium-lead concordia ages of zircons; Rb-Sr, rubidium-strontium whole-rock ages; all values have been recalculated to λ $^{87}Rb = 1.42 \times 10^{-11}$ yr^{-1}.

*The reported age is probably a minimum value.

†The sampled geologic unit may be a volcanic neck that is intrusive into the Albemarle Group.

zones, possible multiple or compound arcs, and a crustal environment ranging from oceanic to continental.

In this volume, J.W.W. Rogers reviews the relationship of geochemistry to tectonic setting, and using published data, he suggests a continental margin arc environment for the Carolina slate belt. Green and others, in a study of sedimentary characteristics of the volcanic deposits southwest of Chapel Hill, North Carolina, also report on the bimodal chemistry of these rocks. This is one of the few sedimentological studies published on rocks of the slate belt. Bourland and Rigby present paleontologic evidence for late Precambrian to Cambrian sponges in the slate belt of South Carolina and suggest a correlation between various parts of the slate belt. Carpenter and others present new radiometric age data from the Little River Series of Georgia, confirm a Cambrian age for these rocks, and suggest a broad similarity of volcanic history from Georgia into central North Carolina. McConnell and Glover present a study of the metaigneous rocks in the Durham, North Carolina, area and postulate a shallow, possibly surface-breaking, intrusion into a submarine volcanic pile. They also present new U-Pb zircon ages for evidence of late Precambrian igneous activity. Geoffrey Feiss gives a short description of several typical ore deposits in the Carolina slate belt and proposed a model relating ore genesis to environmental setting.

Several papers presented at the Carolina slate belt symposium held in 1979 are not represented in this issue. Thurmond and Whitney (1979) discussed the sequence of structural and sedimentological events deduced from mapping of the Chennault quadrangle in Georgia and South Carolina, which contains the Little River series in the slate belt and gneisses and schist in the Charlotte (or Uchee) belt. Significantly, they reported a possible regional uncomformity between the Charlotte and slate belts with no evidence of faulting, and interpreted the Charlotte belt as representing an older cycle of volcanic activity, the younger slate belt rocks resting noncomformably on the Charlotte belt rocks that form the basement in this part of the Piedmont.

Long (1979) suggested a rift basin tectonic setting for slate belt volcanic activity, the axis of the rift being parallel to the strike of the slate belt. His model is based on a positive free air and the Bouguer gravity signature. He proposed that seismic evidence supported this interpretation and indicated a velocity inversion beneath the slate belt. A similar inversion was reported in the Chapel Hill area by Black and others (1975). An examination of refraction data from near Troy, North Carolina (D.H. Warren, 1975, personal commun.), revealed no evidence of a refraction break at 10- to 15-km depth, thus supporting a velocity inversion beneath the slate belt which Long reported.

Maher (1979) reported on the Belair belt of South Carolina and Georgia, which is presumably analogous to the eastern slate belt in North Carolina. The Belair belt was interpreted as the eastern flank of an anticlinorium cored by amphibolite grade rocks of the Kiokee belt. The similarities between rocks of the Belair and slate belts were taken as a strong indication of depositionally correlative parts of a single island-arc sequence.

Snoke and Secor (1979) reported on quartz-rich metasedimentary rocks that are presumed to represent arc sediments in which quartz was derived from a near-by cratonic mass adjacent to the evolving Cambrian section of the Carolina slate belt. They noted similar lithologies in the higher grade rocks of the Kiokee belt.

Butler (1979), in reviewing the geologic setting of the Carolina slate belt of North Carolina and adjacent areas of South Carolina, suggested a late Precambrian to Cambrian island arc, calc-alkaline in nature, resting on either oceanic or possibly a thin continental crust. Butler noted that the basement is either unexposed or not yet recognized. He suggested that early calc-alkaline volcanism shifted toward a bimodal basalt plus rhyolite suite in which rhyolite predominates. This shift of the volcanic chemistry may have been due to crustal thickening.

Pavlides (1979), in a review of the Chopawamsic Formation of Virginia, discussed the possible relationship of this Cambrian age volcanic suite to the slate belt. At the conclusion of the symposium, Blackburn (1979) discussed a compilation study of volcanic rocks in the southern Appalachians, which was sponsored by the International Geological Correlation Project.

Unresolved Problems

Adequate mapping of the slate belt exists only in isolated

areas such as the Virgilina area near the Virginia–North Carolina state line; the Chapel Hill and Albemarle areas of North Carolina; thc arca along thc North Carlina–South Carolina border; in the vicinity of Columbia, South Carolina; and in the vicinity of the South Carolina–Georgia border (Little River Series). This leaves extensive areas unmapped, or at best, mapped on a reconnaissance level.

A number of especially critical areas are unmapped, most notably the area between Chapel Hill and Albemarle in North Carolina where the younger, central and southern slate belt abuts the older, northern slate belt. Another critical area is the border between the Carolina slate belt and the Charlotte belt. In the area of Virgilina, Tobisch and Glover (1969) suggested a conformable sequence with the Charlotte belt at higher metamorphic grade; in other words, the boundary between the belts is merely an isograd in a loose sense of the term. However, along strike to the southwest, Best and others (1973) pointed to a significant gravity gradient along boundary of about 100 mgals, which they suggested represents abutting crustal blocks of a considerable density contrast. Farther along strike, in southern North Carolina, the boundary is highly faulted, possibly an important factor in the mineralization of the Gold Hill area; however, this area lacks the significant gravity gradient found to the northeast. Detailed mapping along this belt boundary would be a significant contribution to establish a correlation between slate belt stratigraphy and that of the western Piedmont. This relationship is essential to an understanding of the southern Appalachian Piedmont, especially in view of the several mutually exclusive environments that have been suggested for the formation of the slate belt.

Paleomagnetic studies may be useful to establish the original location and orientation of the Carolina slate belt and to shed light on the relationship of the slate belt to other Appalachian volcanic terranes of roughly the same age and environment which are found northward along the Atlantic seaboard.

A fundamental issue that must be resolved by further geophysical and geochemical work is the question of what lies beneath the slate belt's 10- to 15-km thickness of volcanic rocks—older continental crust or oceanic crust? Interpretation of deep seismic reflection profiles in the Carolina slate belt (Black and others, 1975; Cook and others, 1979) suggests that much of the Piedmont, including the slate belt, may be an allochthonous sheet. Obviously, if this is the case, the rocks at great depth may not be the original crust over which the slate belt arc formed.

REFERENCES CITED

Best, D. M., Geddes, W. H., Watkins, J. S., 1973, Gravity investigation of the deep source of the Piedmont gravity gradient in Davidson County, North Carolina: Geological Society of America Bulletin, v. 84, p. 1213–1216.

Black, W. W., 1977, The geochronology and geochemistry of the Carolina slate belt of north-central North Carolina [Ph.D. dissert.]: Chapel Hill, University of North Carolina, 118 p.

1978, Chemical characteristics and Rb/Sr ages of metavolcanics from the Carolina slate belt of North Carolina: Geological Society of America Abstracts with Programs, v. 10, p. 162–163.

1980, Chemical characteristics of metavolcanics from the Carolina slate belt of North Carolina, *in* Wones, D. R., ed., The Caledonides in the USA: Department of Geological Sciences, Virginia Polytechnic Institute and State University Memoir no. 2, p. 271–278.

Black, W. W., Fergusion, J. F., and Stewart, D. M., 1975, Reflection seismic survey of a portion of S. Appalachian slate belt near Chapel Hill, N.C.: Geological Society of America Abstracts with Programs, v. 7, p. 470.

Blackburn, W. H., 1979, Compilation of igneous geology in the Appalachian-Caledonide orogen: Geological Society of America Abstracts with Programs, v. 11, p. 171.

Butler, J. R., 1979, The Carolina slate belt in North Carolina and northeastern South Carolina: A review: Geological Society of America Abstracts with Programs, v. 11, p. 172.

Butler, J. R., and Fullagar, P. D., 1975, Lilesville and Pageland plutons and associated meta-rhyolites, eastern Carolina slate belt: Geological Society of America Abstracts with Programs, v. 7, p. 475.

Carpenter, R. H., Odom, A. L., and Hartley, M. E., III, 1982, A geochronological investigation of the Lincolnton metadacite, Georgia and South Carolina, *in* Bearce, D. N., Kish, S. A., Black, W. W., and Tull, J. F., eds., Tectonic studies in the Carolina and the Talladega slate belts in the southern Appalachian orogen: Geological Society of America Special Paper 191, (this volume).

Cook, F. A., Albaugh, D. S., Brown, L. D., Kaufman, S., Oliver, J. E., and Hatcher, R. D., 1979, Thin-skinned tectonics in the crystalline southern Appalachians: COCORP seismic-reflection profiling of the Blue Ridge and Piedmont: Geology, v. 7, p. 563–567.

Fullagar, P. D., 1971, Age and origin of plutonic intrusions in the Piedmont of the southeastern Appalachians: Geological Society of America Bulletin, v. 82, p. 2845–2862.

Glover, L., III, and Sinha, A. K., 1973, The Virgilina deformation, a late Precambrian to early Cambrian orogenic event in the central Piedmont of Virginia and North Carolina: American Journal of Science, v. 273A, p. 234–251.

Hatcher, R. D., Jr., 1972, Development model for the southern Appalachians: Geological Society of America Bulletin, v. 83, p. 2735–2760.

Hills, F. A., and Butler, J. R., 1969, Rubidium-strontium dates for some rhyolites from the Carolina slate belt of the North Carolina Piedmont: Geological Society of America, Abstracts for 1968, Special Paper 21, p. 445.

King, P. B., 1955, A geologic cross-section across the southern Appalachians: an outline of the geology in the segment in Tennessee, North Carolina, and South Carolina, *in* Russell, R. J., ed., Guides to southeastern geology: Geological Society of America, p. 332–373.

Long, L. T., 1979, The Carolina slate belt may be a continental rift: Geological Society of America Abstracts with Programs, v. 11, p. 187.

Maher, H. D., Jr., 1979, The Belair belt of South Carolina and Georgia: Stratigraphy and depositional regime as compared to the Carolina slate belt: Geological Society of America Abstracts with Programs, v. 11, p. 187.

Maher, H. D., Jr., Palmer, A. R., Secor, D. T., and Snoke, A. W., 1981, New trilobite locality in the Piedmont of South Carolina and its regional implications: Geology, v. 9, p. 34–36.

McConnell, K. I., and Glover, Lynn, III, 1982, Age and emplacement of the Flat River Complex, an Eocambrian sub-volcanic pluton near Durham, North Carolina, *in* Bearce, D. N., Kish, S. A., Black, W. W., and Tull, J. F., eds., Tectonic studies in the Carolina and the Talladega slate belts in the southern Appalachian orogen: Geological Society of America Special Paper 191, (this volume).

Odom, A. L., and Fullagar, P. D., 1973, Geochronologic and tectonic relationships between the Inner Piedmont, Brevard zone and Blue Ridge belts, North Carolina: American Journal of Science, v. 273-A, p. 133–149.

Pavlides, L., 1979, The Chopawamsic formation of Virginia and its regional relationships: Geological Society of America Abstracts with Programs, v. 11, p. 207.

Rankin, D. W., 1975, The continental margin of eastern North America in the southern Appalachians: The opening and closing of the proto-Atlantic Ocean: American Journal of Science, v. 275-A, p. 298–336.

Seiders, V. M., 1978, a chemically bimodal, calc-alkalic suite of volcanic rocks, Carolina volcanic slate belt, central North Carolina: Southeastern Geology, v. 19, p. 241–265.

Snoke, A. W., and Secor, D. T., Jr. 1979, The significance of quartz-rich metasedimentary rocks in the Carolina slate belt, South Carolina Piedmont: Geological Society of America Abstracts with Programs, v. 11, p. 213.

St. Jean, J., Jr., 1973, A new Cambrian trilobite from the Piedmont of North Carolina: American Journal of Science, v. 273A, p. 196–216.

Sundelius, H. W., 1970, The Carolina slate belt, *in* Fisher, G. W., Pettijohn, F. J., Reed, J. C., Jr., and Weaver, K. N., eds., Studies of Appalachian geology: Central and southern: New York, Interscience, p. 351–367.

Thurmond, C. J., and Whitney, J. A., 1979, Structural stratigraphic development of the Charlotte and northern slate belts within the Chennault quadrangle, Georgia and South Carolina: Geological Society of America Abstracts with Programs, v. 11, p. 215.

Tobisch, O. T., and Glover, L., III, 1969, Metamorphic changes across part of the Carolina slate belt–Charlotte belt boundary, North Carolina and Virginia: U.S. Geological Survey Professional Paper 650-C, p. C-1 to C-7.

Wright, J. E., and Seiders, V. M., 1980, Age of zircon from volcanic rocks of the central North Carolina Piedmont and tectonic implications for the Carolina volcanic slate belt: Geological Society of America Bulletin, v. 91, part 1, p. 287–294.

Manuscript Accepted by the Society July 8, 1982

Printed in U.S.A.

Geological Society of America
Special Paper 191
1982

Criteria for recognizing environments of formation of volcanic suites; Application of these criteria to volcanic suites in the Carolina Slate Belt

John J. W. Rogers
Department of Geology
University of North Carolina at Chapel Hill
Mitchell Hall 029A
Chapel Hill, N. C. 27514

INTRODUCTION

The Carolina slate belt contains a wide variety of volcanic rocks. As a contribution to deciphering the history of the belt, this paper attempts to determine the environments of formation of a number of individual volcanic suites. For this paper, the term "environment" means not only local conditions (subaqueous, subaerial, etc.) but also broad tectonic features such as the presence or absence of continental crust, subduction zones, and other major characteristics.

The variety of environments is clearly enormous. For the purpose of simplification, this paper subdivides them into six major types. The first section of the paper describes some characteristic properties of each type. The second part discusses criteria for recognizing environments gradational between oceanic island arcs and continental margins. The third part briefly applies the various characteristics to volcanic suites in the Carolina slate belt.

RECOGNITION OF ENVIRONMENTS

This paper emphasizes characteristics of volcanic suites that can be used in the field and on moderately altered rocks. Chemical properties are given for each suite with the understanding that metamorphism, weathering, and so forth may change some primary features. Some effort is made to distinguish chemical features most or least susceptible to alteration.

The number of papers that describe volcanic suites is almost limitless, and no effort is made to refer to many of them. References cited are generally those papers that summarize previous information on one or more suites. General references to a variety of volcanic suites include Kuno (1968), Miyashiro (1974), Condie (1976), and Donnelly and Rogers (1978, 1980).

Generalized characteristics of idealized volcanic suites are shown in Tables 1 to 6. This paper concentrates on the properties of suites rather than on the characteristics of individual rocks. Single samples commonly should not be used for tectonic inferences because of the possibility of secondary modification of primary features or, simply, the chance that they are not representative of the assemblage.

Mid-Ocean Ridge Basalt (MORB) and Back-Arc Basin (Table 1)

Although clearly variable in detail, this suite is characterized by broad-scale homogeneity (Miyashiro and others, 1969; Kay and others, 1970; Donnelly and Rogers, 1980). The dominant characteristic is the monotonous sequence of basalts, many of which show evidence of subaqueous eruption such as pillows.

Some efforts have been made to distinguish basalts of small ocean basins (including back-arc basins) from MORB produced at major spreading ridges. The data of Hawkins (1977), however, mostly from the Lau Basin, show that basalts from the two environments have such an overlap-

TABLE 1. MID-OCEAN RIDGE BASALT (MORB) AND BACK-ARC BASIN

SiO_2 range	- restricted (46 to 52%)	
eruptive style	- subaqueous; some pillows, no pyroclastics	
mineralogy	- olivine and plagioclase phenocrysts	
K_2O	0.1-0.2%	
Rb	2 ppm	lower than in all
Ba	15 ppm	other basalt types
Sr	100 ppm	
TiO_2	1.5%	
Zr	100 ppm	higher than in
Cr	300 ppm	subduction-zone basalts
Ni	100 ppm	
REE pattern	- light REE depletion	
$Zr-TiO_2$	- linear relationship	
FeO*/MgO	- < 1.5 (lower than in arc or continental-margin basalts)	

ping range of features that they are not clearly distinguishable in the geologic section.

In addition to oceanic ridges and back-arc basins, thick sequences of basalt occur in continental plateau areas and in young intra-oceanic island arcs. Plateau basalts are commonly distinguishable on the basis of subaerial eruption (forming soil zones, etc.); in some cases, minor silicic pyroclastic components are also present. Large ion lithophile (LIL) elements (K, Rb, etc) are more abundant in continental plateau basalts than in MORB, although such elements are potentially affected by metamorphism.

Distinction between MORB and basalts from primitive island arcs can be made in most instances, although both are generally erupted subaqueously. As summarized in Table 3, primitive island-arc basalts commonly occur in association with silicic eruptives, which are very rare in MORB suites. Primitive arc suites contain less TiO_2 and Zr than MORB, both elements being regarded as relatively unaffected by metamorphism or weathering. Furthermore, the flat REE pattern of the island-arc basalts is distinct from the light REE depletion of MORB, and these patterns may also be unaffected by secondary processes. Pyroxene phenocrysts (mostly clinopyroxene) are much more common in arc basalts than in MORB.

Oceanic Island (Table 2)

Volcanic rocks formed in oceanic islands are quantitatively less abundant than those formed in other environments and probably do not constitute a large proportion of the geologic record. They should not be particularly difficult to recognize provided that the entire suite is reasonably well represented in outcrop. Complete geochemical studies of volcanic islands or seamounts are not numerous, and the properties discussed here are taken largely from the work of Aumento (1968), Schilling (1975), and Zielinski (1975).

The major, and generally oldest, volcanic assemblage in oceanic islands is tholeiitic (subalkaline) basalt. This basalt is mostly subaqueously erupted, like MORB, but some portions may be subaerial. The island basalts are moderately enriched in LIL elements relative to MORB and, in this regard, are similar to the basalts of primitive island arcs. Among the possibly preservable chemical features of oceanic-island basalts are a relatively high TiO_2 content and a flat or light-enriched REE pattern (contrasted with the light-depleted REE pattern of MORB).

TABLE 2. OCEANIC ISLAND

Two suites (possibly transitional)		
Subalkaline suite		
basalt similar to MORB but with higher:		
TiO_2	1.5%	
K_2O	0.4%	
Ba	50-100 ppm	
REE pattern shows light REE enrichment		
Alkaline suite		
SiO_2 range	- mostly basalt; some rocks up to 65%	
mineralogy	- alkaline; normative, and some modal, nepheline	
Alkali olivine basalts		
higher than other basalts in:		
TiO_2	2.7%	
K_2O	1.2%	higher in most trace elements
Ba	300 ppm	than other basalts
Sr	400 ppm	

The most useful distinguishing property of island suites, however, is the occurrence of small amounts of alkaline rocks, particularly in the later stages of island evolution. Chemically, these rocks would be recognized by high ratios of alkali/SiO_2 or alkali/MgO, where this chemistry is preserved. Alkaline mineralogy, shown by minerals such as aegerine or titanaugite, is more likely to be preserved during metamorphism and weathering than the chemical ratios involving alkali elements. Rocks such as nephelinites, phonolites, and trachytes also occur in environments other than oceanic islands, but the association of these alkaline varieties with large amounts of tholeiitic basalt should be a good indication of an island assemblage.

Suites of rocks commonly referred to as "alkali olivine basalts" (AOB) can be recognized on the basis of (1) a mineralogy consisting primarily of olivine, clinopyroxene, and plagioclase; (2) nepheline in the norm; (3) absence of modal or normative orthopyroxene; and (4) high TiO_2 contents. These rocks occur in such a wide variety of environments that they apparently cannot be used to identify tectonic conditions. Schwarzer and Rogers (1974) have shown the AOB from oceanic islands, island arcs, and continental interiors all have virtually identical compositions.

Early Intra-Oceanic Island Arc (Table 3)

The term "intra-oceanic" is used here to refer to island arcs formed during subduction of oceanic lithosphere under other oceanic lithosphere. During the early stages of this subduction process, volcanic rocks generally form subaqueously, building initial island piles from the ocean floor. These rocks have been referred to by a variety of terms, including "island arc tholeiite" (Jakes and Gill, 1970) and "primitive island arc" (Donnelly and Rogers, 1978, 1980). The discussion of these rocks is taken from various summaries, including Gill (1970), Donnelly and others (1971), Miyashiro (1974), Donnelly and Rogers (1978, 1980), and Garcia (1978).

A characteristic and generally preserved feature of primitive suites is the frequency distribution of SiO_2. Rocks in the range of about 55% to 65% SiO_2 (largely andesites) are rare or absent. This absence contrasts strongly with the abundance of andesites in other subduction-zone assemblages (later stages of island arcs or active continental margins). The primitive suite, therefore, shows a bimodality of the SiO_2 distribution, but the relative abundance of the mafic

TABLE 3. EARLY INTRA-OCEANIC ISLAND ARC

SiO_2 range	- 50% to 75%; andesite rare; may be bimodal (basalt-dacite or spilite-keratophyre)	
eruptive style	- subaqueous; some pillows; some non-explosive pyroclastics	
mineralogy	- clinopyroxene, plagioclase, and some olivine phenocrysts in mafic rocks	
Mafic rocks		
K_2O	0.5%	
Rb	5 ppm	higher than in MORB; similar
Ba	100 ppm	to calcalkaline arc basalts
Sr	200 ppm	
TiO_2	0.8%	
Zr	50 ppm	lower than in MORB; similar
Cr	50 ppm	to calcalkaline arc basalts
Ni	50 ppm	
REE pattern	- flat	
Zr-TiO	- linear relationship	
FeO*/ MgO	- >1.5	
Silicic rocks		
impoverished in lithophilic elements		

and silicic members is not clear. In some primitive suites (e.g., northeastern Caribbean; Donnelly and Rogers, 1980), silicic and mafic rocks occur in subequal abundance. In other areas (e.g., Viti Levu, Fiji; Gill, 1970), silicic rocks are only rarely reported. Whether all primitive suites show bimodal SiO_2 distributions or whether some are almost exclusively mafic is uncertain based on data currently available. Bimodal primitive suites are commonly referred to as "spilite-keratophyre" or "basalt-dacite" assemblages; in some cases, both assemblages are present.

The SiO_2 abundances in primitive arc rocks must be compared with those of other suites. Keratophyric rocks have been identified in mid-oceanic-ridge dredge and core samples, but the amounts are extremely small. Certainly, no thick sequence of subaqueous keratophyres or dacites has been identified as having formed at an oceanic spreading center. Rhyodacites occur on some oceanic islands (summary by McBirney and Gass, 1967), but the relative abundance is small. Bimodal SiO_2 frequencies are found in some continental-interior suites, such as the younger Tertiary rocks of the Basin and Range province of the western United States (McKee, 1971). The continental-interior bimodal suites should be distinguishable from island-arc suites on the basis of a dominantly subaerial eruptive style, including abundance of ignimbritic debris.

The silicic rocks of the primitive arc assemblage are compositionally distinct. They contain very low amounts of LIL elements even in rocks with as much as 75% SiO_2. This criterion, where preserved, permits distinction of these rocks from the more LIL-enriched silicic assemblages of continental margins and continental interiors.

The mafic rocks in primitive arc suites have a variety of characteristic properties. A typical ferromagnesian mineral is clinopyroxene, which commonly occurs as phenocrysts; phenocryst clinopyroxene is rare in MORB. Abundances of LIL elements are higher than in MORB but generally lower than in continental plateau basalts. The flat REE pattern is distinctive and probably preservable through secondary alterations. For some reason currently unexplained, all subduction-zone assemblages have low abundances of TiO_2 and Zr relative to MORB; Cr and Ni are similarly in low abundance in subduction-generated basalts.

Mature Island Arc (Calcalkaline) (Table 4)

Primitive arc assemblages are commonly succeeded upward by rocks that can be termed "calcalkaline" by a large variety of criteria. These criteria include (1) differentiation trend showing low iron enrichment on an AFM diagram (Kuno, 1968); (2) moderate K_2O enrichment with increasing SiO_2 (Gill, 1970); and (3) abundance of andesites and pyroxene andesites. Properties of these suites have been summarized by Gill (1970), Donnelly and others (1971), Miyashiro (1974), Condie (1976), Donnelly and Rogers (1978, 1980), and Garcia (1978).

Some efforts have been made to separate typical calcalkaline suites from more potassic (shoshonitic) assemblages. The stratigraphic intermingling of volcanic rocks of widely different K_2O contents, however, caused Donnelly and Rogers (1980) to regard highly potassic (HK) rocks as a subgroup within the typical calcalkaline suite.

The abundance of andesites, generally subaerially erupted, distinguishes calcalkaline arc suites from those of early arcs, intra-oceanic islands, or spreading centers. The distinction between intra-oceanic-arc and continental-margin andesites is difficult because of the gradational character of the two groups; the presence of intermediate suites is discussed in the next section.

Basalts of calcalkaline island arcs can be distinguished from those of some other environments. The principal characteristic of calcalkaline arc suites is the low ratio of basalt

TABLE 4. MATURE ISLAND ARC (CALCALKALINE)

SiO_2 range	- unimodal; andesite common; >68% rare	
eruptive style	- subaerial; flows and pyroclastics	
mineralogy	- clinopyroxene in andesites and basalts	
Mafic rocks		
similar to early arc basalts except: light REE enriched (contrasting to flat pattern in early basalts); no linear relationship between Zr and TiO_2		
Andesites		
K_2O	1.0%	
Rb	15 ppm	lower than in continental-
Ba	300 ppm	margin andesites
Sr	400 ppm	
TiO_2	0.7%	
Zr	100 ppm	lower than in continental-
Cr	30 ppm	margin andesites
Ni	5 ppm	

Note: Calcalkaline suite shows steady increase in lithophilic element content toward the more silicic rocks.

to andesite and the subaerial, commonly pyroclastic, style of eruption. Chemically, calcalkaline arc basalts are very similar to those of primitive arcs except for greater light REE enrichment in the calcalkaline basalts. Also, calcalkaline mafic rocks commonly do not show the linear Zr-TiO_2 relationships that characterize MORB and primitive arc basalts.

Silicic calcalkaline rocks (dacites and rhyolites) of intra-oceanic arcs contain much higher LIL concentrations than keratophyres or dacites of primitive arcs. The best criterion for distinguishing them from the silicic rocks of calcalkaline continental margins is relative abundance of rock types; the percentage of rocks with more than about 68% SiO_2 is much higher in continental-margin suites than in those of island arcs.

Continental Margin (Calcalkaline) (Table 5)

A great deal of information on continental-margin calcalkaline suites has been obtained from studies in the Andes. Because of its exceptionally thick crust, however, the Andes may not be a typical example of active continental margins. Information on various continental-margin suites has been summarized by Miyashiro (1974), Dickinson (1975), Rogers and Novitsky-Evans (1977), and Rogers and Ragland (1980).

Continental-margin, subduction-zone suites have a number of characteristic features. Principal identifying features are (1) calcalkaline differentiation trends; (2) amphibole as a common varietal mineral (instead of clinopyroxene in island-arc suites); (3) pronounced LIL enrichment toward more differentiated rocks; (4) subaerial eruption, with a high proportion of pyroclastic rocks; and (5) scarcity of rocks with SiO_2 less than 60%. The LIL enrichment is shown not only by such typical elements as K, Rb, and Ba but also by considerable enrichment in light REE. The proportion of high-SiO_2 rocks apparently increases with increasing crustal thickness (Miyashiro, 1974).

Compositional differences occur between andesites of island arcs and continental margins. In general, LIL elements are more enriched in continental-margin suites. Also, continental-margin andesites contain slightly higher concentrations of Cr, Ni, and Zr than oceanic-arc suites.

TABLE 5. CONTINENTAL MARGIN (CALCALKALINE)

SiO_2 range	- mostly 60% to 75%; minor basalts	
eruptive style	- subaerial; commonly pyroclastic, some flows	
mineralogy	- amphibole in intermediate rocks	
Mafic rocks		
similar to intra-oceanic-arc calcalkaline basalts		
Andesites		
K_2O	2.0%	higher than in intra-oceanic arc andesites
Rb	50 ppm	
Ba	500 ppm	
Sr	600 ppm	
TiO_2	0.9%	higher than in intra-oceanic arc andesites
Zr	200 ppm	
Cr	60 ppm	
Ni	30 ppm	

Note: Rate of increase in abundances of lithophilic elements with increasing SiO_2 is higher than in intra-oceanic arc suites.

Continental (Table 6)

An enormous variety of volcanic rocks occurs in continental areas. The most abundant suites are plateau basalts and rift assemblages.

Most plateau basalts are subalkaline; Wright and others (1973) have described the Columbia River basalt as a typical example. The principal identifying characteristics in the field are subaerial eruption (columnar joints, soil zones, etc.) and minor siliceous pyroclastic horizons. The chemical distinction of plateau basalts from MORB and primitive arc basalts is discussed in the section on MORB. Some plateau areas contain mildly alkalic basalts, recognized by being nepheline normative.

The volcanic assemblages of rift zones are very diverse. The variety of rock types is briefly discussed by Mutschler and others (1978). There appears to be no particular type that exclusively indicates a rift environment. Peralkaline rocks are common in rift zones but also occur in both continental and oceanic areas not obviously involved in rifting. The basalts of rift zones include varieties similar to MORB, AOB, and continental plateaus. Alkaline rocks are present in rifts but also occur in oceanic islands and isolated continental areas. In short, there seems to be no special criterion currently available for the recognition of rift-zone environments.

TABLE 6. CONTINENTAL

great diversity of rock types:	
mafic to silicic	
flows and pyroclastics	
subalkaline and alkaline	
peraluminous and peralkaline	
Plateau basalts	
SiO_2 range	- 48% to 52%; rare silicic horizons
eruptive style	- subaerial; flows, rare pyroclastics
mineralogy	- quartz-normative to nepheline-normative basalts
compositionally similar to calcalkaline arc basalts except higher in:	
TiO_2	2.0%
Cr	200 ppm
Ni	30 ppm
Rift zones	
great variety of rock types, including:	
plateau basalts	
alkaline suite (basalts, phonolites, etc.)	
peralkaline suite	
ignimbrites (peralkaline and peraluminous)	

AREAS INTERMEDIATE BETWEEN INTRA-OCEANIC ISLAND ARCS AND CONTINENTAL MARGINS

Rocks from typical intra-oceanic island arcs contrast strongly with those from Andean-type continental margins. Some suites of volcanic rocks, however, seem to have characteristics intermediate between these two ideal types. For example, Dickinson (1975) distinguished peninsular areas, island arcs fringing continental margins, and so forth as having K_2O contents intermediate between those of completely intra-oceanic arcs and those of true continental margins; at equivalent SiO_2 concentrations and depths to underlying subduction zones, K_2O contents are lowest in arc rocks, highest in continental-margin rocks, and occupy an intermediate range in the "intermediate" suites.

A suite of rocks regarded as having formed along a thin continental margin (20- to 30-km thickness) has been described by Rogers and Novitsky-Evans (1977) and Rogers and Ragland (1980). The Eocene to lower Oligocene Clarno Formation of central Oregon has characteristics of rocks from both mature island arcs and continental margins. Island-arc features include (1) preponderance of andesites, with a small percentage of silicic rocks; (2) basalts with a content of LIL elements virtually identical to those of calcalkaline arcs (Table 7); and (3) clinopyroxene as the dominant varietal mineral in the andesites. Continental-margin features include (1) subaerial eruption with interbedded lake sediments and (2) andesites somewhat richer in LIL elements that those of intra-oceanic arcs (Table 7). Regional geologic relationships also show that the Clarno Formation was formed during subduction of oceanic lithosphere under the western margin of North America in mid-Tertiary time.

Table 7 summarizes data from a variety of sources on the typical andesites and basalts of mature (calcalkaline) island arcs and continental margins (references and data in Rogers and Novitsky-Evans, 1977, and Rogers and Ragland, 1980). The "intermediate" character of the Clarno Formation is most clearly shown by LIL abundances in the andesites, with the Clarno basalts very similar to those of island arcs. Rogers and Novitsky-Evans (1977) concluded that formations such as the Clarno, with intermediate properties, could best be explained as having formed on a crust that was "intermediate" between continental and oceanic. This type of crust presumably has a thickness less than that of continents and greater than that of ocean basins: furthermore, the composition of the crust may also be intermediate between continental and oceanic.

TABLE 7. COMPOSITIONS OF BASALTS AND ANDESITES FROM INTRA-OCEANIC ISLAND ARCS, CONTINENTAL MARGINS, AND INTERMEDIATE ENVIRONMENTS

	Basalts			Andesites		
	continental margin	inter-mediate (Clarno)	intra-oceanic arc	continental margin	inter-mediate (Clarno)	intra-oceanic arc
SiO_2	51.6	50.0	51.5	60.3	60.0	59.4
TiO_2	1.29	1.20	0.87	0.89	1.14	0.64
Al_2O_3	18.2	14.9	18.0	16.8	15.6	16.5
Fe_2O_3	9.4	10.5	10.1	6.2	7.8	7.9
MgO	6.1	8.1	5.4	3.7	2.7	3.0
CaO	8.9	10.5	11.2	5.7	6.4	7.4
Na_2O	3.5	2.8	2.7	3.9	3.7	3.3
K_2O	0.85	0.48	0.51	2.2	1.3	1.0
Rb	20	10	7	62	43	17
Sr	501	311	344	576	345	406
Y		28	12	13	25	22
Zr		124	49	203	217	81
Nb		16			17	
Ba	274	201	167	568	409	338
V		232	271	136	127	147
Cr	197	432	95	74	86	37
Ni	88	135	52	40	35	12

Note: Fe_2O_3 is total iron as Fe_2O_3. Blank spaces indicate insufficient data. Major-element abundances are in weight percent; trace-element abundances in parts per million.

Continental-margin and intra-oceanic-arc values are averages of individual suites compiled by Rogers and Ragland (1980), who also cite original references. Basalt data are averages of rocks with SiO_2 contents between about 48% and 52%. Andesite data are averages of rocks with SiO_2 contents between about 58% and 62%.

Clarno data are from Rogers and Ragland (1980). Basalts are a suite of 12 samples with SiO_2 contents between 48.0% and 52%. Andesites are a suite of 46 samples with SiO_2 contents between 58.0% and 62.0%.

ENVIRONMENTS IN THE SLATE BELT

The slate belt is one of the northeast-southwest-trending linear belts of the Appalachian orogen in the southeastern United States. The complete extent of the belt is unclear because of the possible correlation of the slate belt of the Carolinas with similar lithologic sequences both to the northeast and the southwest. Lithologically, the belt consists of volcanic, volcaniclastic, and clastic rocks spanning a period of time probably extending from the late Precambrian to the early Paleozoic. Rocks of the belt commonly show low-grade (greenschist facies) metamorphism, and most areas have undergone mild deformation. The belt is approximately 50 miles wide, both margins, where exposed, being represented by rapid transitions to more highly metamorphosed rocks of adjoining belts. The nature of the margins (depositional, tectonic, etc.) is not clear. Thus, much of the slate belt escaped the most intense deformation in the Appalachian Piedmont during Paleozoic time. The belt has, however, been intruded by plutons of various Paleozoic ages, culminating in the final tectonic/magmatic activity in the area about 300 m.y. ago (Fullagar and Butler, 1979).

The criteria discussed in previous sections for the recognition of various tectonic environments can be applied to volcanic assemblages in the slate belt. The conclusions drawn about each assemblage are discussed below. Sundelius (1970) summarized the geology of the slate belt and compiled previous geochemical data; his data have been used in addition to the specific studies of individual suites referenced below.

Uwharrie Formation, North Carolina

The Uwharrie Formation of south-central North Carolina is largely a series of fine-grained, felsic pyroclastic rocks with some mafic pyroclastics, felsic flows, and mudstones. Lithologic and stratigraphic evidence has led to controversy about whether the pyroclastic rocks were deposited subaerially or subaqueously.

Chemical information on the Uwharrie Formation is provided mostly by Seiders (1978). Only major-element data are available, and pyroclastic rocks are particularly susceptible to secondary alteration. Sodium contents in the suite, however, are reasonably uniform (3% to 4% Na_2O), and this constancy is commonly not found in areas of major metasomatism. Assuming no excessive alteration, the high K_2O contents (2% to 3%) indicate a continental-margin or continental-interior environment. Certainly this dominantly pyroclastic suite is not characteristic of ocean basins or early arcs.

Albemarle Group, North Carolina

The Albemarle Group (including the Tillery Formation) overlies the Uwharrie Formation in south-central North Carolina. It consists of mafic and felsic pyroclastic rocks and minor flows in a thick sequence of marine clastic sedimentary rocks.

Major-element chemical information has been provided by Butler and Ragland (1969), Black (1977, 1980), and Seiders (1978); Butler and Ragland (1969) also obtained Rb values, and Black (1977, 1980) determined Rb, Sr, Y, Zr, and Ba.

The following summary observations can be made:

1. The suite is clearly bimodal in SiO_2 content, with a scarcity of andesites and an abundance of both felsic and mafic rocks.
2. Metasomatism has probably been extensive, as shown by Na_2O contents variable from less than 1% to more than 5%; K_2O, Rb, and Ba contents are also highly variable in rocks of equivalent SiO_2 content.
3. In the mafic rocks, TiO_2 contents are generally less than 1%, possibly indicating a subduction-zone assemblage.

Chemical and lithologic data place the Albemarle Group as either an early island-arc or a continental-interior assemblage. Interbedded marine sediments clearly indicate that the environment was not a continental interior. A totally oceanic environment for the Albemarle Group, however, is unlikely in view of the "continental" character of the immediately underlying Uwharrie Formation. Furthermore, the bimodal assemblages of intra-oceanic arcs commonly form massive volcanic piles rather than pyroclastic interbeds with clastic sediments. A most likely environment for the Albemarle Group, therefore, seems to be subduction under some type of crust intermediate between that of typical oceans or oceanic arcs and that of typical continents.

Upper Volcanic Assemblage of Albemarle area, North Carolina

This group of rocks either overlies or at least represents upper parts of the Albemarle Group. It was referred to by Butler and Ragland (1969) as the Tater Top Group. The assemblage consists of basaltic to andesitic tuffs and flows plus the Morrow Mountain rhyolite. Butler and Ragland (1969) and Black (1977, 1980) provided major-element and some trace-element information.

Rocks of this upper assemblage show the same mobility of Na_2O and other alkali elements as in the underlying Albemarle Group. Subduction-zone activity is implied by the presence of andesites and the relatively low (less than 1%) TiO_2 contents of the basalts; Zr contents are also low (less than 100 ppm). The Morrow Mountain rhyolite is not a typical continental or continental-margin rock. Although MgO contents in the rhyolite are low (less than 0.5%), abundances of alkali elements are lower than might be expected for a continental or continental-margin felsic rock; K_2O contents are less than 3.0% and Rb less than 100 ppm. Barium contents are variable in the felsic rocks.

As with the Albemarle Group, the most likely environment of formation of this upper volcanic assemblage is an area intermediate between oceanic and continental structure. The group was probably formed in response to subduction-zone activity. If secondary alteration had not modified original rock composition, there might have been considerable similarity between the upper volcanic assemblage in the Albemarle area and the Clarno Formation of Oregon, discussed above.

Chapel Hill Area, North Carolina

Black (1977, 1980) summarized information on volcanic-sedimentary sequences in the Chapel Hill area. The rocks consist of felsic to mafic flows and pyroclastic rocks with associated argillites and other clastic sedimentary rocks.

Black (1977) provided data on 30 individual samples. Only 1 sample has Na_2O less than 1% and none have more than 5% Na_2O. Other elements seem to follow typical variation trends with SiO_2. Therefore, it seems likely that metasomatism has been less intense in the Chapel Hill area than in other parts of the slate belt. The dominant characteristics of the suite are as follows:

1. Rocks with high SiO_2 contents (65% or more) are common; the suite is also probably bimodal in SiO_2 contents.

2. The composition of the basalts is similar to that of continental-margin suites (compare Tables 7 and 8).

3. The composition of rhyodacites (Table 8) is consistent with continental-margin, calcalkaline differentiation trends (compare with Table 7).

4. The TiO_2 contents of the basalts (Table 8) are characteristic of subduction-zone environments.

These chemical properties, plus the mixed flow and pyroclastic rock types, indicate that the rocks of the Chapel Hill area were formed by subduction-zone activity along a continental margin. As wth the Albemarle Group, it seems likely that the margin was comparatively thin. The nature of the lithosphere subducted to form the suite is not clear.

Virgilina Area, North Carolina

An extensive sequence of volcanic and epiclastic rocks occurs in North Carolina near the Virginia state border. Lithologic information is summarized by Glover and Sinha (1973), and Bland (1978) reported chemical data.

The Virgilina area contains a thick series of deformed volcanogenic rocks overlying gneiss of uncertain age. Most of the volcanic suite is pyroclastic or epiclastic, ranging in composition from mafic to felsic. The suite appears to be unimodal in SiO_2 contents (Black, 1980) and at least partly andesitic. A sequence of metabasalts (both flow and pyroclastic rocks) in the upper part of the section is commonly referred to as the Virgilina greenstone. Analyses of this greenstone (from Bland, 1978) are shown in Table 8.

Lithologically, the Virgilina sequence appears to be typical of subduction in mature island arcs or thin continental margins. Although the extent of metasomatism is unknown, the composition of the greenstone is consistent with this interpretation (compare Tables 7 and 8). In particular, subduction is indicated by the low TiO_2 and Zr contents. The relatively high K_2O abundance (1.45%) would not be found in intra-oceanic-arc basalts, but it may simply be the result of secondary processes.

Pavlides (1979) has studied the Chopawamsic Formation of Virginia, a possible slate belt correlative. He reports a unimodal SiO_2 distribution, which is consistent with observations in the Virgilina area.

Lancaster Area, South Carolina

A suite of pyroclastic and flow rocks from Lancaster County, South Carolina, has been discussed and analyzed by Black (1977), These rocks range from felsic to mafic and are overlain by a sequence of argillites.

Only six samples have been analyzed, all from the felsic assemblage, and two have clearly undergone alkali metasomatism (data from Black, 1977). The four samples that show "normal" and consistent alkali contents have high K_2O (2% to 5%) and high Ba (600 to 1,400 ppm) abundances. Not enough information is available to draw extensive conclusions, but the rocks seem to be somewhat analogous to the Uwharrie Formation, as discussed above. No typical oceanic crust seems to have been present in the Lancaster area when these rocks were formed.

TABLE 8. SOME BASALTS AND RHYODACITES OF SLATE BELT

	Chapel Hill Area		Virgilina Area
	Basalt	Rhyodacite	Greenstone
SiO_2	50.6	65.9	49.1
TiO_2	1.05	0.68	0.86
Al_2O_3	18.0	15.2	14.2
Fe_2O_3	10.3	5.1	12.0
MgO	5.4	1.1	8.6
CaO	8.0	2.3	8.7
Na_2O	3.4	3.7	2.4
K_2O	0.67	2.8	0.54
Rb	16	63	14
Sr	571	259	407
Y	13	32	21
Zr	91	241	90
Ba	396	750	--

Note: Fe_2O_3 is total iron as Fe_2O_3. Major-element abundances are in weight percent; trace-element abundances in parts per million. For Chapel Hill Area: basalt average is based on 4 samples with SiO_2 contents between 50.0% and 52.0%; rhyodacite average is based on 7 samples (6 for Ba) with SiO_2 contents between 63.0% and 67.0%; all data from Black (1977). Virgilina data are from Bland (1978), based on 12 basalts.

Little River Series, South Carolina

A southwestern extension of the slate belts occurs near Lincolnton, Georgia, and McCormick, South Carolina. Rocks in this area are referred to as the Little River Series. The assemblage consists of felsic volcanic and volcaniclastic rocks with minor mafic flows, tuffs, and dikes. The predominantly volcanic section is overlain by a sedimentary sequence consisting of argillites and graywackes, partly volcanically derived; this sequence shows abundant structures characteristic of subaqueous deposition. The assemblage has been described and analyzed by Whitney and others (1978).

Whitney and others (1978) concluded that the Little River Series was formed as part of a primitive, intra-oceanic, island arc. Chemically, their conclusion is based on bimodality of the SiO_2 distribution, lack of variation in Al_2O_3 from mafic to felsic rocks, and low K_2O contents; (rocks with 70% to 75% SiO_2 generally contain only 0.5% to 1.5% K_2O). The rocks are also interbedded with marine sediments.

There are two arguments against the possibility of formation of the Little River Series in a primitive island arc. First, only two of the eight dacites analyzed contain less than 4.5% Na_2O. Concentrations of Na_2O greater than 4.5% commonly indicate metasomatic alteration. Whether this alteration could cause the very low K_2O contents of the suite is uncertain. A second argument is the very limited abundance of mafic rocks. The only mafic rocks analyzed by Whitney and others (1978) are dikes. Significant amounts of basalt or spilite-flow rocks are apparently absent. In terms of quantities of mafic and felsic rocks, the Little River Series compares more closely with the bimodal sequences of the Basin and Range area (McKee, 1971) than with primitive island arcs; no suggestion is made here, however, that the series formed in a continental interior.

As pointed out in the preceding discussion, the Little River Series is similar to some other rocks of the slate belt in being primarily a suite of felsic tuffs interbedded with marine sedimentary rocks. Water depth and other environmental parameters are not available for the sedimentary sequence. Nevertheless, as at other localities in the slate belt, the assemblage of sedimentary rocks and felsic tuffs does not seem to be an adequate indication of formation on an oceanic crust or in a primitive island arc. As elsewhere, a plausible environment of formation has both continental and oceanic characteristics.

SUMMARY

Based on present information, definitive statements can hardly be made about tectonic conditions in the slate belt during formation of the volcanic-sedimentary assemblages. There seems to be evidence for two different types of suites:

1. Andesitic assemblages formed by subduction along continental margins (Chapel Hill and Virgilina areas, possibly Chopawamsic Formation in Virginia).
2. Assemblages consisting largely of interbedded felsic tuff and fine-grained, clastic, marine sediments. A most probable tectonic environment includes both continental and oceanic features. One possibility is a continental crust thin enough for the accumulation of significant amounts of marine sediments.

There is no evidence of purely oceanic crust in the area of the southeastern Appalachian slate belt during formation of its volcanic-sedimentary sequences. Whether the entire area was underlain by a continental crust of variable thickness is not certain. A full understanding of the belt obviously depends on much more information than is currently available.

ACKNOWLEDGMENTS

Preparation of the criteria for recognizing volcanic environments has been greatly aided by my association, throughout many years, with T. W. Donnelly, J. M. Novitsky-Evans, P. C. Ragland, and R. R. Schwarzer. Comments on volcanic rocks in the slate belt could not have been made without the cooperation of W. W. Black, J. R. Butler, and P. D. Fullagar. The entire paper has been made possible through the efforts of S. A. Kish.

REFERENCES CITED

Aumento, F., 1968, The mid-Atlantic ridge near 45° N, II—Basalts from the area of Confederation Peak: Canadian Journal of Earth Sciences, v. 5, p. 1–21.

Black, W. W., 1977, The geochronology and geochemistry of the Carolina slate belt of north-central North Carolina [Ph.D. Thesis]: Chapel Hill, University of North Carolina at Chapel Hill, 118 p.

1980, Chemical characteristics of metavolcanics in the central Carolina slate belt, *in* Wones, D. R., ed., Proceedings of "The Caledonides in the USA," Blacksburg, Virginia Polytechnic Institute and State University, Dept. of Geological Sciences, Memoir 2, p. 271–278.

Bland, A. E., 1978, Trace element geochemistry of volcanic sequences of Maryland, Virginia, and North Carolina, and its bearing on the tectonic evolution of the central Appalachians [Ph.D. Thesis]: Lexington, University of Kentucky, 328 p.

Butler, J. R., and Ragland, P. C., 1969, Petrology and chemistry of meta-igneous rocks in the Albemarle area, North Carolina slate belt: American Journal of Science, v. 267, p. 700–726.

Condie, K. C., 1976, Trace-element geochemistry of Archean greenstone belts: Earth Science Reviews, v. 12, p. 393–417.

Dickinson, W. R., 1975, Potash-depth (K-h) relations in continental margin and intra-oceanic magmatic arcs: Geology, v. 3, p. 53–56.

Donnelly, T. W., and Rogers, J.J.W., 1978, The distribution of igneous rock suites throughout the Caribbean, *in* MacGillavry, H. J., and Beets, D. J., eds., The 8th Caribbean Geological Conference, Willemstad, 1977: Geologie en Mijnbouw, v. 57, p. 151–162.

——1980, Igneous series in island arcs—The northeastern Caribbean compared with worldwide island-arc assemblages: Bulletin Volcanologique, v. 43, p. 347–382.

Donnelly, T. W., Rogers, J.J.W., Pushkar, P., and Armstrong, R. L., 1971, Chemical evolution of the igneous rocks of the eastern West Indies—An investigation of thorium, uranium, and potassium distributions, and lead and strontium isotopic ratios, *in* Donnelly, T. W., ed., Caribbean geophysical, tectonic, and petrologic studies: Geological Society of America Memoir 130, p. 181–224.

Fullagar, P. D., and Butler, J. R., 1979, 325- to 265-m.y.-old granitic plutons in the Piedmont of the southeastern Appalachians: American Journal of Science, v. 279, p. 161–185.

Garcia, M. D., 1978, Criteria for the identification of ancient volcanic arcs: Earth Science Reviews, v. 14, p. 147–165.

Gill, J. B., 1970, Geochemistry of Viti Levu, Fiji, and its evolution as an island arc: Contributions to Mineralogy and Petrology, v. 27, p. 179–203.

Glover, L., III, and Sinha, A. K., 1973, The Virgilina deformation, a late Precambrian to Early Cambrian(?) orogenic event in the central Piedmont of Virginia and North Carolina, *in* Glover, L., III, and Ribbe, P. H., The Byron N. Cooper Volume: American Journal of Science, v. 273-A, p. 234–251.

Hawkins, J. W., Jr., 1977, Petrologic and geochemical characteristics of marginal basin basalts, *in* Talwani, M., and Pitman, W. C., III, eds., Island arcs, deep sea trenches, and back-arc basins: American Geophysical Union, Maurice Ewing Series, v. 1, p. 355–365.

Jakes, P., and Gill, J. B., 1970, Rare earth elements and the island arc tholeiitic series: Earth and Planetary Science Letters, v. 9, p. 17–28.

Kay, R., Hubbard, N. J., and Gast, P. W., 1970, Chemical characteristics and origin of oceanic ridge volcanic rocks: Journal of Geophysical Research, v. 75, p. 1585–1613.

Kuno, H., 1968, Differentiation of basalt magmas, *in* Hess, H. H., and Poldervaart, A., eds., Basalts—The Poldervaart treatise on rocks of basaltic composition: New York, Interscience, v. 2, p. 623–688.

McBirney, A. R., and Gass, I. G., 1967, Relations of oceanic volcanic rocks in mid-oceanic rises and heat flow: Earth and Planetary Science Letters, v. 2, p. 265–276.

McKee, E. H., 1971, Tertiary igneous chronology of the Great Basin of western United States—Implications for tectonic models: Geological Society of America Bulletin, v. 82, p. 3497–3502.

Miyashiro, A., 1974, Volcanic rock series in island arcs and active continental margins: American Journal of Science, v. 274, p. 321–355.

Miyashiro, A., Shido, F., and Ewing, M., 1969, Diversity and origin of abyssal tholeiite from the mid-Atlantic ridge near 24° and 30° north latitude: Contributions to Mineralogy and Petrology, v. 23, p. 38–52.

Mutschler, F. E., Finn, D. D., and Ludington, S., 1978, Magmatism and related ore deposits of extensile continental environments—Computer exercises in petrochemical pattern recognition and prediction: Program and Abstracts of 1978 International Symposium on the Rio Grande Rift, Santa Fe, New Mexico, Los Alamos Scientific Laboratory, p. 64–65.

Pavlides, L., 1979, The Chopawamsic Formation of Virginia and its regional relationships: Geological Society of America Abstracts with Programs, v. 11, p. 207–208.

Rogers, J.J.W., and Novitsky-Evans, J. M., 1977, The Clarno Formation of central Oregon, U.S.A.—Volcanism on a thin continental margin: Earth and Planetary Science Letters, v. 34, p. 56–66.

Rogers, J.J.W., and Ragland, P. C., 1980, Trace elements in continental-margin magmatism, Part I—Trace elements in the Clarno Formation of central Oregon and the nature of the continental margin on which eruption occurred: Geological Society of America Bulletin, Part I, v. 91, p. 196–198; Part II, v. 91, p. 1217–1292.

Schilling, J. G., 1975, Azores mantle blob—Rare earth evidence: Earth and Planetary Science Letters, v. 25, p. 103–115.

Schwarzer, R. R., and Rogers, J.J.W., 1974, A worldwide comparison of alkali olivine basalts and their differentiation trends: Earth and Planetary Science Letters, v. 23, p. 286–296.

Seiders, V. M., 1978, A chemically bimodal calc-alkalic suite of volcanic rocks, Carolina volcanic slate belt, central North Carolina: Southeastern Geology, v. 19, p. 241–265.

Sundelius, H. W., 1970, The Carolina slate belt, *in* Fisher, G. W., Pettijohn, F. J., Reed, J. C., Jr., and Weaver, K. M., eds., Studies of Appalachian geology, Central and Southern: New York, Interscience, p. 351–367.

Whitney, J. A., Parks, T. A., Carpenter, R. H., and Hartley, M. E., III, 1978, Volcanic evolution of the southern slate belt of Georgia and South Carolina—A primitive island arc: Journal of Geology, v. 86, p. 173–192.

Wright, T. L., Grolier, M. J., and Swanson, D. A., 1973, Chemical variation related to the stratigraphy of the Columbia River basalt: Geological Society of America Bulletin, v. 84, p. 371–386.

Zielinski, R. A., 1975, Trace element evaluation of a suite of rocks from Reunion Island, Indian Ocean: Geochimica et Cosmochimica Acta, v. 39, p. 713–734.

MANUSCRIPT ACCEPTED BY THE SOCIETY JULY 8, 1982

Printed in U.S.A.

Geological Society of America
Special Paper 191
1982

Volcanic and volcaniclastic facies in a part of the slate belt of North Carolina

George B. Green*
Victor V. Cavaroc
Edward F. Stoddard
Abdelzahir M. Abdelzahir*
Department of Geosciences
North Carolina State University
Raleigh, North Carolina 27650

ABSTRACT

Slate belt rocks of southern Chatham and northern Moore Counties, North Carolina, constitute an overlapping series of metamorphosed volcanic and volcaniclastic strata reflecting three distinct sites of volcanic activity separated in time, space, and composition. At the base of the section, felsic volcanic and epiclastic rocks had a northeastern source. Subsequently, deposition was dominated by a center of relatively mafic (50 to 57 wt% SiO_2) volcanism to the northwest. Accretionary lapilli and sedimentary structures indicate subaerial to shallow subaqueous conditions in the north; southward decrease in grain size and changes in structures suggest deposition on the deeper-water slopes of the volcanic edifice. A still younger site of felsic (68 to 75 wt% SiO_2) volcanism abruptly developed on the southern flank of the waning mafic eruptive complex; local depositional paleoslope reversed, as felsic flows and tuffs interfingered northward with banded mudstones.

The bimodal igneous rock geochemistry is consistent with the stratigraphic model. Low K/Na ($K_2O/Na_2O < 0.2$) may reflect alteration and metamorphism; some (more stable) minor and trace elements suggest calc-alkalic affinities.

INTRODUCTION

Purpose of the Study

The volcanic and volcaniclastic rocks that constitute the slate belt of the Piedmont province in the eastern United States have stood for many decades as classic examples of eugeosynclinal facies rocks (e.g., King, 1959). Since the early 1960s their implications for global tectonic models as well as their known association with sulfide mineralization have generated renewed interest in these volcanic belt rocks. As a result, both regional studies and investigations of specific features of these rocks have been common. Conversely, detailed local studies of the component lithofacies and their variations and spatial relationships are few in nuber. This is in part due to the low relief and deep weathering profiles of the region.

This paper presents the results of a detailed study conducted periodically during the period 1973 to 1979 in a small (~390 km^2) area of slate belt rocks from North Carolina. The specific goals of this paper are to characterize (1) the lithofacies present, (2) their stratigraphic relationships, and (3) their geologic framework in terms of both the character of volcanic activity and the depositional environments of the resultant strata.

Geologic Setting

Slate belt rocks form a zone of greenschist facies metavol-

*Present addresses: Green, 1435 Sedwick Road, Durham, North Carolina 27713; Abdelzahir, Geology and Mineral Resources Department, P.O. Box 410, Khartoum, Sudan.

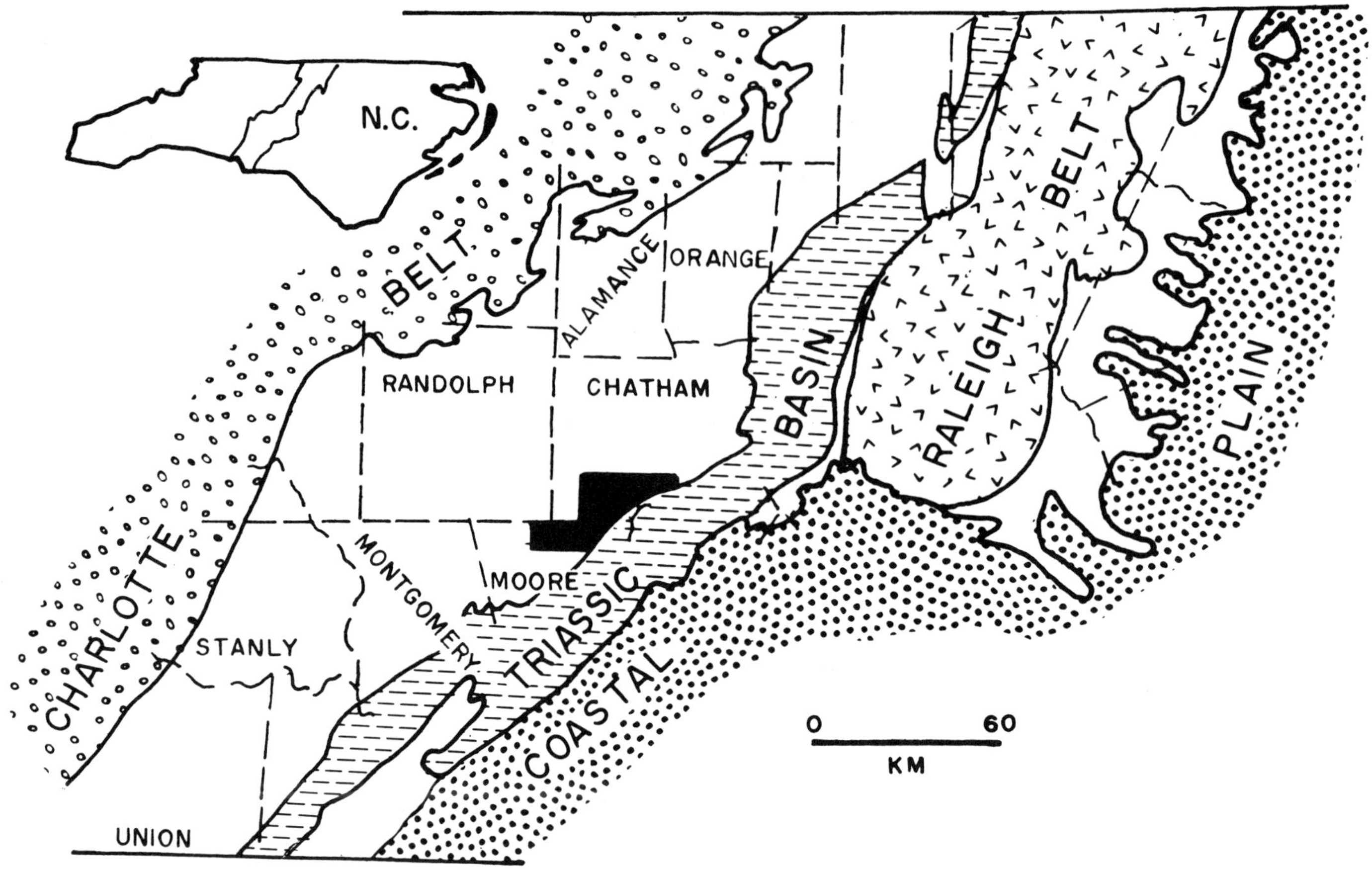

Figure 1. General geologic map of central North Carolina showing the location of North Carolina slate belt rocks in white. Counties listed are those referred to in the text; the study area is shown in black.

canic and metasedimentary rocks that extends in a southwesterly direction from southeastern Virginia into central Georgia. In North Carolina, the slate belt lithofacies (Fig. 1) is divided into an eastern and a western part by the gneisses, schists, and granitic intrusives of the Raleigh belt and the unmetamorphosed sedimentary rocks of the Durham and Wadesboro Triassic basins. The contact between rocks of the slate belt and the amphibolite facies metamorphic rocks of the Raleigh belt has been variously interpreted as conformable (Parker, 1968) or as tectonic in nature (Hatcher and others, 1977; Farrar, 1980). The eastern part of the slate belt covers a smaller area and has received less study than the western part. The eastern slate belt is overlapped by Mesozoic-Cenozoic age Coastal Plain sediments to the east, but deep wells have encountered rocks lithologically similar to those of the slate belt as far as 100 km to the east (Stuckey, 1965, p. 90). The more extensive western slate belt is bounded on the east by downfaulted Triassic sediments and on the west by the plutonic and high-grade metamorphic rocks of the Charlotte belt.

Trilobites (St. Jean, 1973) from slate belt rocks located about 60 km to the west of the study area suggest an Early to Middle Cambrian age for those rocks; nearby metarhyolites studied geochronologically (Rb/Sr) by Hills and Butler (1969) yield revised ages (Wright and Seiders, 1980) also in the Cambrian. However, other parts of the slate belt have yielded radiometric ages of late Precambrian (Glover and Sinha, 1973; Black and Fullagar, 1976; Wright and Seiders, 1977).

The area investigated in the present study (Fig. 1) lies along the eastern margin of the western slate belt. It includes the Goldston and Bear Creek 7.5-minute quadrangles in southern Chatham County, together with the northern parts of the Putnam and Robbins 7.5-minute quadrangles of northern Moore County.

Previous Work

Lithology, Stratigraphy, and Depositional Model

Conley (1962a) mapped the Albemarle 15-minute quadrangle, located about 50 km southwest of the study area along the Stanly-Montgomery county line. He divided the rocks into a lower volcanic sequence, a middle volcanic-sedimentary sequence, and an upper volcanic sequence, the upper volcanic sequence separated from the two older units by an unconformity. In reconnaissance fashion, he also mapped Moore County (Conley, 1962b), which overlaps the southern edge of the study area. He considered the rocks of Moore County to be stratigraphically equivalent to the two lower sequences of the Albemarle quadrangle. On the basis of Conley's mapping, Conley and Bain (1965) defined several formations and projected them over most of the western slate belt.

Subsequent studies in the vicinity of the study area include Stuckey's (1967) description of several of the pyrophyllite mines and prospects and Bain's (1966) geology and groundwater report, which differentiates argillite from non-argillite rocks. A general reconnaissance map by the North Carolina Department of Mineral and Economic Resources has been published (Wilson and Carpenter, 1975). In addition, McDaniel (1976) reevaluated pyrophyllite deposits occurring within the study area.

Butler and Ragland (1969a) postulated an island arc setting for the deposition of rocks in the Albemarle area and suggested Unalaska Island in the Aleutians as a modern analogue. Randazzo (1972) suggested that the rocks in part of Union County resemble the "back deep" and "inner volcanic arc" of a volcanic arc system and proposed the Andaman Sea, as described by Bemmelen (1949; 1954), as an analogue. Still more recent studies of the slate belt (e.g., Black, 1977; Whitney and others, 1978) have also argued for a volcanic arc origin.

Structural Framework

The structure of the western slate belt is characterized by folds whose axes are subparallel to the northeastward outcrop trend; axial planes generally dip to the northwest. Conley (1962b) stated that the dominant structure in the central western slate belt is the southwestward-plunging Troy anticlinorium. This fold is over 40 km wide; the present study area lies in a part of its southeastern limb (see Fig. 3). The axial zone is occupied by Conley's lower volcanic sequence, while overlying andesitic tuffs and slates define the southeastern flank of the structure. A series of smaller, generally doubly plunging anticlines and synclines, which vary in wavelength from 1 to 5 km, occurs along the southeast flank of the anticlinorium. These folds are overturned to the southeast, and axial plane cleavage dips monotonously to the northwest at 55° to 70° (Conley, 1962b). While the minor folds are easily observed in the field, extension of the Troy anticlinorium into the present study area is contingent upon the regional correlation of the four major stratigraphic units of Conley and Bain (1965) across the slate belt. The continuity of these units into the study area, however, is questionable due in part to the lack of detailed mapping in intervening areas. A significant argument against such long-distance correlation is recognition that the slate belt rocks appear to represent a very complex interfingering of similar lithofacies from multiple sources. This observation is compatible with current interpretations of slate belt rocks as the remnants of an ancient volcanic and volcanogenic sedimentary complex.

Faults are not a pronounced feature on geologic maps of the slate belt; perhaps this is due in part to the lack of detailed mapping. In his study of Moore County, Conley (1962b) divided faults into two groups. The first consists of northeast-trending, Paleozoic longitudinal faults, intruded by quartz veins, and often silicified and mineralized. The other group consists of northwest-trending cross-faults that offset the longitudinal faults and commonly are intruded by Mesozoic diabase dikes.

Metamorphism

Pervasive Paleozoic regional metamorphism has affected the rocks of the Carolina slate belt, developing mineral assemblages indicative of P-T conditions spanning the greenschist facies; still, primary volcanic and sedimentary features commonly are preserved. For this reason, the prefix "meta" will be omitted from rock names in this paper. The age of regional metamorphism can be bracketed with certainty between Early Cambrian and Carboniferous (Butler, 1972; Fullagar and Butler, 1979). K/Ar dating of slate belt metasedimentary rocks (Kish and others, 1979) suggests a metamorphic age of 483 ± 15 m.y. ($\sim$Taconic).

Techniques

Several hundred exposures were studied in the area. These ranged from creek beds to a freshly dug grave site. Many were saprolite or partially weathered rock; however, numerous others, including prospect pits and quarries, provided fresh samples for subsequent analysis. In a few instances, a portable diamond-bit coring tool was used for fresh sample collection.

Petrographic mineral identification was corroborated by X-ray diffraction scans of randomly oriented whole-rock powders; X-ray fluorescence analysis was used for determination of major and minor element whole-rock chemistry.

VOLCANOGENIC STRATIGRAPHY

General Stratigraphic Sequence

The slate belt rocks of the area are lithologically similar to those found in other areas of the western slate belt of North Carolina and reflect a chemically bimodal, probably calc-alkalic suite such as described by Seiders (1978). Repetitious rock types, poor outcrop continuity, rapid facies changes, and degree of weathering precluded attempts to trace individual rock units. The stratigraphic divisions of Figure 2 therefore were selected as mappable facies of similar internal lithology which also reflect consistency of depositional conditions. Nearly all contacts between these subdivisions in the field appear to be vertically gradational. No attempt has been made to correlate these stratigraphic divisions to the rock units defined by Conley (1962a). The authors believe the current low level of understanding of the rates of lateral variations in these rocks, combined with the paucity of detailed mapping of the intervening areas, would make such correlations tenuous at best.

The five stratigraphic units depicted in Figure 2 are summarized as follows (from oldest to youngest):

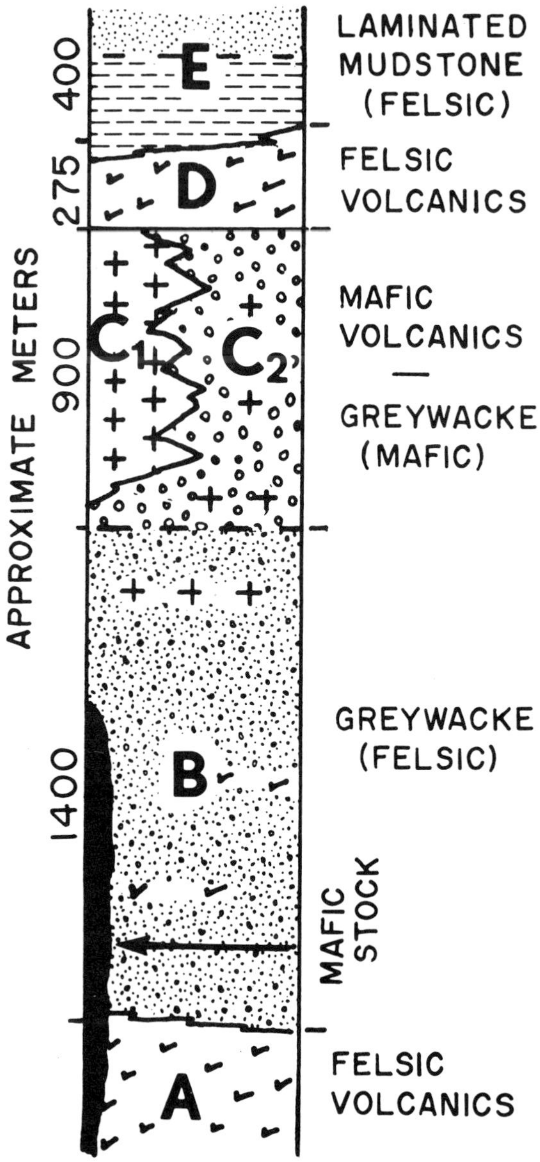

Figure 2. General stratigraphic column; alphabetical unit designations are referred to in the text. Unit thicknesses are approximate and based on map outcrop widths corrected for average structural dip.

Unit A. Felsic ash-flow tuff and interbedded epiclastic rocks; limited exposure, base not observed; gradational upward into Unit B.

Unit B. Predominantly felsic volcanic graywacke sandstone interbedded with thinly laminated mudstone; local conglomerate and felsic tuff; 1,300 m minimum thickness; gradational upward into Unit C-2.

Unit C. Intermediate to mafic volcanic and sedimentary rocks; consists of two laterally equivalent subunits; 900 m minimum thickness; abrupt contact with Unit D. ***C-1.*** Interbedded, apparently non-pillowed, intermediate to mafic lava flows, volcanic breccia, and tuff. ***C-2.*** Volcanic graywacke sandstone interbedded with laminated mudstone and local andesitic tuff.

Unit D. Predominantly light-colored felsic volcanic rocks; contains flows, pyroclastic rocks, and epiclastic rocks; 275 m minimum thickness; intercalates northward with base of Unit E.

Unit E. Predominantly thinly laminated, blue-gray claystone to mudstone; coarsens upward; minimum thickness 400 m (top eroded); upper 120 m with thinly laminated red to red-purple sandstone.

Structural Setting

The dominant structure of the area (Fig. 3) is a large, doubly plunging syncline with a northeasterly axial trend and is overturned to the southeast. Except in the vicinity of the axial plane, both bedding and well-developed axial-plane cleavage dip northwesterly 35° to 50°. At the northwestern edge of the map area, two large anticlines (10 to 12 km wavelength) with axial planes approximately parallel to the syncline plunge into the area from opposite directions.

The southwestern nose of the syncline is offset by northwest-trending cross-faults, while a major northeast-striking longitudinal fault has removed the adjacent anticlinal axis to the south. The northeast-striking fault exhibits high-angle reverse displacement (Conley, 1962b), is associated with vein quartz, and locally is mineralized by pyrophyllite, sericite, chloritoid, and pyrite (e.g., McDaniel, 1976). A band of similar mineralization in the vicinity of the anticlines at the northwestern edge of the map suggests similar faulting, but this possibility has not been documented. Faults of the second set of northwest-trending cross-faults commonly are associated with diabase dikes and offset both major northeast-striking faults and Triassic strata. Minor faults that cannot be traced also occur within the study area.

Lithologic Properties and Variations

Unit A: Lower Felsic Volcanics

The oldest unit exposed in the area contains felsic vol-

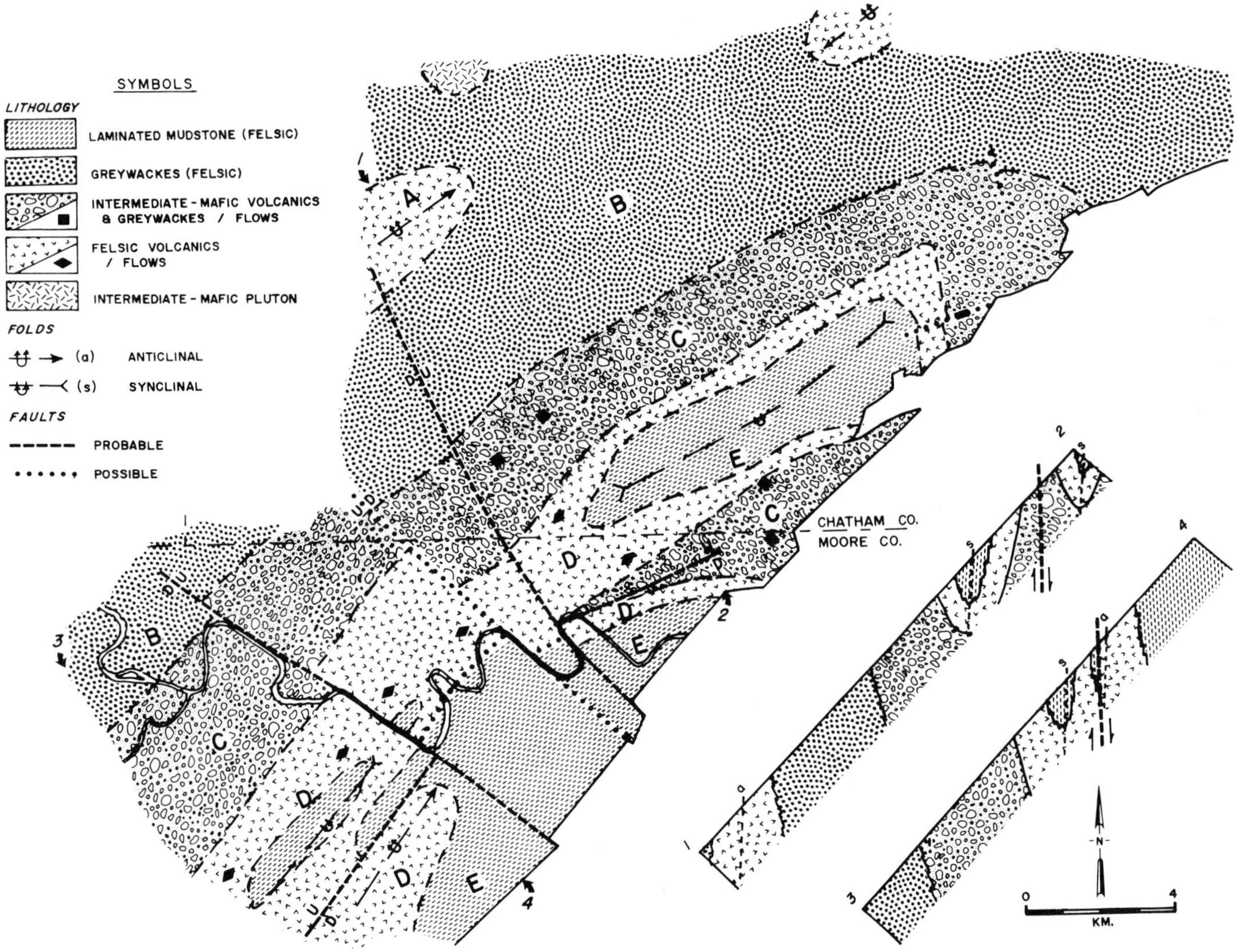

Figure 3. Geologic map of the study area. Alphabetical unit designations are those in the stratigraphic section of Figure 2.

canic rocks, at least some of which were deposited subaerially. These volcanic rocks consist of ash-flow tuffs, probable lava flows, and airfall pyroclastic accumulations. Coarse-grained volcanic conglomerate, breccia, and fine-grained, thinly bedded sedimentary rocks also are included in the unit. Unit A is best exposed in the core of the northeast-plunging anticline; its thickness cannot be estimated because its base is not exposed in the area.

The rocks inferred to be ash-flow tuffs or lava flows are very dense, tough, blue-gray to white flow-banded rocks with aphanitic groundmasses and angular, white to pinkish sodic plagioclase phenocrysts, some of which display zonation in thin section. Fine-grained, more strongly foliated and more easily weathered tuffs are interbedded with these deposits and may have resulted from airfall between flows. Alternatively, both may be components of the same ash flows, representing more welded and indurated areas where the ash flow was hottest alternating with cooler parts of the ash flows that were not as well indurated (E. Burt, 1977, personal communication). Chemical analysis of a sample from this unit (Table 1, analysis 1) indicates that the rock is dacitic in composition.

Discontinuous units of epiclastic volcanic cobble and pebble conglomerate are found in this unit. These conglomerates are moderately to well sorted and contain subangular to rounded rock fragments mineralogically similar to adjacent felsic volcanics.

Along the nose of the southwest-plunging anticline, the oldest rocks are porphyritic (plagioclase phenocrysts) felsic lava flows and breccias. At this location the apparently monolithic breccias are younger than the flows and contain what appear to be kaolinized rock fragments, locally altered to pyrophyllite. The flow material and breccias, as well as some of the overlying sediments, have been hydrothermally

TABLE 1. MAJOR OXIDE CHEMISTRY OF SELECTED VOLCANIC AND RELATED ROCKS, CHATHAM AND MOORE COUNTIES, NORTH CAROLINA

	Unit A	Unit B					Unit C				
	1	2	3	4	5	6	7	8	9	10	11
SiO_2	73.9	68.2	47.5	48.7	49.7	50.5	48.1	54.9	56.3	50.1	52.5
(SiO_2)*	(75.3)	(69.4)	(50.4)	(50.8)	(52.9)	(53.3)	(50.3)	(57.0)	(57.3)	(52.5)	(53.6)
Al_2O_3	13.1	16.1	16.7	17.6	17.3	17.7	17.6	14.8	16.8	18.7	19.1
TiO_2	0.34	0.38	0.69	0.73	0.65	0.64	0.72	0.70	0.77	0.93	0.90
Fe_2O_3†	1.62	3.13	10.30	9.97	8.58	8.55	8.91	9.55	8.50	10.48	9.94
MgO	0.2	1.1	7.7	7.3	4.9	5.1	7.3	9.9	6.1	5.2	3.3
CaO	0.64	3.20	7.70	7.16	6.93	6.17	10.05	2.43	5.47	4.85	6.41
K_2O	1.43	2.33	0.10	0.13	0.43	0.46	0.04	0.01	0.20	0.04	0.05
Na_2O	6.8	3.8	3.4	4.1	5.3	5.4	2.7	4.0	3.9	4.9	5.5
P_2O_5	0.02	0.08	0.21	0.25	0.19	0.19	0.15	0.12	0.13	0.18	0.15
Ign. loss	0.8	2.6	5.5	5.1	5.2	4.9	4.0	4.1	3.4	3.3	3.2
Total	98.9	100.9	99.8	101.0	99.2	99.6	99.6	100.5	101.6	98.7	101.1

	Unit C			Unit D						Pluton	
	12	13	14	15	16	17	18	19	20	21	22
SiO_2	45.7	53.4	55.5	73.8	74.4	64.7	85.7	68.6	67.6	48.0	52.0
(SiO_2)*	(51.9)	(57.7)	(57.9)	(74.5)	(76.2)	(67.1)	(85.7)	(71.3)	(69.1)	(48.6)	(54.0)
Al_2O_3	12.3	15.7	17.6	13.5	12.5	15.2	7.3	13.1	17.5	14.8	16.2
TiO_2	0.48	0.69	0.45	0.39	0.25	0.52	0.41	0.34	0.23	1.73	0.63
Fe_2O_3†	7.81	6.23	5.20	2.10	1.60	4.88	0.90	2.87	1.78	14.28	8.03
MgO	1.6	4.3	4.2	0.4	0.3	1.7	0.1	0.2	0.4	6.0	6.9
CaO	15.4	8.09	5.80	0.35	0.01	4.10	0.95	6.13	0.20	9.65	8.41
K_2O	1.13	1.05	0.40	5.21	2.81	0.74	0.14	0.34	0.08	0.25	0.52
Na_2O	3.6	2.8	6.5	3.3	5.8	4.4	4.5	4.8	10.1	3.7	3.5
P_2O_5	0.12	0.27	0.22	na	na	0.13	na	0.12	0.01	0.28	0.08
Ign. loss	13.2	8.5	5.7	3.1	2.5	3.2	1.7	1.2	1.0	1.1	2.9
Total	101.3	101.0	101.6	102.2	100.2	99.6	101.7	97.4	98.9	99.8	99.2

Key to analyzed samples

1. Flow-banded porphyritic dacite[§], Unit A, Chatham County.
2. Dacitic crystal tuff, Unit B, Deep River, Moore County.
3,4. Porphyritic hornblende-bearing tuff, upper Unit B, N Chatham area.
5,6. Andesitic flow(?) interbeds, upper Unit B, SW Chatham area.
7. Amygdaloidal basalt, Unit C-1, SE Chatham area.
8. Andesite flow, Unit C-1, NE Moore area.
9. Andesitic tuff(?), Unit C-2, NE Moore area.
10,11. Mafic-intermediate flow or tuff, Unit C, Deep River, NE Moore.
12-14. Volcanic clasts, Unit C-2 graywacke, SW Moore.
15. Flow-banded rhyolite, Unit D, Moore County.
16. Flow-banded dacite, Unit D, Moore County.
17. Dacitic flow, Unit D, near county line.
18. Silicified felsic porphyry, Unit D, Moore County.
19,20. Altered felsic flow(?) rocks, Unit D, county line.
21. Medium-grained augite-hornblende gabbro, pluton, NW Chatham area.
22. Porphyritic hornblende diorite from same pluton as 21.

Note: X-ray fluorescence analyses, NCSU; na, not analyzed.

*Figures in parentheses represent SiO_2 recalculated on a volatile-free basis.

†Total Fe as Fe_2O_3.

§Rock names based on normative mineralogy, using the IUGS classification (Streckeisen, 1979).

altered. Immediately overlying these volcanic rocks is a sedimentary sequence consisting mainly of fine-grained, thinly bedded sediments with graded bedding, scour and fill structures, and local interbedded poorly sorted, medium to coarse-grained graywacke sandstones. A well-sorted, well-rounded, clast-supported polylithic volcanic conglomerate (Fig. 4A) occurs near the contact with the overlying unit.

Unit B: Felsic Graywacke Unit

A sedimentary sequence consisting predominantly of interbedded, tan to light-brown, medium-grained volcanic graywacke sandstone and siltstone occurs stratigraphically above and interfingers with the lower felsic unit. Through much of Chatham County these sandstones have flat to gently curved, scoured bases with coarse-grained sand just above the scour and are interbedded with thinly laminated siltstones. The thickness of each sandstone and associated siltstone varies, but averages between 2 and 3 m. The sandstones commonly are pebbly, texturally submature, and consist predominantly of rock fragments. The rock fragments include angular volcanic types as well as small, rounded to subrounded and angular sedimentary fragments up to several centimeters long of the same lithology as the fine-grained sediments below the scour. The sands generally contain abundant broken but fresh feldspar crystals, together with deeply weathered feldspar grains. The grain size, color, and fine laminations of the fresh siltstones resemble the argillites of Unite E near the center of the syncline (see below). The siltstones differ, however, in that they are interbedded in cyclic fashion with sandstones and the laminations are often thicker and not as sharply delineated as in the younger argillite, Unit E. The sandstone mineralogy of Unit B apparently reflects a felsic source area (Table 2; MC-29). In thin section, groundmass quartz grains penetrate the boundaries of small mudstone clasts, suggesting that these were poorly lithified at the time of deposition.

At irregular intervals, resistant detrital volcanic conglomeratic lenses scour through these sediments. These conglomerates are usually only 1.5 to 2 m thick, not noticeably graded, and form resistant outcrops that can be traced 100 m or more. Grain size of the pebbles varies from 0.5 to 5.0 cm or slightly more. Sand-size matrix ranges from 5% to 30% of the rock. The fairly well sorted pebbles and cobbles are usually of varied volcanic types; however, well-rounded quartz pebbles about 2 cm in diameter do occur in some of the conglomerates. These conglomerates usually appear quite massive in outcrop, and primary structures are difficult to detect. In one of these conglomerates, however, internal festoon cross-bedding is indicated by straight but erratic, cross-cutting laminations.

Southward into Moore County, the interval becomes thicker and grain size averages fine to medium. Clasts are smaller (1 to 2 cm) and are composed of felsic volcanic rock fragments plus graywacke and argillite fragments. Here, sedimentary structures consist primarily of graded bedding with low-angle, cross-cutting laminations.

A localized unit of sedimentary graywacke breccia (Fig. 4B) is exposed near the northwestern corner of the area flanking a gabbro-diorite composite pluton (Table 1, analyses 21, 22). In hand specimen, the breccia has a rough, greenish-gray appearance and lacks foliation. Bedding is expressed in outcrop by the alternation of coarser grained beds with slightly thinner, fine-grained beds. The rock itself is a moderately well sorted, immature, angular to subangular graywacke containing large angular sedimentary rock fragments in a sand-sized matrix. The matrix consists of rounded volcanic rock fragments of several types, broken plagioclase crystals (some of which appear fresh; others of which may have undergone weathering prior to deposition), quartz, chlorite, and minor amounts of epidote and magnetite. The large sedimentary rock fragments range in size from 15 cm down to the sand size of the matrix and were derived from fine-grained, thinly bedded sandstone or siltstone. The fine bedding is accented on the weathered surfaces of the fragments, and many fragments show well-preserved cross-bedding. Graded bedding is not apparent. Trough-type cross-bedding can be seen in the coarser materials. Seiders and Wright (1977) reported the occurrence of a sedimentary breccia surrounding a felsic volcanic neck in the Asheboro quadrangle. They suggested that this rock was related to volcanic emplacement possibly by penecontemporaneous doming of the strata into the zone of wave action.

Local thin pyroclastic rocks are interbedded with the sediments. Rhyodacitic or dacitic crystal tuffs (Table 1, analysis 2; Table 2, MC-25) are the most common type. Stratigraphically upward, more mafic interbeds (Table 1, analyses 3 to 6) become common, so that the boundary with the overlying Unit C is extremely gradational.

Unit C: Intermediate to Mafic Volcanics (C-1) and Graywacke (C-2)

The mafic to intermediate volcanic subunit consists of tuffs, lava flows, and breccias (Unit C-1); it occupies the same stratigraphic level as, and interfingers with, the mafic graywacke (Unit C-2) sandstone-siltstone subunit (Fig. 2).

C-1. The intermediate to mafic volcanic subunit differs from the equivalent graywacke sandstone-siltstone subunit (C-2) mainly in the lack of laminated mudstone and the presence of lava flows. Volcanic rocks include an amygdaloidal basalt flow (Table 1, analysis 7; Fig. 4C; Table 2, CCB) and mixed flows and tuffs. Interbedded with the flows are purplish to gray breccias with one or two types of poorly sorted, angular volcanic rock fragments that attain a size of 2.5 cm or more. One well-exposed andesitic flow (Table 1, analysis 8) is at least 20 m thick, massive green, and locally

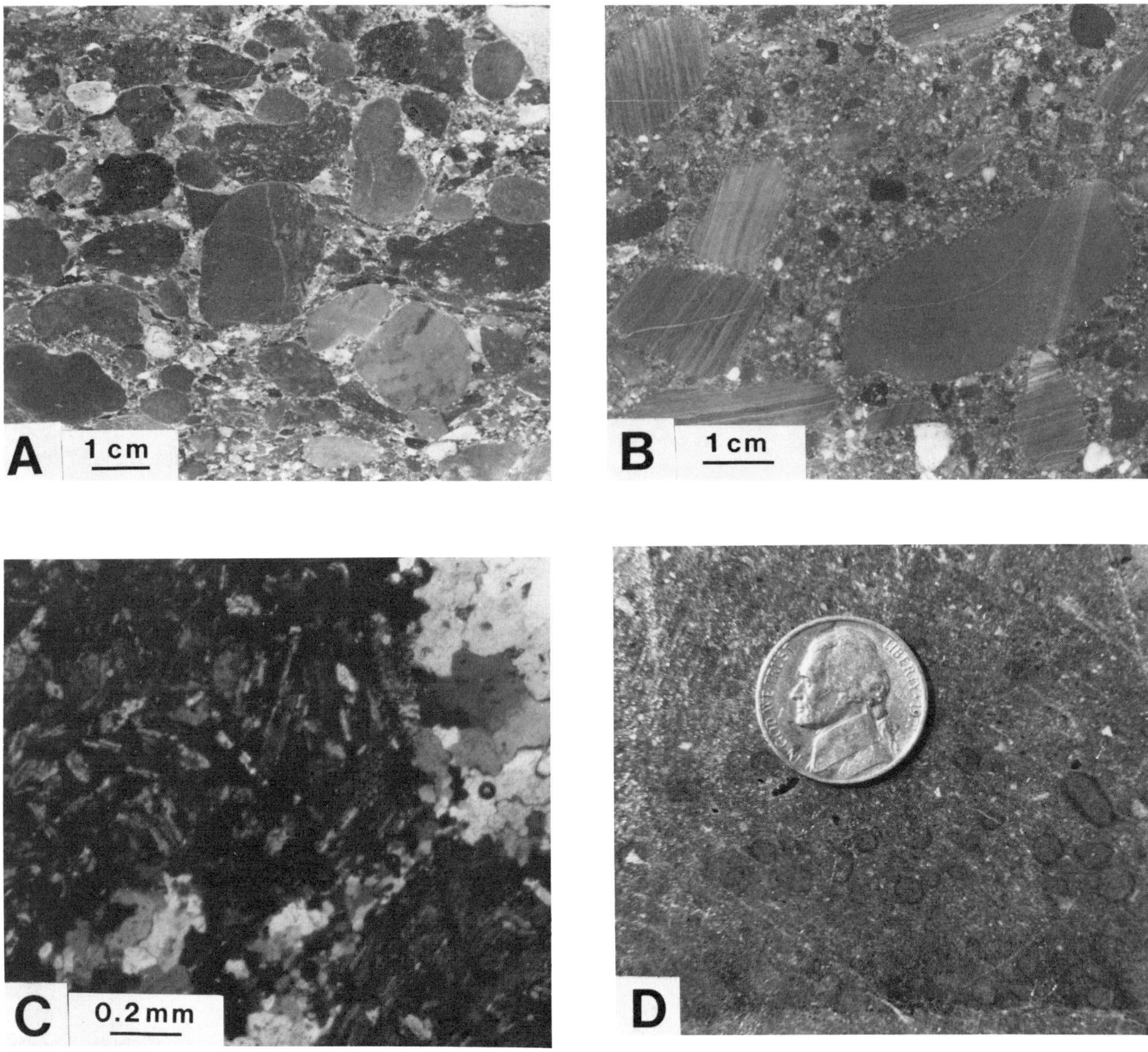

Figure 4 (This and facing page). Selected photographs. A. Clast-supported conglomerate containing rounded, polylithic but largely volcanic clasts, from the top of Unit A. B. Matrix-supported angular siltstone clasts in the sedimentary breccia of Unit B near the gabbro-diorite pluton. C. Photomicrograph of amygdaloidal basalt from Unit C-1 under crossed polars. Amygdules are filled by mosaic of quartz rimmed by epidote; matrix of chlorite+albite+epidote preserves relict ophitic texture. D. Sawed surface of thinly bedded tuff containing accretionary lapilli from Unit C in Chatham County. E. Lahar-like deposit within Unit C-2 containing large, monolithic volcanic blocks in a fine-grained matrix. F. Photomicrograph of flow-banded rhyodacite from Unit D, crossed polars. Note euhedral plagioclase phenocrysts and sheaves of metamorphic stilpnomelane. G. Autobrecciated rhyodacite flow from Unit D, northernmost Moore County. H. "Argillite" with fine, regular laminations, Unit E.

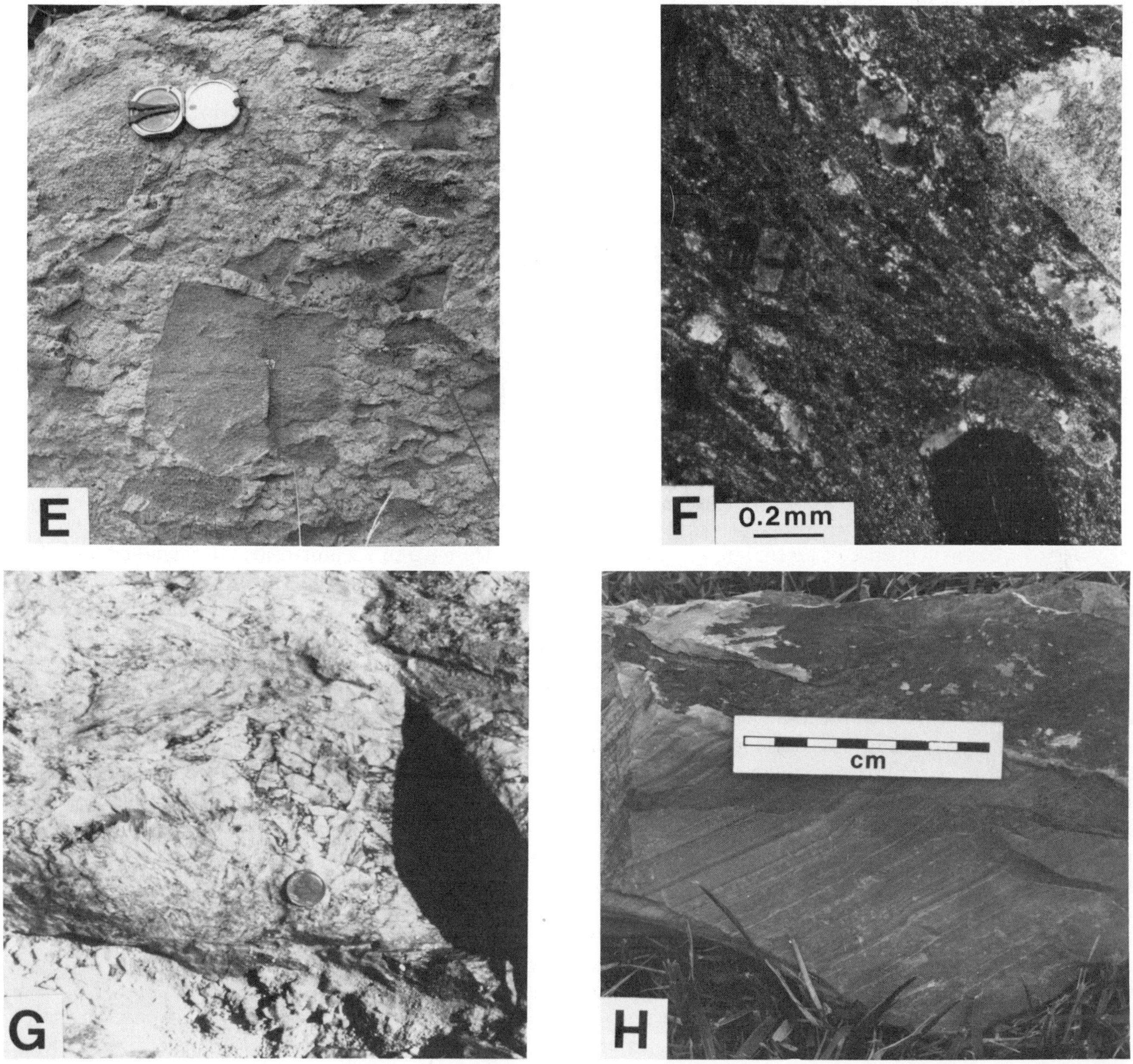

amygdaloidal. Below this flow, a purple breccia is exposed. This breccia, containing particles up to 8 cm in size, is definitely bedded, and many of its beds are size-graded, suggesting possible water transport or deposition.

Epiclastic volcanic conglomerates also occur. These contain well-rounded, moderately well sorted, polylithic pebbles and cobbles, again suggesting water transport. One such conglomerate bed is about 3 m thick and is exposed for a distance of about 20 m.

A final form of clastic rock contained in this subunit is a very poorly sorted volcanic breccia with compositionally varied angular fragments, many of which are very large (over 1 m). This breccia was mapped by Stuckey (1928).

Accretionary lapilli were found at one locality within Unit C-1. The lapilli are found as spheres of fine-grained ash approximately 2 to 8 mm in diameter and surrounded by rims of finer grained ash (Figure 4D). They constitute several thin beds less than 5 cm thick in a thinly bedded sequence of red mudstones with layers of angular volcanic rock fragments (probably airfall debris).

C-2. To both north and south, the volcanic subunit interfingers with graywacke sandstone and siltstone in some respects very similar to those of the underlying graywacke, Unit B. However, the volcanic detritus in Unit B is largely felsic, whereas that in C-2 is intermediate to mafic. These latter sandstones also tend to exhibit poorer size sorting and higher matrix content than do the underlying graywackes. Nevertheless, the rocks of these two units probably were formed in similar sedimentary environments.

In Chatham County, this graywacke unit is predomi-

TABLE 2. MINERALOGY OF SELECTED ROCKS FROM THE CHATHAM-MOORE COUNTY AREA, NORTH CAROLINA

Unit	Lithology	Sample	Quartz	Sodic plag.	Calcite	Epidote	Chlorite	Muscovite
B	Fine graywacke	MC29	2	2	..	3	3	..
	Felsic tuff	MC25	1	3	..	..	..	3
C	Amygdaloidal basalt	CCB	3	1	..	2	1	..
	Coarse Graywacke	MC1	2	2	3	3	2	3
	Mudstone	MC11	3	2	..	3	2	3
	Clasts in graywacke	BC	3	1	3	..	3	4
		GC	2	2	3	..	2	4
		PC	2	2	3	..	3	4
	Andesitic tuff	MC20	2	2	..	3	2	..
D	Rhyodacite	MC47	1	2	..	3	..	..
		MC50	1	2	..	3	..	3
		MC55	1	2	..	..	..	..
E	Laminated mudstone	CCA	1	3	..	..	4	2
		MC7	1	3	..	..	3	..
		MC40	1	..	..	..	3	3

Note: Mineralogy determined by thin-section petrography and X-ray diffraction; relative mineral abundances (1 = greatest) based upon comparison of X-ray peak heights with those from rocks of known modal mineralogy.

nantly a monotonous sequence of green, immature to submature, coarse- to medium-grained graywacke sandstones interbedded with purple, thinly bedded siltstones and claystones. The thickness of each couplet of sandstone and associated finer sediment, while variable, averages about 2 to 3 m. The sandstones consist of mainly angular polylithic volcanic rock fragments and invariably contain angular fragments of the underlying, finer grained purple mudstones. Although rare, volcanic conglomerates similar to those of Unit B also occur in this unit. A unique feature of this unit is a lahar-like volcanic breccia (Figure 4E) consisting of poorly sorted, angular to subrounded, monolithologic volcanic boulders (up to 60 cm in diameter) in a sandy matrix.

In Moore County, very thick (up to 20 m), greenish-gray graywacke sandstones exhibit very crude normal and reverse size grading. Sandstone-supported rounded and flattened volcanic clasts are up to 12 cm in length but appear to become smaller and less common southward. In the extreme south, clasts occur in uncommon bands suspended within coarse (medium sand to fine pebble) sandstones; rarely, isolated clasts are suspended within the sandstone. Cross-bedding was not observed, except in the extreme south where asymmetric ripple cross-beds (up to 2 cm in height) occur in thin zones of better sorted, medium-grained sandstone. Local zones of penecontemporaneous contorted bedding are present. Interbedded locally silty or sandy claystones in the extreme south range from a few centimeters to more than 15 m thick. These thick claystones form the upper portion of the crude cycles, with their tops scoured by subsequent sandstones. The thinner of the claystones form bands within the sandstones and commonly are truncated laterally by gently curved scour surfaces.

The bulk composition of these graywackes appears to be intermediate to mafic. This compositional range is reflected in the mineralogy of both sandstones (Table 2, MC-1) and the interbedded claystones (Table 2, MC-11). Three volcanic clasts from the sandstone indicate andesitic to basaltic source rocks (Table 1, analyses 12-14; Table 2, BC, GC, PC). Andesitic tuffs (Table 1, analysis 9; Table 2, MC-20) locally are interbedded within this unit, indicating coeval intermediate-composition volcanic activity.

Unit D: Upper Felsic Volcanics

The upward change from intermediate to the felsic composition rocks of Unit D is very abrupt. Pyroclastic rocks compose a large part of Unit D to the north, while flows (Table 2; MC-47, 50) are more common to the south. The flows are aphanitic to sparsely porphyritic in hand specimen, although one flow is spherulitic. These most commonly are flow-banded rhyolites or dacites (Fig. 4F; Table 1, analyses 15, 16), with one flow exhibiting autobrecciation (Fig. 4G). Porphyritic flows contain euhedral albite phenocrysts from 0.5 to 1.5 mm in size. In a detailed investigation in the southernmost part of the study area, Moore (1980) has recently described a volcanogenic textural sequence for a thick rhyodacitic flow at the top of Unit D which is similar in detail to those seen in modern felsic flows.

An exposure of finely laminated and locally brecciated rock consisting of greater than 90% SiO_2 was found in southern Chatham County. This unit has been described as a metasiliceous sinter (Spence, 1975; McDaniel, 1976) and may be related to the adjacent, fault-controlled pyrophyllite bodies. Other rocks with high silica content in adjacent Moore County (Table 1, analysis 18; Table 2, MC-55) show albite or oligoclase (up to 1.5 mm) as subhedral to euhedral phenocrysts. These rocks may have originated as felsic flows or tuffs, subsequently silicified by hydrothermal solutions associated with the development of the pyrophyllite deposits.

Tuffs constitute the other major rock type in the southern part of the study area. Finely laminated, slightly silty argillite with minor interbeds of epiclastic sandstone also occurs. Pyroclastic rocks dominate the interval to the north.

The pyroclastic rocks are of three types based on grain size, composition, and physical characteristics. The first are coarse-grained lapilli tuffs which contain angular to rounded volcanic and sedimentary fragments (up to 20 cm) in a finer matrix. The second are medium-grained pyroclastic units which are most widespread and consist of felsic rock fragments and broken but fresh plagioclase crystals in a sericitic groundmass. Although they generally appear massive in outcrop, these medium-grained pyroclastic rocks show hints of bedding, and locally contain thin layers of rounded volcanic pebbles and cobbles that suggest reworking. The third type of pyroclastic rock consists of bedded fine-grained material with very regular, thin (average of less than 1 cm) laminations. The medium- and fine-grained types are fairly evenly distributed, while coarse-grained types are more predominant in the southern areas that contain felsic flow material.

In the south, the top of the felsic volcanic unit gradationally intercalates upward with the laminated mudstone of Unit E. In the north, the transition is rapid with a single 10- to 12-m thick, fine-grained tuff occurring about 16 m above the base of Unit E. Southward, the mudstones become coarser grained and banding becomes less well developed as they interbed with the felsic tuffs near the top of Unit D.

Unit E: Laminated Mudstones

This unit consists of a blue to bluish-gray laminated mudstone (argillite of previous workers) that weathers olive green. It coarsens upward, ultimately grading into a purple sandstone. In Chatham County, approximately 500 m are exposed. Individual ultra-thin beds in the mudstone average a little over 1 cm (Fig. 4H) in thickness. They typically show well-developed size grading from silt up to clay, although fine- or even medium-grained graywacke sand is present, especially higher in the unit. The bedding is very regular and can be traced for long distances within outcrops. Wide but shallow cut-and-fill features, indicative of some bottom currents, are common; slump features are present but uncommon. Ripple marks occur in the upper coarser parts of the unit. These mudstones reflect a predominantly felsic source (Table 2, CCA, MC-7, MC-40). Limonite pseudomorphs after pyrite are disseminated throughout.

The grain size of the mudstone increases upward, and it grades into purple to reddish, finely laminated siltstones and fine-grained purple graywacke sandstones. The sandstones are texturally immature and exhibit rippled to graded bedding. Individual laminations of the volcanic sandstones average about 5 mm in thickness and display graded bedding from fine sand up to silt, with a thin layer of opaque minerals above basal scours. Bedding is much less regular than in the underlying mudstones, and individual laminations in the sandstones tend to be truncated by younger laminations within 1 m or less. Unlike the underlying mudstones, the bedding in the upper purple facies commonly exhibits penecontemporaneous deformation. Individual beds may be contorted, sometimes two or three beds at a time, while beds directly above them lack similar contortions.

EVOLUTION OF VOLCANIC ACTIVITY

Compositions of Igneous Rocks

Major element chemical analyses of samples of volcanic rocks from Units A, B, C, and D, and two analyses of plutonic rocks possibly associated with Unit C are listed in Table 1. These rocks have been subjected to greenschist facies metamorphic conditions, possibly in association with marine pore waters, and have undergone varying degrees of metasomatic alteration. The rocks have been altered locally by hydrothermal fluids.

Although the freshest, least altered appearing samples were chosen for analysis, the effects of alteration are nonetheless apparent in several samples. These include silicified felsic rock (e.g., Unit D, analysis 18); rocks with SiO_2 exceeding 90 wt%, interpreted as siliceous sinter by Spence (1975) and McDaniel (1976); calcite-rich mafic to intermediate rocks (Unit C, analyses 12 to 14); and felsic rocks from Unit D with high CaO (analysis 19), and Na_2O (analysis 20). The latter may have undergone alkali exchange with seawater (see also Butler and Ragland, 1969b).

The remainder of the analyzed samples appear not to have been so modified as to obscure the original volcanic rock types. However, oxide contents (especially silica) of these rocks may appear low because of hydration; silica contents recalculated on a volatile-free basis provide a better indication of the original rock type. In Table 1, these are listed in parentheses immediately following the measured silica contents. Their bimodal distribution on the histogram of Figure 5 is consistent with the stratigraphy and evolution of the area: it is dominated by three distinct volcanic sources, one mafic (in Unit C), and two felsic (Units A and D). Seiders (1978) reported a similar distribution of silica contents from volcanic rocks of other areas of the slate belt; a simplistic view of the slate belt as a province of intermediate volcanics appears to be unwarranted, at least in some areas.

The overall volcanic evolution of the Chatham-Moore area (discussed in detail above) begins with Unit A: felsic epiclastics with subordinate, poorly exposed felsic flows (analysis 1). Unit B time was dominated by sedimentary processes; volcanic rocks are rare but appear to represent a transition from earlier felsic (e.g., dacitic tuff, analysis 2) to more mafic activity (e.g., mafic hornblende-bearing porphyritic tuff, analyses 3 and 4; and andesitic flows(?), analyses 5 and 6). Unit C was dominated by the center of mafic activity in the north; it is represented by the amygdaloidal basalt (analysis 7) and by other flows and tuffs of basaltic to

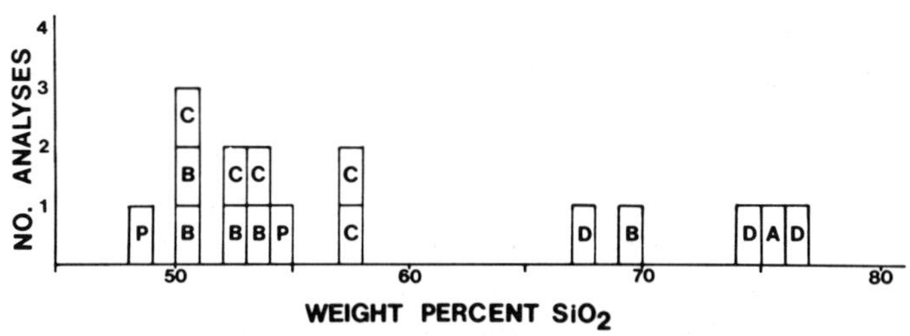

Figure 5. Histogram of silica contents (recalculated on an anhydrous basis) of the least-altered volcanic rocks from the study area. Letters correspond to the stratigraphic units of Figure 2; samples from gabbro-diorite pluton indicated by the letter P.

andesitic composition (analyses 8 to 11). In addition, the small stock composed of gabbro (analysis 21) and porphyritic diorite (analysis 22) in the northwestern corner of the map area may represent plutonic equivalents of the volcanics of Unit C. When mafic activity waned, a center of felsic volcanism began to dominate deposition in the area; flows and tuffs from Unit D are dacitic to rhyolitic (analyses 15 to 17). Local volcanism apparently ceased after deposition of Unit D, as the remainder of the stratigraphy is represented by sedimentary rocks, with but a few thin tuffs.

Although chemical variation diagrams such as the Peacock alkali-lime diagram (Fig. 6) may be generally suggestive of calc-alkaline affinities, the data are so meager and the scatter and uncertainty introduced by alteration are so great that any such conclusion would be unwarranted. However, the predominantly fragmental textures of the volcanic rocks of the area are consistent with the explosive eruptive character of volcanic arcs, in contrast to the more effusive nature of oceanic ridge and island volcanics (e.g., Macdonald, 1972; Garcia, 1978).

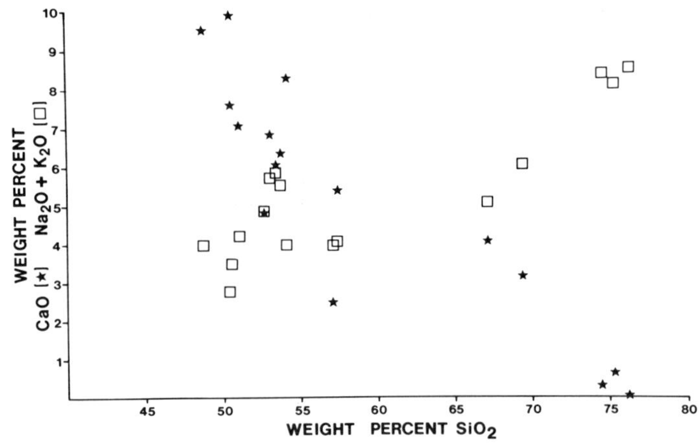

Figure 6. Alkali-lime variation diagram for the least-altered volcanic rocks from the study area. SiO_2 recalculated on an anhydrous basis.

Volcanogenic Facies

The rock facies and their lateral variations indicate that only portions of the areal extent of each volcanogenic unit are present within the study area. This is compatible with the scale of the study area as roughly equivalent to a portion of the slope of a major volcanic cone. It also is reasonable to assume these rocks were derived directly or indirectly from surrounding contemporaneous volcanic activity, because there is no evidence of an external source of sediment.

Within this area, Units A and B record an initial episode of felsic eruptive activity subsequently resulting in deposition of volcaniclastic graywacke sandstones of a similar composition. Shallow water depths to local subaerial conditions are suggested for Unit A by the repetitive ash-flow tuffs and the localized zones of subangular to rounded monolithologic conglomerates within the unit. The well-rounded, clast-supported, polylithologic conglomerate (Fig. 4A) found consistently near the contact between Units A and B further suggests the likelihood of reworked, near shore deposits. Such volcanic conglomerates have been recently described from subaerial and near-shore environments in the modern Lesser Antilles arc by Sigurdsson and others (1980).

The graywacke sandstones of Unit B in Chatham County are coarse-grained and locally conglomeratic, have scoured bases, contain sedimentary rip-clasts, and locally exhibit festoon cross-bedding, all indicative of high-energy current transport. Mixed fresh and weathered feldspar grains in these sandstones additionally suggest subaerial exposure within their source area (Folk, 1974). The laminated greenish mudstone interbedded with the sandstones indicates the coarser detritus was introduced as pulses into an overall quiet-water bottom, reducing to weakly oxidizing environment. Southward in Moore County, Unit B is much thicker. Limited exposures show the sandstones to be finer grained with local graded beds. Pebble-size clasts are very rare and cross-bedding was not observed. These features suggest a transition to lower energy, deeper waters, and increased distance from the detrital source in a generally southerly to southwesterly direction.

Regional mapping by Wilson and Carpenter (1975) showed felsic composition rocks in a position analogous to Unit A (Fig. 7) to the east and northeast of the study area and Babiker (1978) reported rhyolites common within this interval about 8 km east of the study area. In the same general directions, the volcaniclastic rocks of Unit B appear to wedge out (Fig. 7); reconnaissance has disclosed additional thick (up to 7 m), discontinuous, presumed near-shore deposits of polylithologic, clast-supported conglomerates along this wedge-out boundary. These associations, together with the direct overlap of Unit C on Unit A in these areas (Fig. 7), strongly argue that Unit B represents a subaqueous apron of erosional debris developed on the flanks of the felsic eruptive complex represented by Unit A. These regional relationships further support higher elevations of the felsic cone slopes in the north and/or east.

The felsic graywacke sandstone of Unit B gradually changes upward into the intermediate to mafic sandstones and maroon to greenish mudstones of Unit C-2. These in turn interfinger laterally into andesitic to basaltic flows and epiclastic rocks of Unit C-1. In the northern part of the study area, the presence of transported epiclastics interbedded with these flows combined with the red mudstones and the apparent absence of pillow structures, found elsewhere in the slate belt (e.g., Wilson and Allen, 1968), suggests possible subaerial conditions. Such an inference is reinforced in this case by the presence of accretionary lapilli (previously reported from other areas in the North Carolina slate belt by

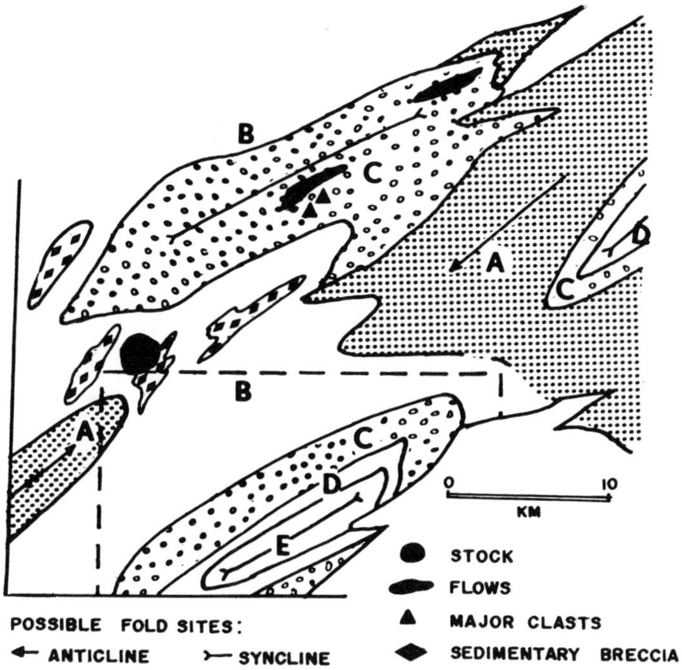

Figure 7. Map of southeastern Moore and Chatham Counties modified from Wilson and Carpenter (1975). The portion of Figure 3 included by this map is delineated by the dashed line. Alphabetical unit designations are those of Figure 2.

Sundelius (1963) and Randazzo (1969)). Accretionary lapilli require a subaerial tephra cloud and would have disintegrated with substantial immersion in water (Moore and Peck, 1962; Lowe and Knauth, 1978).

As with the underlying units, paleoslope appears to have been generally southward. To the north in Chatham County, the sandstones are thinner, coarser grained, more poorly sorted, and contain structures indicative of moderate-velocity, confined current transport (runoff channels?). Interbedded mudstones are maroon to reddish-purple and a very coarse lahar-like slump occurs in this area. Southward into Moore County, the equivalent but thicker sandstones become finer, indicating increased distance from their source. Concurrently, the interbedded sandy, reddish-purple mudstones become thicker, greenish-gray muddy claystones. The sandstones in the south are characterized by sand-supported normal and reverse crudely size-graded clasts, thin beds of suspended clasts, contorted zones, gentle internal scours, and a general lack of current cross-lamination. These features are more similar to the laminar flow of subaqueous sand debris flows (Hampton, 1972; Padgett and others, 1977) than to either classical confined channel or turbidity flow deposits. The Roseau pyroclastic debris flow deposit in the Grenada Basin, recently discussed by Sigurdsson and others (1980), appears to be similar in many respects and may be a good modern analogue for these sandstones.

The preceding association of features in Unit C-2 is interpreted as deposition along a southward-sloping volcanic flank, possibly in the form of vertical sedimentation of suspended muds, punctuated by occasional subaqueous sand-debris flows, perhaps initiated from subaerial to shallow-water lahars.

The roughly circular (1.5-km-diameter) gabbro-diorite pluton which cuts Unit B in the northwestern portion of the study area may represent the remnants of a volcanic neck. The halo of strongly indurated sedimentary breccia surrounding this stock (Fig. 7), suggests pre-eruptive intrusion and doming, resulting in the formation of sedimentary breccia near the top of Unit B, possibly within the zone of wave action as suggested by the large-scale cross-beds. In the syncline immediately to the north of the study area (Fig. 7), Wilson and Carpenter (1975) have mapped extensive mafic tuffs and andesitic flows with subordinate basalt flows and coarse wacke conglomerate. A cursory field reconnaissance encountered intermediate composition flows interbedded with lithic tuffs containing angular clasts in excess of 25 cm. Near the northern end of this syncline, multiple conglomerates up to 25 m thick contain angular to subrounded clasts several centimeters in diameter. The spatial relationships of these rocks, the flows and coarse angular clasts of the study area, and the sedimentary breccia combine to place a significant intermediate to mafic volcanic vent at or somewhat north of the recognized intrusive stock.

The transition from Unit C upward into Unit D represents an abrupt change to more explosive flesic eruptive activity concurrent with a shift in the site of volcanism. Throughout the study area, the transition between these rock units is rapid and exhibits no evidence of unit intercalation. In Moore County, the fine- to coarse-grained felsic pyroclastic rocks commonly contain aphanitic to sparsely porphyritic flows. The same unit in the extreme southern portion of the synclinal outcrop in Chatham County contains an autobrecciated flow. Northward from this position, flows are absent and the unit is dominated by laminated, fine-grained water-lain tuffs interbedded with thick, homogeneous, medium-grained deposits of moderately well sorted felsic volcanic debris. These spatial relationships indicate abrupt development of felsic volcanic activity, perhaps a localized subaerial dome because of the nonbrecciated nature of the flows to the south of the study area. The onset of this felsic activity probably was coincident with initial subsidence of parts of the earlier volcanic edifice in Chatham County. Local faults may reflect this collapse; one such structure occurs within Unit C of Chatham County and is associated with large (up to 1 m) rounded oblong boulders of Unit C red mudstone in a sandy matrix.

The reversal in paleogradient suggested by the change from coarse pyroclastics with flows to water-lain tuffs is underscored by the northward intercalation of Unit D into the quiet-water, laminated mudstones of Unit E. In addition to their physical interfingering, the similarity in composition

TABLE 3. SUMMARY OF INFERRED DEPOSITIONAL FRAMEWORK

	Unit A Lower felsic volcanics	Unit B Felsic graywacke	Unit C Intermediate to mafic C-1: volcanics; C-2: graywacke		Unit D Upper felsic volcanics	Unit E Laminated mudstone
Depositional setting	Shallow water; locally subaerial	Deeper, quiet water; pulses of debris flows	Subaerial to shallow; subaqueous	Subaqueous; moderate depths; debris flows	Subaerial in south; subaqueous in north	Deeper, quiet water
Paleoslope		to S, SW	to S		to N	
Source area		N, NE	N, NW		S	
Volcanic evolution						
Felsic volcanic	Construction	Destruction				
Intermediate-mafic volcanic source to N, NW			Construction; terminated by rapid subsidence			
Felsic volcanic source to S					Construction	Destruction

of these two units and the southward-coarsening grain size of the mudstone support the lateral equivalence of these units.

The inferred sequence of depositional settings is summarized in Table 3.

CONCLUSIONS

Depositional Model

The rocks of southern Chatham and northern Moore Counties exhibit a complex interfingering of volcanic and sedimentary facies derived directly or indirectly from different sources. Although deposition took place in an overall subaqueous marine environment, the association of oxidized muds, accretionary lapilli, clast-supported conglomerates, and cross-bedding indicates substantial areas of subaerial to shallow subaqueous deposition. Tectonic mobility of the area is indicated by changes in paleoslope, occurrences of penecontemporaneous faulting, as well as possible doming about the small stock. In analogy, the overall depositional controls appear similar in many respects to the Tertiary volcanics of the Ohanepecosh Formation of Washington as described by Fiske (1963). Those rocks were described as accumulating in shallow water, with additions of volcanic materials offsetting crustal subsidence. Fiske (1963) concluded that the distribution of topographically higher areas and deeper waters was rearranged constantly as overall subsidence was modified by the building and destruction of individual volcanoes, some of which were at least partially subaerial.

The bimodal composition of the Chatham and Moore County rocks is exhibited in Units C through E by a change from mafic to felsic eruptive activity (Table 1). This is associated with a shift in the site of volcanism and depositional slope, such that at least some of the previously subaerial parts of the intermediate to mafic cone complex were overlain with thick deposits of finely laminated mudstone during the latter stages of felsic eruption. It is possible to model this sequence as independent volcanic events, as in the Lesser Antilles arc where adjacent silicic and mafic volcanic centers are the sources of a great variety of volcanogenic sedimentary deposits (Sigurdsson and others, 1980). However, similar sequences occurring in other calc-alkaline volcanic terranes (Williams, 1932, 1942; Macdonald, 1972), such as the Cascade Range (e.g., Mount Mazama; Brokeoff Volcano) have been interpreted as the results of a single volcanogenic cycle. There, numerous felsic eruptive centers developed along the flanks of intermediate to mafic composite cones preceding cauldron collapse in the final stages of volcanic activity.

Hughes and Bruckner (1971) studied the lithologic facies of a late Precambrian to early Paleozoic volcanic series in southeasternmost Newfoundland. They recognized penecontemporaneous facies analogous to those of this study area and reported vertical crustal movements to be the dominant form of deformation occurring during volcanism. Their facies for a volcanic island complex reflect an initial synvolcanic constructional phase (intercalating vent, alluvial, marine, and plutonic facies) followed by a postvolcanic destructive phase (erosion with alluvial and marine facies). Application of their model to the Chatham and Moore County area (Table 3) for Units A through C would result in Unit A representing the constructional phase of a vent(?) located generally northeast of the study area. Unit B (the destructional phase), thin to absent north of the study area (Fig. 7), grades southward from alluvial and reworked shallow-water facies in the north to more basinward facies in the south. Unit C represents reinstitution of volcanic activity accompanied by a change in magmatic composition and likely in location of the volcanic center. Constructional vent and alluvial facies are well represented in Chatham County and pass into lower subaqueous slope facies in Moore County. Except for the single mafic stock, plutonic facies are not present, probably because of insufficient depth of erosion. The absence of any record of a destructional phase following Unit C supports the earlier inference of subsidence and/or collapse coincident with the new constructional phase of Unit D. The more southerly felsic vent facies, perhaps developed as a felsic dome, is represented by the rhyodacites and ash beds in the south of the study region grading northward into probable airfall tuffs and finely banded mudstones (Unit E) derived from these felsic materials. Although the top of Unit E is truncated by erosion, the upward passage from gray laminated mudstone to cross-bedded maroon volcaniclastic sandstones preserved in the center of the Chatham County syncline probably reflects development of the subsequent destructive phase.

Stratigraphic Implications

The stratigraphic implications of the rocks of the study area appear to be twofold. First, individual beds are usually not practical mapping units in the slate belt because of the poor exposures, the likelihood of concurrent but spatially separated sources, and the rapid lateral changes in depositional conditions. The most effective mapping units instead appear to be thick units of mixed volcanics and clastics characterized by a dominant bulk composition. Second, at least two discrete volcanic sources existed within this small study area, creating a complex interrelationship of lithic facies. By analogy, numerous such point sources with similar volcanogenic cycles can be inferred to have developed within the slate belt, separated by both time and space. Under these conditions, previous attempts at long-distance correlation based solely on lithologic similarities of succession appear to be somewhat tenuous, as do slate belt models dependent upon such correlations.

ACKNOWLEDGMENTS

The authors wish to acknowledge W. H. Spence (American Selco, Inc.) and E. R. Burt (North Carolina Geological Survey) for freely contributing substantial field time and many significant ideas and criticisms during the period of this study. This paper was based in large part on unpublished thesis research by Green (1977) and Abdelzahir (1978). Partial field support was provided by the North Carolina Department of Natural Resources and Community Development. We also thank J. R. Butler, M. O. Garcia, V. M. Seiders, and A. F. Randazzo for their thoughtful reviews.

REFERENCES CITED

Abdelzahir, A. M., 1978, The geology of the Carolina slate belt, northern Moore County, North Carolina [M.S. thesis]: Raleigh, North Carolina State University, 67 p.

Babiker, H. M., 1978, Geology of the Clegg Copper Mine and Vicinity, Lee and Chatham Counties, North Carolina [M.S. thesis]: Raleigh, North Carolina State University, 67 p.

Bain, G. L., 1966, Geology and ground-water in the Durham area, North Carolina: North Carolina Department of Water Resources Bulletin, v. 7, 147 p.

Bemmelen, R. W. van, 1949, The geology of Indonesia: General Geology, v. 1A: The Hague, Netherland Printing Office, 732 p.

——1954, Mountain Building: The Hague, Martinus Nijhoff, 1977 p.

Black, W. W., 1977, The geochronology and geochemistry of the Carolina slate belt of north-central North Carolina [Ph.D. thesis]: Chapel Hill, University of North Carolina, 118 p.

Black, W. W., and Fullagar, P. D., 1976, Avalonian ages of metavolcanics and plutons of the Carolina slate belt near Chapel Hill, North Carolina: Geological Society of America Abstracts with Programs, v. 8, p. 136.

Butler, J. R., 1972, Age of Paleozoic regional metamorphism in the Carolinas, Georgia, and Tennessee, southern Appalachians: American Journal of Science, v. 272, p. 319–333.

Butler, J. R., and Ragland, P. C., 1969a, Petrology and chemistry of meta-igneous rocks in the Albemarle area, North Carolina slate belt: American Journal of Science, v. 267, p. 700–726.

——1969b, A petrochemical survey of plutonic intrusions in the Piedmont, southeastern Appalachians, U.S.A.: Contributions to Mineralogy and Petrology, v. 24, p. 164–190.

Conley, J. F., 1962a, Geology of the Albemarle quadrangle, North Carolina: North Carolina Department of Conservation and Development, Division of Mineral Resources Bulletin, v . 75, 26 p.

——1962b, Geology and mineral resources of Moore County, North Carolina: North Carolina Department of Conservation and Development, Division of Mineral Resources Bulletin, v. 76, 40 p.

Conley, J. F., and Bain, G. L., 1965, Geology of the Carolina slate belt west of the Deep River Wadesboro Triassic basin, North Carolina: Southeastern Geology, v. 6, p. 117–138.

Farrar, S. S., 1980, Geology of the Raleigh Block and the adjacent Piedmont of North Carolina, *in* Costain, J. K., and others, Evaluation and targeting of geothermal energy resources in the Southeastern United States, Progress Report, January 1–June 30, 1980: National Technical Information Service, U.S. Department of Commerce, Springfield, Virginia 22161, Document No. VPI+SU-78ET-27001-8, p. A-1–A-53.

Fiske, R. S., 1963, Subaqueous pyroclastic flows in the Ohanepecosh Formation, Washington: Geological Society of America Bulletin, v. 74, p. 391–406.

Folk, R. L., 1974, Petrology of sedimentary rocks: Austin, Hemphill Publishing Co., 182 p.

Fullagar, P. D., and Butler, J. R., 1979, 325- to 265-m.y.-old granitic plutons in the Piedmont of the southeastern Appalachians: American Journal of Science, v. 279, p. 161–186.

Garcia, M. O., 1978, Criteria for the identification of ancient volcanic arcs: Earth Science Reviews, v. 14, p. 147–165.

Glover, L., III, and Sinha, A. K., 1973, The Virgilina deformation, a late Precambrian to Early Cambrian(?) orogenic event in the central Piedmont of Virginia and North Carolina: American Journal of Science, v. 273-A, p. 234–251.

Green, G. B., 1977, The geology of the slate belt rocks of the Goldston and Bear Creek quadrangles, North Carolina [M.S. thesis]: Raleigh, North Carolina State University, 68 p.

Hampton, M. A., 1972, The role of subaqueous debris flows in generating turbidity currents: Journal of Sedimentary Petrology, v. 42, p. 775–793.

Hatcher, R. D., Jr., Howell, D. E., and Talwani, P., 1977, Eastern Piedmont fault system: Speculations on its extent: Geology, v. 5, p. 636–640.

Hills, F. A., and Butler, J. R., 1969, Rubidium-strontium dates from some rhyolites from the Carolina slate belt of the North Carolina Piedmont [abs.]: Geological Society of America Special Paper 131, p. 445.

Hughes, C. J., and Bruckner, W. D., 1971, Late Precambrian rocks of eastern Avalon peninsula, Newfoundland—a volcanic island complex: Canadian Journal of Earth Sciences, v. 8, p. 899–915.

King, P. B., 1959, The evolution of North America: Princeton, Princeton University Press, 189 p.

Kish, S. A., Butler, J. R., and Fullagar, P. D., 1979, The timing of metamorphism and deformation in the central and eastern Piedmont of North Carolina: Geological Society of America Abstracts with Programs, v. 11, p. 184–185.

Lowe, D. R., and Knauth, L. P., 1978, The oldest marine carbonate ooids reinterpreted as volcanic accretionary lapilli, Onverwacht Group, South Africa: Journal of Sedimentary Petrology, v. 48, p. 709–722.

Macdonald, G. A., 1972, Volcanoes: Englewood Cliffs, New Jersey, Prentice-Hall, 510 p.

McDaniel, R. D., 1976, Application of a hot spring-furmarole alteration model to the genesis of the pyrophyllite deposits of the Carolina slate belt [M.S. thesis]: Raleigh, North Carolina State University, 75 p.

Moore, B. R., 1980, Investigation of a felsic flow, northern Moore County, North Carolina [M.S. thesis]: Raleigh, North Carolina State University, 70 p.

Moore, J. G., and Peck, D. L., 1962, Accretionary lapilli in volcanic rocks of the western continental United States: Journal of Geology, v. 70, p. 182–193.

Padgett, G., Ehrlich, R., and Moody, M., 1977, Submarine debris flow deposits in an extensional setting—Upper Devonian of western Morocco: Journal of Sedimentary Petrology, v. 47, p. 811–818.

Parker, J. M., III, 1968, Structure of the easternmost North Carolina Piedmont: Southeastern Geology, v. 9, p. 117–131.

Randazzo, A. F., 1969, Crystal molds in argillites of the Carolina slate belt, North Carolina: Journal of Sedimentary Petrology, v. 39, p. 820–821.

——1972, Petrography and stratigraphy of the Carolina slate belt, Union County, North Carolina: North Carolina Department of Natural and Economic Resources Special Publication 4, 38 p.

Seiders, V. M., 1978, A chemically bimodal, calc-alkalic suite of volcanic rocks, Carolina volcanic slate belt, central North Carolina: Southeastern Geology, v. 19, p. 241–266.

Seiders, V. M., and Wright, J. E., 1977, Geology of the Carolina volcanic slate belt in the Asheboro, North Carolina area, *in* Burt, E. R., ed., Field guides for Geological Society of America, southeastern Section Meeting, Winston-Salem, North Carolina: Geological Society of America, p. 1–34.

Sigurdsson, H., Sparks, R.S.J., Carey, S. N., and Huang, T. C., 1980, Volcanogenic sedimentation in the Lesser Antilles arc: Journal of Geology, v. 88, p. 523–540.

Spence, W. H., 1975, A model for the origin of the pyrophyllite deposits in the Carolina slate belt: Geological Society of America Abstracts with Programs, v. 7, p. 536.

St. Jean, J., 1973, A new cambrian trilobite from the Piedmont of North Carolina: American Journal of Science, v. 273-A, p. 196–216.

Streckeisen, A., 1979, Classification and nomenclature of volcanic rocks, lamprophyres, carbonatites, and melilitic rocks: Recommendations and suggestions of the IUGS Subcommission on the systematics of Igneous Rocks: Geology, v. 7, p. 331–335.

Stuckey, J. L., 1928, The pyrophyllite deposits of North Carolina: North Carolina Department of Conservation and Development, Division of Mineral Resources Bulletin, v. 37, 62 p.

——1965, North Carolina: Its geology and mineral resources: Raleigh, North Carolina Department of Conservation and Development, Division of Mineral Resources, 550 p.

——1967, Pyrophyllite deposits in North Carolina: North Carolina Department of Conservation and Development, Division of Mineral Resources Bulletin, v. 80, 38 p.

Sundelius, H. W., 1963, Accretionary lapilli in rocks of the Carolina slate belt, Stanly County, North Carolina: U.S. Geological Survey Professional Paper 475-B, p. B42–B45.

Whitney, J. A., Paris, T. A., Carpenter, R. H., and Hartley, M. E., 1978, Volcanic evolution of the southern slate belt of Georgia and South Carolina: A primitive oceanic island arc: Journal of Geology, v. 86, p. 173–192.

Williams, H., 1932, Geology of the Lassen Volcanic National Park, California: University of California Publications in the Geological Sciences, v. 21, p. 195–385.

——1942, The geology of Crater Lake National Park, Oregon, with a reconnaissance of the Cascade Range southward to Mount Shasta: Carnegie Institution of Washington Publication 540, 162 p.

Wilson, W. F., and Allen, E. P., 1968, Spilitic amygdaloidal basalt flow rocks and associated pillow structure in Orange County, North Carolina: Southeastern Geology, v. 9, p. 133–142.

Wilson, W. F., and Carpenter, P. A., 1975, Region J. geology: A guide for North Carolina mineral resource development and land use planning: North Carolina Department of Natural and Economic Resources, Division of Resource Planning and Evaluation, Mineral Resources Section, Regional Geology series No. 1, 76 p.

Wright, J. E., and Seiders, V. M., 1977, U-Pb dating of zircons from the Carolina volcanic slate belt, central North Carolina: Geological Society of America Abstracts with Programs, v. 9, p. 197–198.

——1980, Age of zircon from volcanic rocks of the central North Carolina Piedmont and tectonic implications for the Carolina volcanic slate belt: Geological Society of America Bulletin, Part I, v. 91, p. 287–294.

MANUSCRIPT ACCEPTED BY THE SOCIETY JULY 8, 1982

Printed in U.S.A.

Geological Society of America
Special Paper 191
1982

Sponge spicules from early Paleozoic rocks of the Carolina slate belt

W. C. Bourland
Counsulting Geologist
2308 Warner Road
Fort Worth, Texas 76110

J. K. Rigby
Department of Geology
Brigham Young University
Provo, Utah 84601

ABSTRACT

A phyllite unit from the Carolina slate belt of South Carolina contains sponge spicules. The sponge spicules, which occur along bedding planes, are of three types: hexactines, tetraclads, and monactines. Even though the overall morphology of the spicules is very simple, they suggest a Middle Cambrian or younger age. The discovery of sponge spicules and recent discoveries of trilobites and metazoan trace fossils demonstrate that the Carolina slate belt is probably comprised of late Precambrain to Middle Ordovician metasediments.

INTRODUCTION

South of the Pageland pluton near the North Carolina–South Carolina border are two isolated outcrops of phyllite (Figs. 1 and 2). Most of the phyllitic rocks in this area have been metamorphosed by emplacement of the Pageland pluton or altered during hydrothermal events that formed the Hale-Brewer mining area. As shown in Figure 2, the fossil locality is in phyllites south of the Hale-Brewer mine. Phyllites in this area are lower greenschist facies and outside the contact aureole of the pluton and the hydrothermal aureoles of the mine (Butler and Howell, 1977).

The fossiliferous phyllite was collected from about the center of the outcrop shown in Figure 3. Along the exposed bedding-plane surface, 20 to 30 elongate impressions about 0.5 mm in length were observed. Initially these were thought to be weathered impressions of a mineral formed during one of the metamorphic events. After scanning the surface with an electron microscope, it was observed that the impressions had high levels of organization and identity similar to modern sponge spicules. The impressions were later observed to be either monactine sponge spicules or the broken fragments of hexactine and tetraclad sponge spicules (Pl. 1).

In hand specimen, the rock is a weathered light green to tan metasiltstone with thin (1 to 7 mm), alternating, discon-

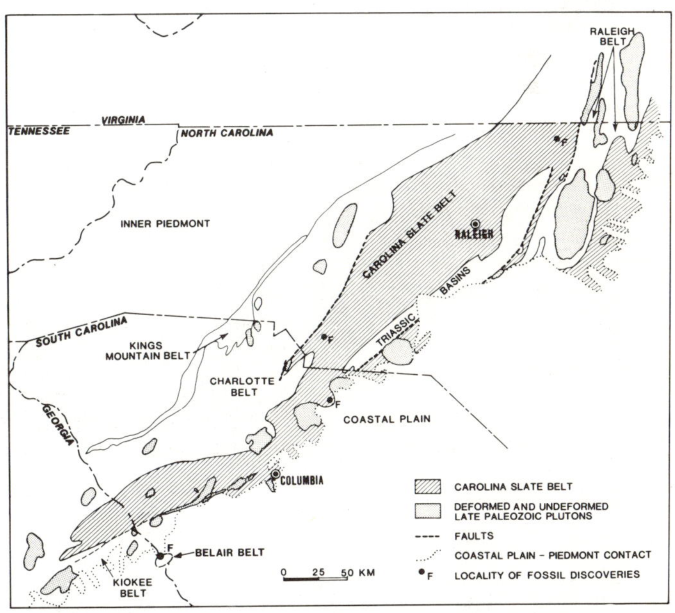

Figure 1. Geological sketch map of the metamorphic belts of the Eastern Piedmont of the United States. All the known fossil localities are designated by the letter "F." The sponge spicules (reported here) are from the outcrop in South Carolina near the South Carolina–North Carolina border.

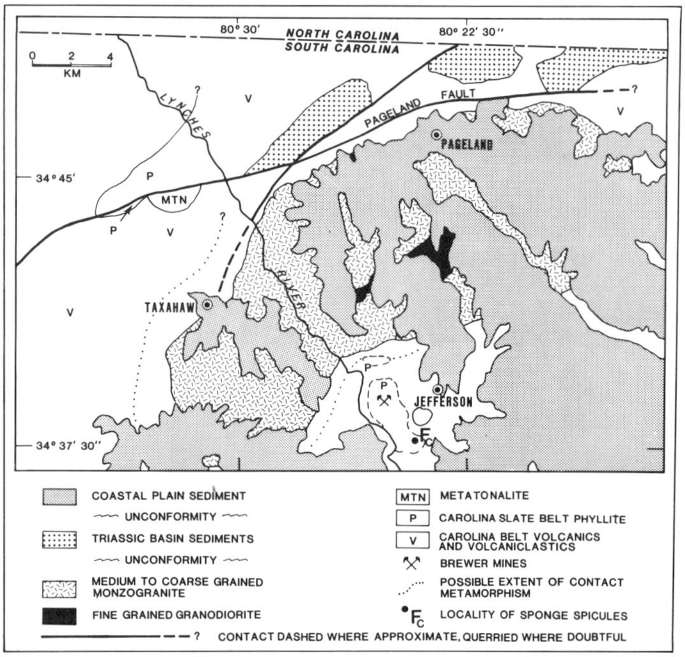

Figure 2. Generalized geologic map of the area surrounding the Pageland pluton, northern South Carolina (after Butler and Howell, 1977).

tinuous laminae of mud and sand-sized grains (Fig. 4). Bedding is irregular and disrupted by burrowing and/or soft sediment deformation.

This sample has only one foliation along the bedding, unlike the phyllitic rocks in the same outcrop that apparently have two penetrative foliations, varying from parallel to normal to the bedding.

The lithology is mainly silt- and sand-sized grains in a mud matrix (Fig. 5). Subrounded quartz, plagioclase, glauconite, and limonite with rare muscovite fragments compose the granular fraction of the rock. Two small amphibolite(?) grains were found in the coarse fraction. The mud matrix is a combination of carbonate and silica.

DESCRIPTION OF THE SPONGE SPICULES

Scanning electron photomicrographs of chips of fossiliferous phyllite show three kinds of spicules: several moderately well preserved hexactines, one four-rayed tetraclad, and numerous fragments of probable monactine spicules (Pl. 1).

One of the best preserved hexactines is specimen 477 (Pl. 1, fig. 1) in which the moderately stubby, sharply tapering

Figure 3. Outcrop containing the fossiliferous rock. The outcrop is 2.5 miles south of Jefferson, South Carolina, on State Road 151 and 1.2 miles west on an unmarked county road that runs along the southern border of the Jefferson, South Carolina, 7½-minute quadrangle map.

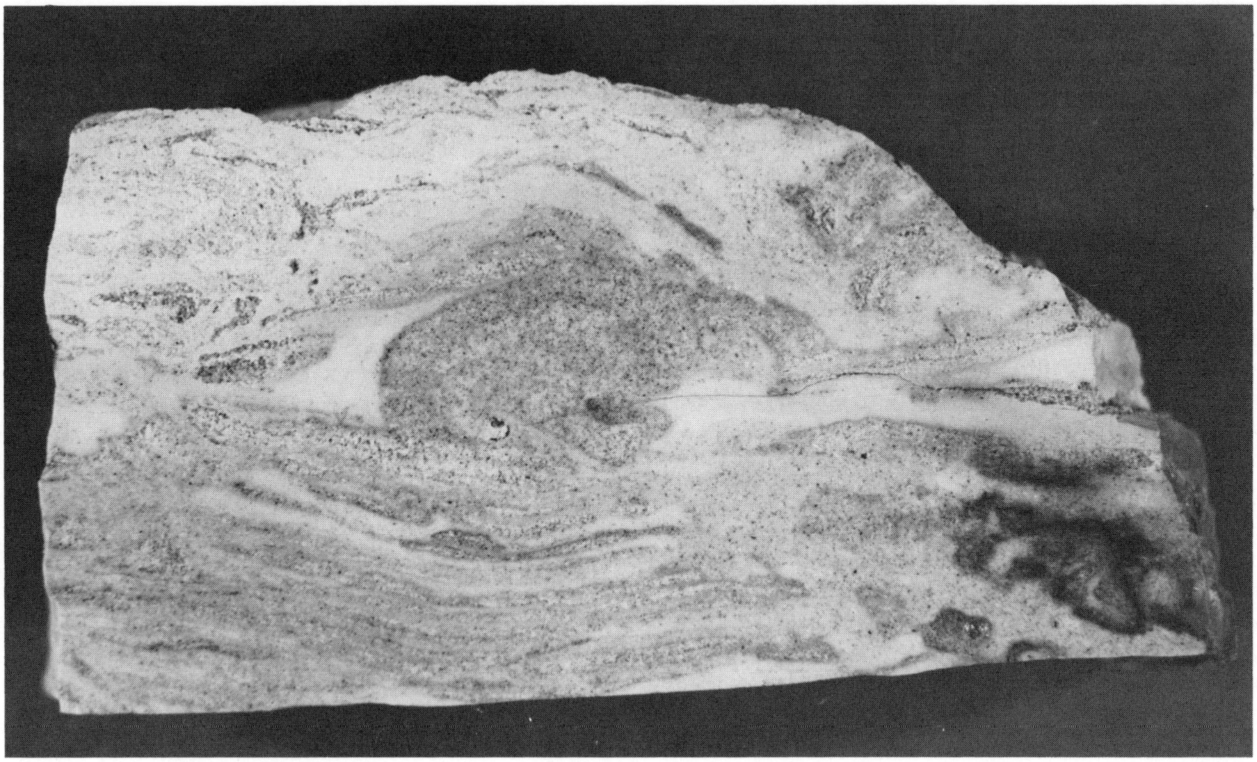

Figure 4. Slab section of fossiliferous rock from the Carolina slate belt of northern South Carolina (magnified × 2).

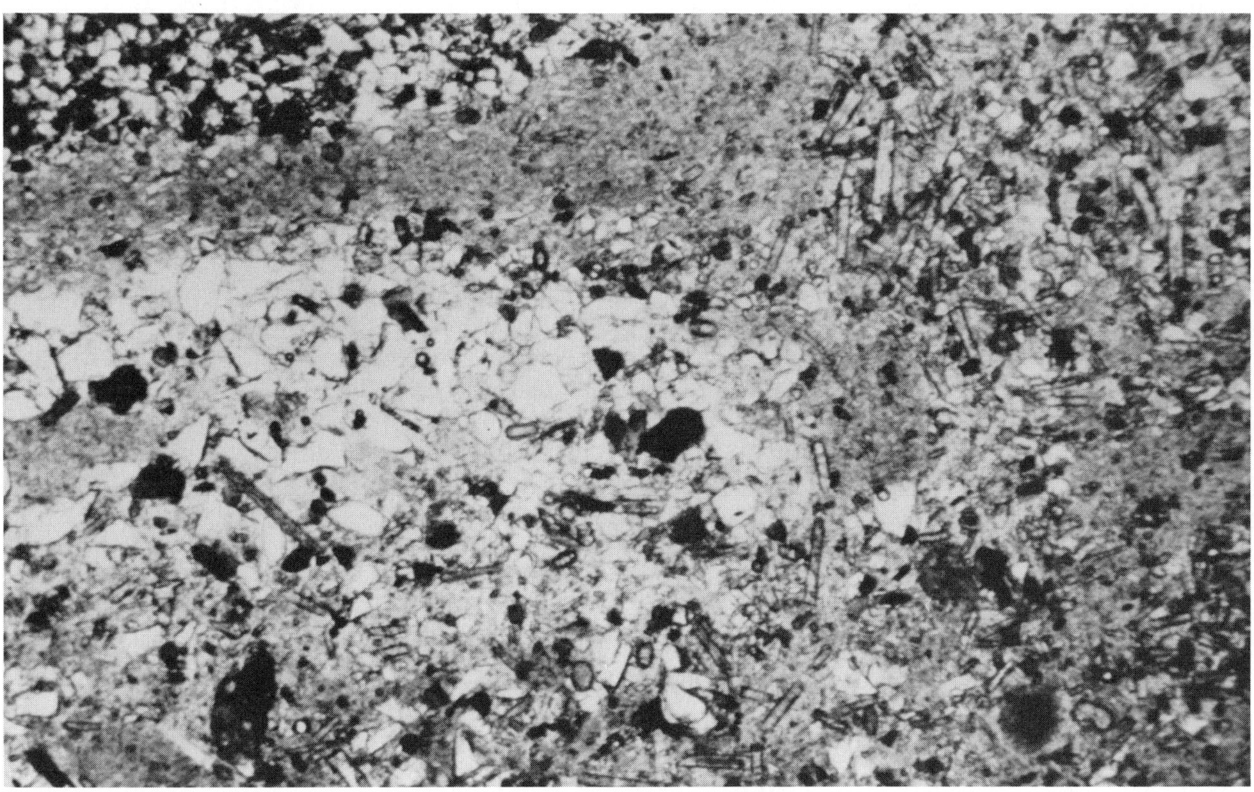

Figure 5. Thin section of fossiliferous rock that shows grain-size differences and the presence of sponge spicules in the fine-grained layers (magnified × 60).

rays are arranged at mutual angles of approximately 90°. One of the rays is markedly swollen with a subcylindrical section near the ray junction. The other rays, however, appear to taper quickly from the ray junctions. The large ray has a maximum basal diameter of approximately 0.025 mm and may be as much as twice that long, although the outer end is broken. The short, stubby rays have basal ray diameters of 0.017 and 0.019 mm and where best preserved, they taper abruptly at a distance of 0.03 mm to diameters of approximately 0.01 mm. The tips of the rays are broken.

A similarly well-preserved hexactine, specimen 502 (Pl. 1, fig. 2), has basal rays that are slightly smaller, subcylindrical, and arranged mutually at about 90°. It has rays 0.006 to 0.007 mm in basal ray diameter. Fragments of the rays are as long as 0.025 mm, but tips are buried in the rosette-strewn surface. Ray fragments taper only very gently so that they could be several times as long as the exposed fragments.

A single tetraclad spicule, specimen 486 (Pl. 1, fig. 3), has individual rays that appear to join at angles of approximately 120°, although one ray may rise more or less normal to the plane in which the other three rays occur. It is possible that the spicule may be a modified hexactine with an aborted ray rather than a true tetraclad. The tiny spicule has preserved ray fragments as long as 0.03 mm. They are subcylindrical and taper gently from basal ray diameters of 0.005 to 0.006 mm. The ray tips are either buried or broken.

In addition to spicules that are preserved essentially intact, some spicules are preserved as molds (Pl. 1, fig. 4). The molds appear to be impressions of monactines, or they may be fragments of hexactines. Most of these fragments are strongly tapering, some with outer sharp-pointed ray tips. Some are subcylindrical, abruptly broken fragments. The largest of these rays are 0.12 to 0.15 mm long and taper from maximum diameters of approximately 0.022 mm to sharp tips. Smaller forms taper abruptly from maximum diameters of 0.02 mm to sharp tips. Subcylindrical fragments have sharply broken abrupt ends. In a few impressions both molds and crude casts are preserved, generally separated by curved hairline fractures.

Placement of individual spicules within genera or families, when they are this generalized, is difficult. Overall characters of the spicules are so simple that even the hexactines are difficult to bracket in time. However, hexactine spicules are common in Middle Cambrian and younger rocks, and the morphology of the spicules does suggest an age of Middle Cambrian or younger. Such a date would fit with an early Paleozoic age, as has been assumed for most of the Piedmont. There is nothing in the sponge occurrence that would indicate an age older than Middle Cambrian.

COMPARISON OF THE FOSSIL AGES AND STRATIGRAPHY OF THE CAROLINA SLATE BELT

Geochronology and the lithostratigraphy indicate that the southern Carolina slate belt in South Carolina is no younger than Ordovician, and the sponge spicules (as reported here) indicate that it is no older than Middle Cambrian. Maher and others (1981) reported a similar Middle Cambrian to Ordovician age for a trilobite in southern South Carolina from the Belair belt (Fig. 1; Fig. 6, cross-sections C and D; Fig. 7, cross-sections C and D). Carpenter and others (1982), on the basis of geochronology of the Lincolnton metadacite, derived a late Precambrian to Cambrian age for similar lithologies from Georgia and southern South Carolina. In a general sense, therefore, fossil evidence and geochronology show that the southern Carolina slate belt and the Belair belt were deposited within the time interval from late Precambrian to Ordivician, which corroborates previous works on geochronology and lithostratigraphy from the northern half of the Carolina slate belt (Figs. 6 and 7).

Lithologically, the units in Figure 6 that contain the North Carolina trilobite, the South Carolina trilobite, and the sponge spicules are very similar (St. Jean, 1973; Maher and others, 1981; Bourland and Farrar, 1980). It has been suggested (Secor and Wagener, 1968) that the stratigraphy of the Carolina slate belt in South Carolina is equivalent to the upper part of the stratigraphic section of North Carolina, but there is no detailed mapping that proves this.

The generalized stratigraphic columns in Figure 6 represent the most detailed studies of stratigraphy, geochronology, and fossil occurrences from the Carolina slate belt and the Belair belt. There is a problem with the correlations made in Figure 6. The sponge spicules, Middle Cambrian or younger in age from the southern Carolina slate belt, occur in a lower stratigraphic position (Fig. 1; and Fig. 6, cross-sections B and C) than the *Paradoxides* trilobite, Early to Middle Cambrian in age from the northern Carolina slate belt, even though the spicules should occur stratigraphically equal or higher. The problem may be the biostratigraphic correlation of the trilobite locality. Beds stratigraphically below the trilobite could be either Middle Cambrian (Fig. 6, cross-section B; Fig. 7, cross-sections B1 and B2) (Wright and Seiders, 1980; Hills and Butler, 1969; and White and others, 1963) or Ordovician(?) or older (Stromquist and Sundelius, 1969, reported coalified algae of Ordovician(?) age). Since the trilobite is Early to Middle Cambrian in age and there are Ordovician(?) and Middle Cambrian sediments below the trilobite, then there should be a structural discontinuity, but none has been reported (Seiders and Wright, 1977). If the problem of inconsistencies within the sedimentary section below the trilobite is ignored, then the age of the trilobite in cross-section B and the age of the sponge spicules in cross-section C are roughly correlative. A correlation based on fossil ages, however, has very little lithologic correlation. If the correlation of fossil ages is correct, then the lithologic differences would have to be explained by placing the cross-sections in separate basins or

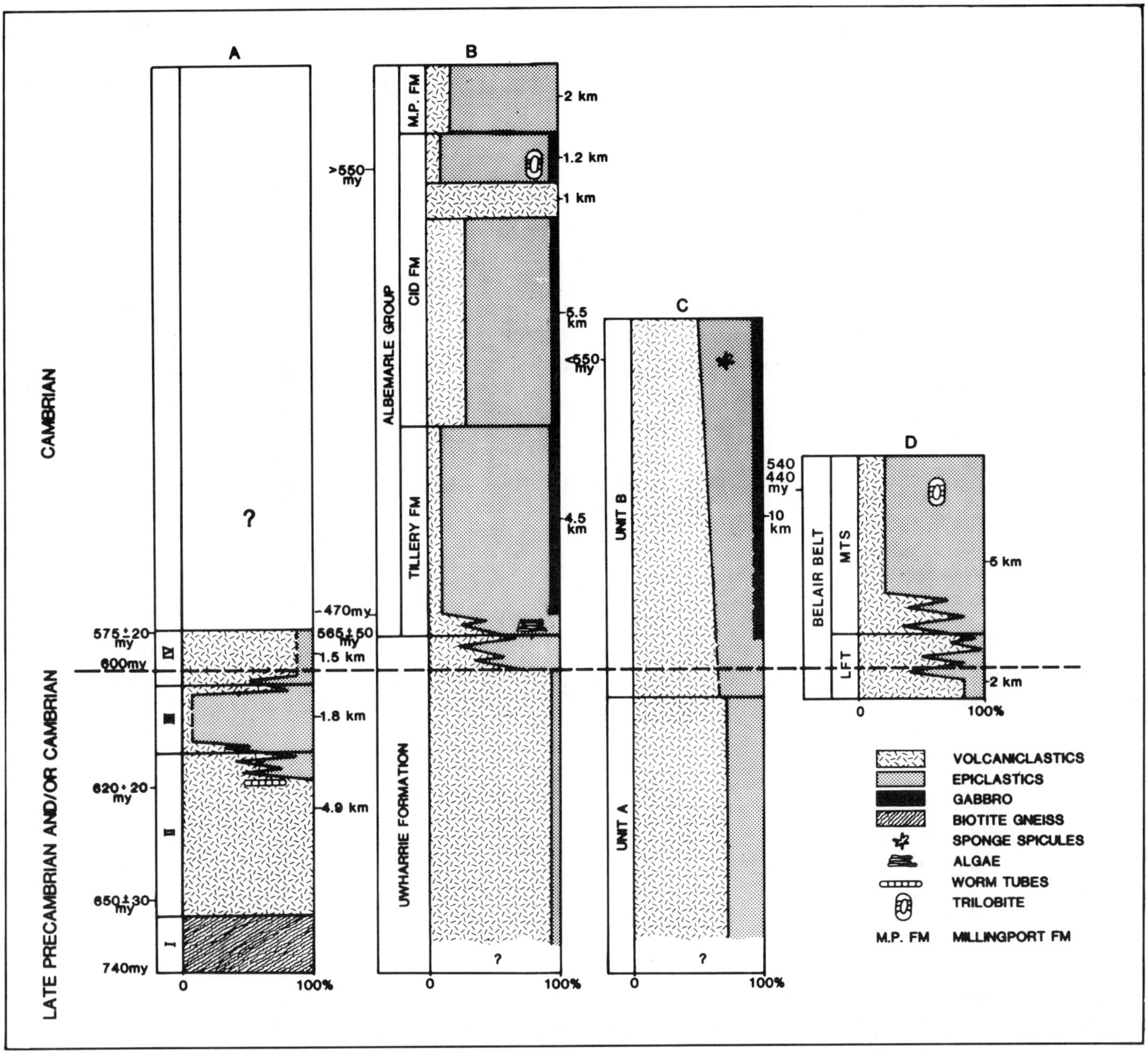

Figure 6. General stratigraphic correlation of all known fossil-bearing sections in the Carolina slate belt (A, B, and C) and the Belair belt (D) compiled from available geochronologic, paleontologic, and lithostratigraphic data: (A) Glover and Sinha (1973); (B) Seiders and Wright (1977); St. Jean (1973); Hills and Butler (1969); Conley and Bain (1965); (C) Bourland and Farrar (1980); and (D) Secor and Snoke (1978); Maher and others (1981). (The Carolina slate belt and Belair belt may not be stratigraphically correlative even though both belts appear to be geochronologically correlative).

by allowing the cross-sections to be part of the same basin that had different sediment-source terrains. Exactly how to correlate the stratigraphic cross-sections B and C in Figure 6 is not very clear.

The northern Carolina slate belt has the best documented age, 650 to 575 m.y., of any of the cross-sections in Figures 6 and 7. The ages shown on cross-section A were established by the combined geochronology of igneous units above and below Precambrian metazoan worm tubes from northern North Carolina (Fig. 1; Figs. 6 and 7, cross-section A) (Glover and Sinha 1973; Cloud and others, 1976; McConnell and Glover, 1982). The age of these units supports other evidence that the Carolina slate belt is younger to the south and might explain why the Virgilina event (620 to 575 m.y.)

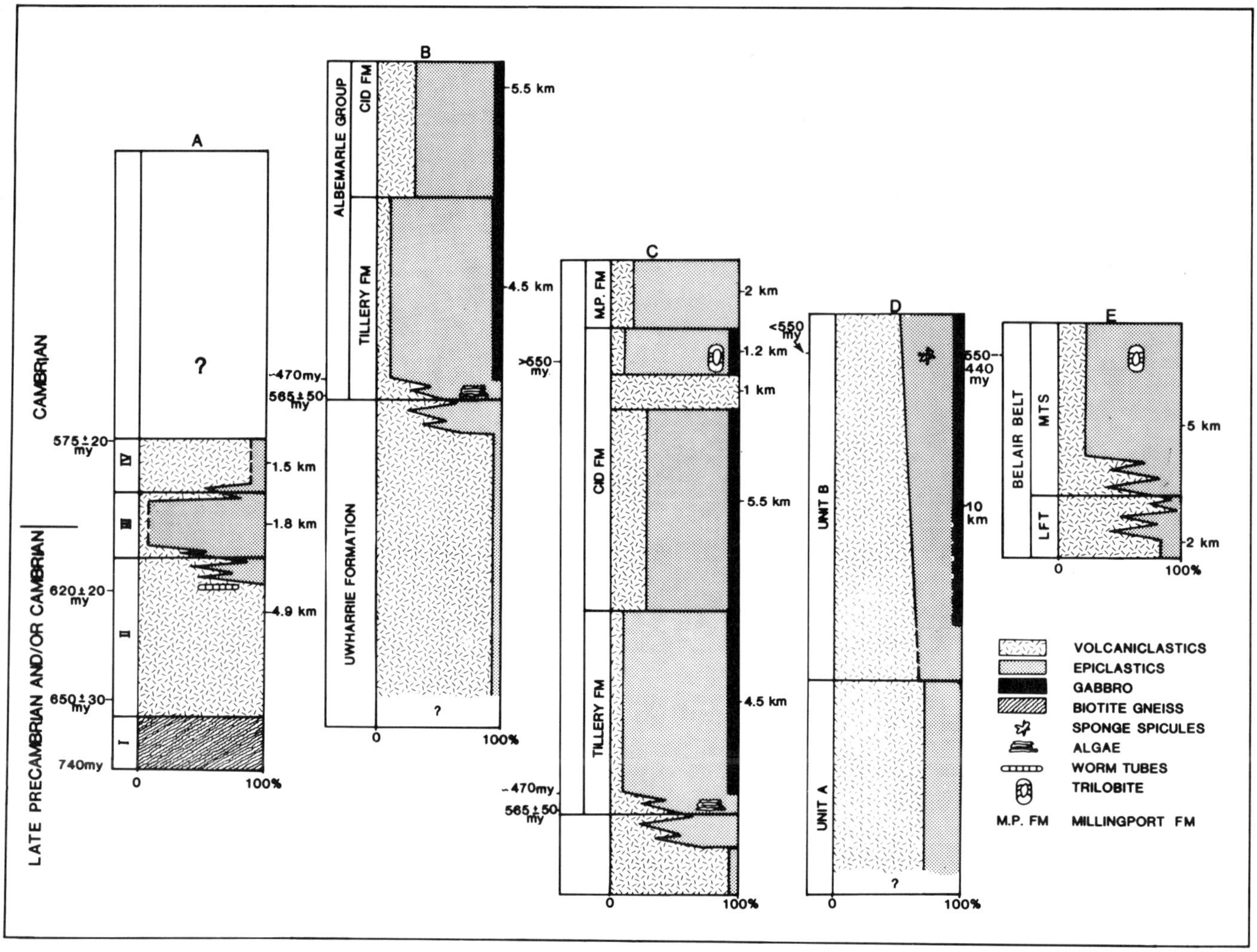

Figure 7. Biostratigraphic correlations based on the same data as Figure 6. Column B1 is correlated using the radiometric ages of Hills and Butler (1969) and Column B2 is correlated on the Early to Middle Cambrian trilobite age (St. Jean, 1973).

has not been observed in the southern Carolina slate belt (Glover and Sinha, 1973; Black, 1978; Black and Fullagar, 1976; Wright and Seiders, 1980). Both biostratigraphically and lithostratigraphically this section is older than the rest of the Carolina slate belt.

An attempt to correlate Carolina slate belt strata from North Carolina and South Carolina based on lithology and fossil occurrence is made in Figures 6 and 7. This is only preliminary because of the lack of clearly related strata, as explained with Figure 6. A more complete biostratigraphic correlation for the entire Piedmont is presented by St. Jean (1973). Lithofacies suggest that the Carolina slate belt is a back-arc basin along an active continental margin (Secor and Snoke, 1978; Glover and Sinha, 1973; Wright and Sciders, 1980) with lateral and vertical facies variations, depending on the nature of the source. A vertical facies change from predominantly volcaniclastic to epiclastic is observed in three of the lithologic sections correlated in Figure 6 (cross-sections B, C, and D). If all three cross-sections were from one continuous basin, then the nearly synchronous timing of a major lithologic change from volcaniclastic to epiclastic would be expected. However, the discrepancies in fossil ages and the lack of any stratigraphic correlation of the Carolina slate belt and Belair belt make the idea of one continuous basin seem implausible. It is more plausible to accept that the three cross-sections are all from different, small basins that may or may not have been synchronous. The similarities between the three cross-sections may then be attributed to a similar tectonic history for each basin, with only local variations in the source terrains. A biostratigraphic correlation is made in Figure 7, with two North Carolina stratigraphic columns, one based on the age of the trilobite (Fig. 7, cross-section B1 and B2) (St. Jean, 1973) and the other based on the geochronology of the sediments below the trilobite (Hills and Butler, 1969). A correlation using the fossilized algae was not included because of the large uncer-

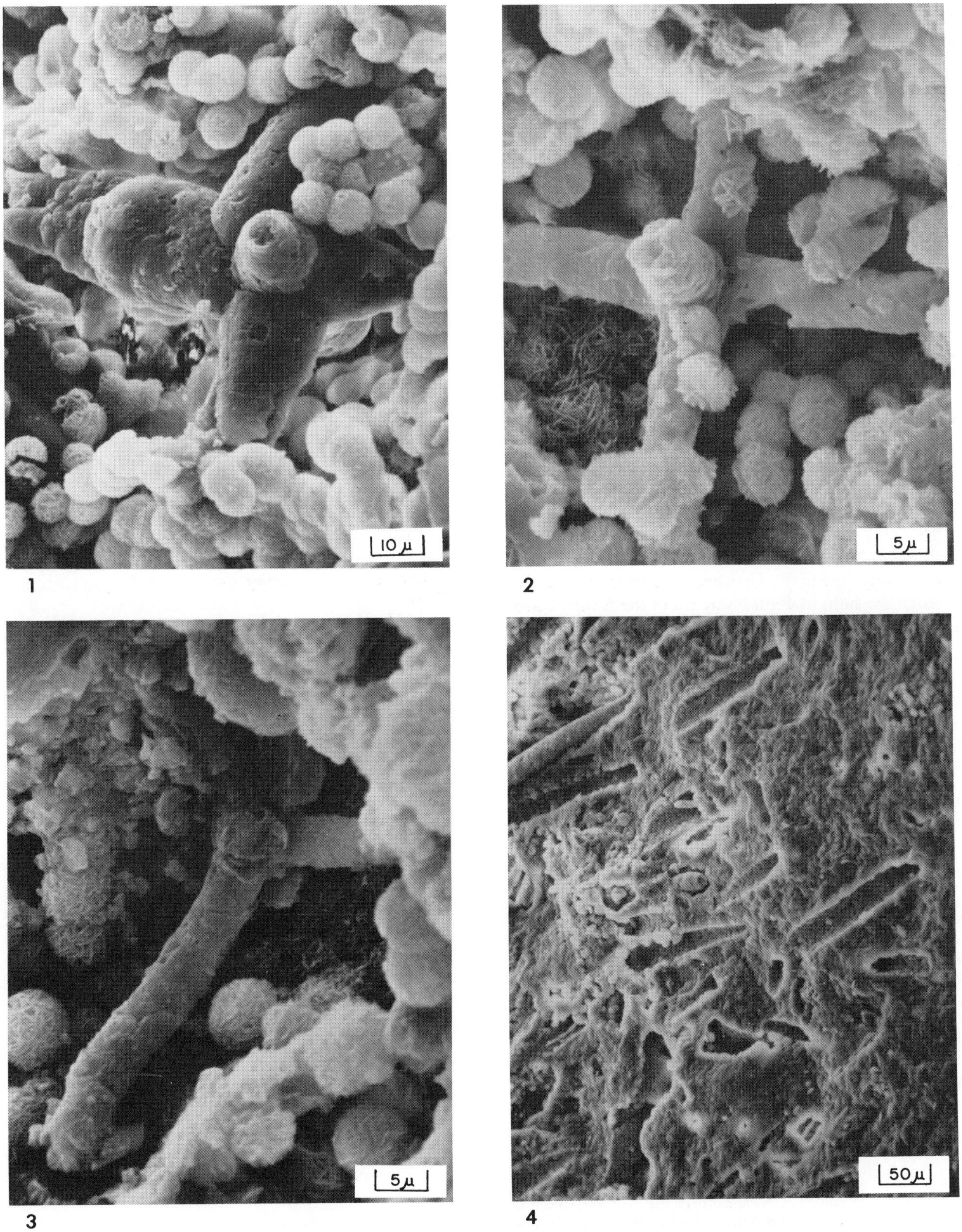

Plate 1. Electron photomicrographs of sponge spicules from metamorphosed rocks of the Carolina slate belt of South Carolina: figure 1, hexactine with somewhat swollen rays, 579-477; figure 2, hexactine with ascending ray broken, 579-502; figure 3, partial tetraclad in which rays appear to join at angles to approximately 120°, with ascending ray broken, 579-486; figure 4, spicule fragments preserved both as molds and positive relief features 579-479.

tainty in their age (Stromquist and Sundelius, 1969). Figure 7 shows a laterally variable set of lithofacies, indicating multiple depositional centers through time.

The correlations in both Figures 6 and 7 show the northern Carolina slate belt is older and different in lithofacies from the southern Carolina slate belt. Although there is only inconclusive evidence, one explanation may be that the northern Carolina slate belt was an older geographically separated basin from the southern Carolina slate belt and the Belair belt, but within approximately the same tectonic regime. The discrepancy in ages and stratigraphy along the Carolina slate belt may indicate a system of small separate basins, with younger basins forming to the south as the tectonic activity along the continental margin migrated southward.

ACKNOWLEDGMENTS

We gratefully acknowledge the suggestions by W. H. Abbott of the South Carolina Geological Survey, who originally thought there were microfossils in the low-grade rocks of the Piedmont. Field work for this study was part of the Virginia Polytechnic Institute and State University–South Carolina Geological Survey Geothermal Program. We also acknowledge the help of Paul Nystrom of the South Carolina Geological Survey, who has done most of the detailed mapping in the area, and Doris Phelps of Phillips Petroleum, Geological Research Branch, who did the scanning electron microscope studies.

Support for this research was provided by Virginia Polytechnic Institute and State University's Orogenic Studies Laboratory and the South Carolina Geological Survey through ERDA Contract EV-76-S-05-5104 and DOE Contract ET-78-C-05-5648.

REFERENCES CITED

Black, W. W., 1978, Chemical characteristics and RB/Sr ages of metavolcanics from the Carolina slate belt of North Carolina: Geological Society of America Abstracts with Programs, v. 10, no. 4, p. 162–163.

Black, W. W., and Fullagar, P. D., 1976, Avalonian ages of metavolcanics and plutons of the Carolina slate belt near Chapel Hill, N.C.: Geological Society of America Bulletin, v. 89, p. 511–521.

Bourland, W. C., and Farrar, Stewart, 1980, Tectogenesis of the rocks surrounding the Winnsboro Intrusive Complex: South Carolina Geology, South Carolina Geological Survey, p. 1–18.

Butler, J. R., and Howell, D. E., 1977, Geology of the Taxahaw quadrangle, Lancaster County, South Carolina: Geologic Notes, Division of Geology, South Carolina State Development Board, p. 133–149.

Carpenter, R. H., Odom, A. L., and Hartley, M. E., III, 1982, A geochronological investigation of the Lincolnton metadacite, Georgia and South Carolina, *in* Bearce, D. N., and others, eds., Tectonic studies in the Talladega and Carolina slate belts, southern Appalachian orogen: Geological Society of America Special Paper 191 (this volume).

Cloud, Preston, Wright, J. E., and Glover, Lynn, III, 1976, Traces of animal life from 620-million-year-old rocks in North Carolina: American Scientist, v. 64, p. 396–406.

Conley, J. F., and Bain, G. L., 1965, Geology of the Carolina slate belt west of Deep River–Wadesboro Triassic Basin, North Carolina: Southeastern Geology, v. 6, p. 117–138.

Glover, Lynn, III, and Sinha, A. K., 1973, The Virgilina deformation, a late Precambrian to Early Cambrian(?) orogenic event in the Central Piedmont of Virginia and North Carolina: American Journal of Science, v. 273-A (Cooper Volume), p. 234–251.

Hills, F. A., and Butler, J. R., 1969, Rubidium-strontium dates for some rhyolites from the Carolina slate belt of the North Carolina Piedmont: Geological Society of America Abstracts for 1968, Special Paper 121, p. 445.

Maher, H. D., Palmer, A. R., Secor, D. T., and Snoke, A. W., 1981, New trilobite locality in the Piedmont of South Carolina and its regional implications: Geology, v. 9, p. 34–36.

McConnell, K. I., and Glover, Lynn, III, 1982, Age and emplacement of the Flat River complex, an Eocambrian sub-volcanic pluton near Durham, North Carolina, *in* Bearce, D. N., and others, eds., Tectonic studies in the Talladega and Carolina slate belts, southern Appalachian orogen: Geological Society of America Special Paper 191 (this volume).

Secor, D. T., and Snoke, A. W., 1978, Stratigraphy, structure, and plutonism in the central South Carolina Piedmont, *in* Snoke, A. W., ed., Geologic investigations of the eastern Piedmont, southern Appalachians: Division of Geology, South Carolina State Development Board, Carolina Geological Society 1978 Guidebook, p. 65–123.

Secor, D. T., and Wagener, H. D., 1968, Stratigraphy, structure, and petrology of the Piedmont in central South Carolina: Carolina Geological Society Field Trip, October 1968, Geological Notes, Division of Geology, South Carolina Development Board, v. 12, no. 4, p. 67–84.

Seiders, V. M., and Wright, J. E., 1977, Geology of the Carolina volcanic slate belt in the Asheboro, North Carolina area: Geological Society of America, Southeastern Section Meeting, 1977, Guidebook, p. 1–34. (Raleigh, North Carolina Department of Natural and Economic Resources).

Stromquist, A. A., and Sundelius, H. W., 1969, Stratigraphy of the Albemarle Group of the Carolina slate belt in central North Carolina: U.S. Geological Survey Bulletin 1274-B, p. 1–22.

St. Jean, Joseph, 1973, A new Cambrian trilobite from the Piedmont of North Carolina: American Journal of Science, v. 273-A (Cooper Volume), p. 196–216.

White, A. M., Stromquist, A. A., and Stern, T. W., 1963, Ordovician age for some rocks of the Carolina slate belt in North Carolina: U.S. Geological Survey Professional Paper 475-C, p. C107–C109.

Wright, J. E., and Seiders, V. M., 1980, Age of zircon from volcanic rocks and tectonic implications from the Carolina volcanic slate belt: Geological Society of America Bulletin, v. 91, p. 287–294.

Manuscript Accepted by the Society July 8, 1982

Printed in U.S.A.

Geological Society of America
Special Paper 191
1982

Age and emplacement of the Flat River complex, an Eocambrian sub-volcanic pluton near Durham, North Carolina

Keith I. McConnell*
Lynn Glover III
Orogenic Studies Laboratory, Department of Geological Sciences
Virginia Polytechnic Institute and State University
Blacksburg, Virginia 24061

ABSTRACT

This study documents an eruptive center, the instrusive Flat River complex and associated Carolina slate belt volcanic rocks near Durham, North Carolina. The zircon Pb/U age of the granite, granodiorite, quartz diorite, gabbro Flat River complex is 650 ± 30 m.y. Local vent breccias, chilled and protoclastically deformed margins, and petrographic comparisons with the experimental results in the qz-ab-or system confirm that the Flat River complex was emplaced at a depth of less than 1 km and was locally surface breaking. Thus, the Flat River can be viewed as a very shallow fossil volcanic magma chamber and may be typical of many granitoid plutons in the Carolina slate belt.

The Flat River was intruded into a pile of coarse, near-vent pyroclastic deposits and lavas of felsic to intermediate composition. Intercalated in these is an exhalative unit of thinly layered to laminated metachert and iron oxide. Shallow to deep (or quiet) probably marine conditions prevailed. Some subaerial pyroclastic rocks may be present in the older parts of this >650-m.y. to 620-m.y.-old sequence of volcanic rocks.

INTRODUCTION

Little is known about the relation between plutons and eruptive volcanic rocks in the crystalline terrain of the central and southern Appalachians. Glover and Sinha (1973, p. 241) suggested that the Moriah pluton in the Carolina slate belt near Durham, North Carolina, may have been eruptive, and this study was designed to test that hypothesis.

The herein-named Flat River complex (composed of the Moriah pluton and the Butner stock) in the Carolina slate belt of North Carolina (Fig. 1) has been spared the severe deformation accompanying the higher grades of metamorphism common to other belts in the Piedmont. Therefore, and important to this study, the rocks retain much of their original igneous texture. Cater (1969), Fiske and others (1963), and Tabor and Crowder (1969) have described the relations between shallow level plutons in the Cascades of Washington and surrounding volcanic material. With this base of information, the well-preserved textures of the Flat River complex permit a similar study in the southern Appalachian Piedmont.

*Present address: Georgia Geologic Survey, Room 400, 19 Martin Luther King, Jr., Drive, SW, Atlanta, Georgia 30334.

METAMORPHIC AND STRUCTURAL HISTORY OF THE STUDY AREA

All pre-Mesozoic rocks in the area have been through one greenschist metamorphic event. Four deformational events have been recognized in the general area of the Flat River complex by Tobisch and Glover (1971). Glover and Sinha (1973) determined the timing of three of these events: (1) Eocambrian-Cambrian(?) (~600 m.y. ago) pre-metamorphic folding and faulting, termed the Virgilina deformation, which formed the Virgilina synclinorium in the study area (Fig. 1); (2) mid-Paleozoic (probably Taconic according to data of Kish and others, 1979; Black, 1977; and Bell and others, 1972) low-grade metamorphism and related cleavage (S_1) formation; and (3) later local folds and a spaced cleav-

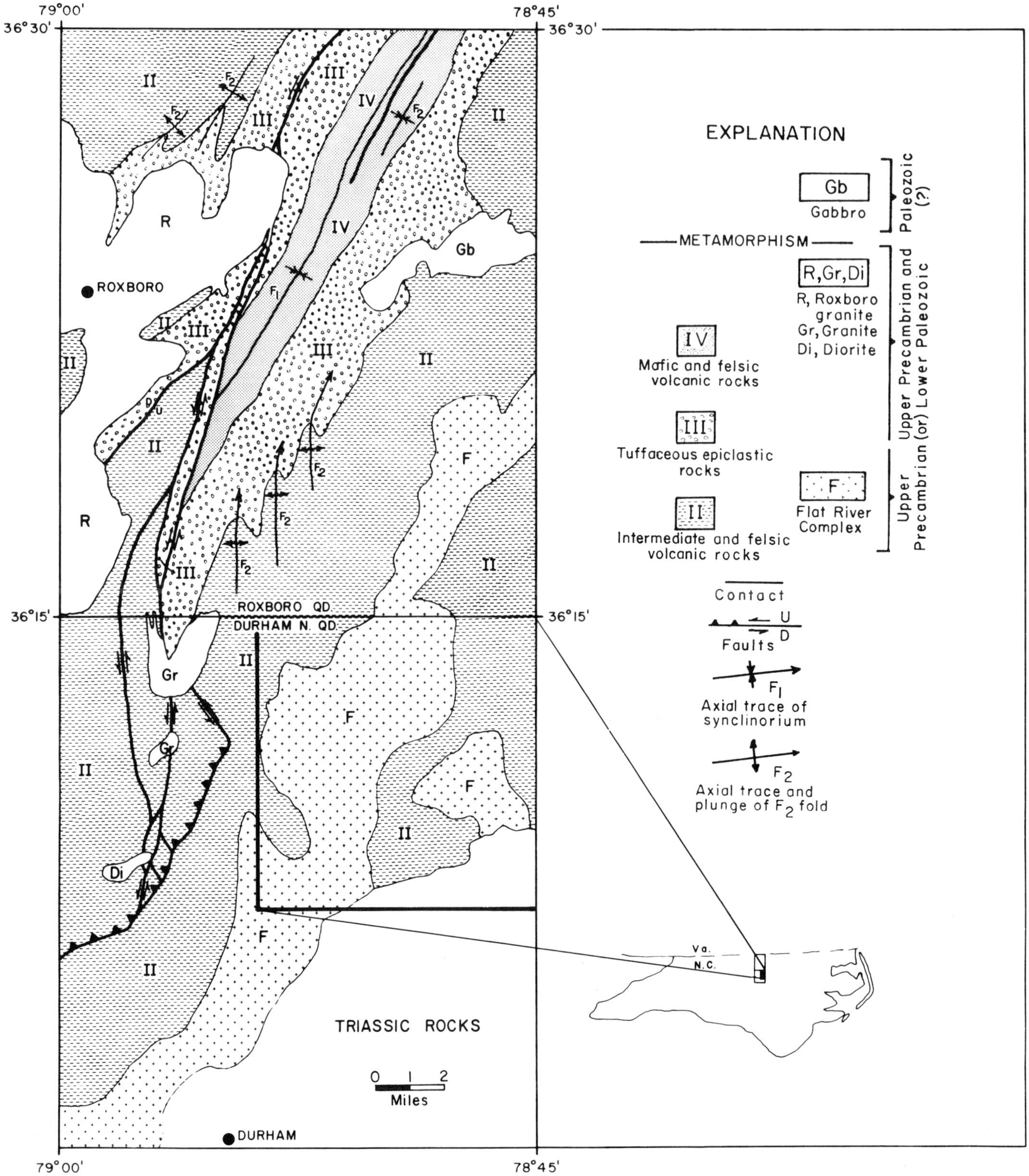

Figure 1. Generalized geology of the Roxboro-Durham area, North Carolina. Geology by Lynn Glover III, K. I. McConnell, S. C. Werner, and J. E. Wright (1967–1974).

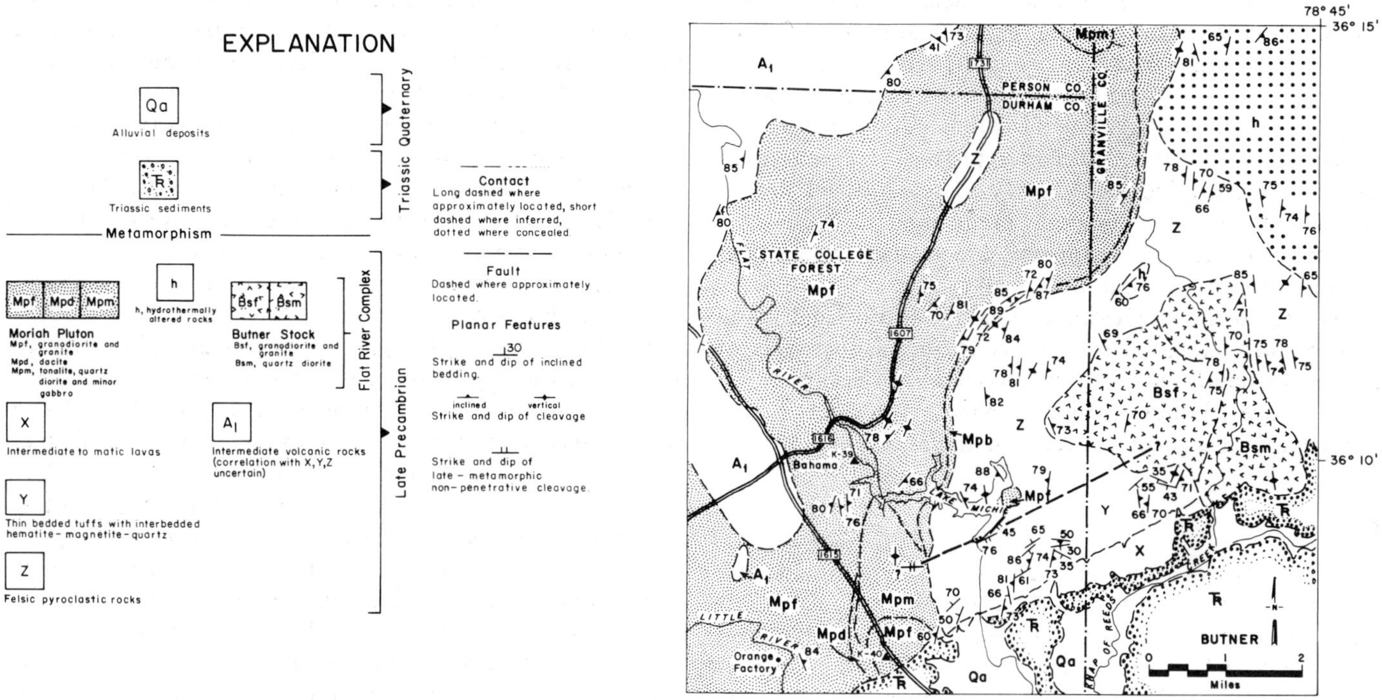

Figure 2. Geologic map of the northeast part of the Durham North, North Carolina, 15-minute quadrangle.

age (S_2). The final event (4) was Mesozoic rifting that formed the Triassic basin in the southeastern part of the map area (Fig. 1).

STRATIGRAPHY

General Statement

Volcanic rocks exposed in the northeast quadrant of the Durham North 15-minute quadrangle represent lower unit II of Glover and Sinha (1973) described in the adjacent Roxboro quadrangle (Fig. 1). They found that unit II on the western limb of the Virgilina synclinorium is about 4,900 m (16,000 ft) thick. East of the Flat River complex (Fig. 2), rocks of the lower part of unit II are herein divided into lithostratigraphic members on the basis of sedimentary structure and texture and are designated from older to younger: Z, Y, and X. Unit A_1 of Wright (1974) west of the Flat River complex is correlative with part of unit II of Glover and Sinha but cannot be directly related to Z, Y, or X. The present structural data and contact relations suggest that units Y and X form a homoclinal structure dipping to the southeast and may represent the southeastern limb of an anticlinorium, adjacent to the Virgilina synclinorium. Unit Z may contain the crest of this anticlinorium. The stratigraphic sequence also includes post-metamorphic rocks of Triassic age. These occur in a basin, part of which is shown in Figure 2.

Member Z

Member Z is a thick sequence of predominantly felsic pyroclastic rocks that lies along the eastern border of the Moriah pluton (Fig. 2). At the western margin of member Z, the volcanic rocks are in contact with a finer-grained chilled margin of porphyritic dacite, part of the Moriah pluton. The contact between member Z and member Y is placed at the change from primarily massive lapilli tuff and coarse tuff of member Z to fine-grained, thinly bedded, and locally laminated tuffs of member Y. This contact has not been observed in outcrop, but it can be seen in float over a distance of about 50 ft. This abrupt change in grain size and bedding thickness and the presence, near the contact, of a cleavage that is not conformable with the regional cleavage suggest that the contact is a fault.

Member Z was dominated originally by vitric lapilli and ash fragments (classification according to Fisher, 1966), less commonly by lithic fragments. Large amounts of pumiceous debris are found locally. Fragment size and lithic component of member Z increase from northeast to southwest. Volcanic textures are well preserved in member Z. Spherulites were found in both the groundmass tuff and in lapilli size fragmental material of one sample from the northeastern part of the member Z. A possible eutaxitic texture, which possibly formed as a result of the collapse of pumice during welding (Ross and Smith, 1961), and round bubble pumice were identified in the rocks from the southwestern part of member Z.

Mid-Paleozoic greenschist facies metamorphism recrystallized these rocks, obscuring the original rock type comprising tephra and other fragments. Quartz is common in the groundmass and in ash- to lapilli-size fragments. Sodic plagioclase, determined by the Michel-Levy method, is abundant and commonly shows broken and bent twin lamellae. The plagioclase crystals were saussuritized during metamorphism. The dominant mafic mineral is chlorite. Metamorphic accessory minerals are calcite, white mica, epidote, pyrite, and stilpnomelane(?).

Features such as spherulites, eutaxitic texture and abundant pumice, common to member Z, are characteristic of subaerially deposited ash-flow tuffs (Ross and Smith, 1961). The absence of these features in parts of member Z may imply local subaqueous pyroclastic deposition.

Unit A_1

Unit A_1 occurs along the western border of the Moriah pluton and was described in detail by Wright (1974). This unit may be correlative with member Z if the Moriah pluton cores an anticlinorium adjacent to the Virgilina synclinorium, but textural and compositional differences (local turbidities and more mafic composition of unit A_1) between the two units do not support this.

The contact between unit A_1 and the Moriah pluton is sharp. Porphyritic dacite, common to the eastern border of the Moriah pluton, is absent at this contact. A somewhat finer-grained, felsic intrusive rock, minor sulfide mineralization, and shearing in the volcanic rock are evident at the contact.

The thickness of unit A_1 has been estimated to be a minimum of 2,500 m (8,100 ft) (Wright, 1974). In this area, it is a heterogeneous mixture of poorly sorted andesitic pyroclastic rocks, andesitic lava, and minor amounts of dacitic pyroclastic rocks (Wright, 1974). Andesitic lapilli tuff predominates in the unit, but varying amounts of tuff breccia, crystal tuff, and lava are also present. Interbedded with these rocks is well-sorted, thin-bedded tuff that commonly shows graded bedding with small-scale cross bedding and soft sediment faulting and slumping. Subaqueous depositional features such as pull-apart structures in the tuff breccias and quench brecciation of the lavas occur in unit A_1 in the area of Figure 2.

Member Y

Member Y is a unit of fine-grained, intermediate to felsic, volcaniclastic rocks that commonly show graded bedding and laminated bedding. As mentioned above, it is probably in fault contact with member Z. The Moriah pluton and the Butner stock (Fig. 2), described below, discordantly intruded the east and west borders of member Y. Minor sulfide mineralization occurs at the contact.

The minimum thickness of member Y is 1,600 m (5,200 ft). The member is composed primarily of well-sorted fine tuff and small amounts of lapilli tuff. A conspicuous subunit of interbedded quartzite (recrystallized chert) and iron oxides occurs within the member. Most of the volcaniclastic material is very fine ash (0.1 mm), although grains range in size to as much as 0.5 mm in the coarser beds. Graded bedding is commonly present and is expressed by thin basal layers of crystals and recrystallized glass fragments overlain by ash that grades vertically into finer sizes. Current indicators such as scour channels, cross-bedding, and ripple marks were not observed.

Bedded iron oxide (magnetite and hematite) interlayered with quartzite occurs in float and outcrop and provides a conspicuous marker unit showing an east-northeast trend for member Y. The iron oxides and quartzite (iron formation) are seen along strike for approximately 4,150 m (15,800 ft). The unit is deformed by soft sediment faulting and folding, and is about 300 m (1,159 ft) thick. The iron oxide grains are commonly euhedral and make up as much as 50% of some samples. Iron oxide layers generally occur as 2 to 5 mm laminae, but are locally as much as 10 cm thick. Quartzite layers and laminae are interleaved with and grade abruptly into the iron oxide layers. The quartzite is commonly very pure and exhibits polygonal (strain free) grain boundaries meeting at 120° angles. We interpret the iron formation to have been a subaqueous chemical precipitate of chert and iron oxides. It may represent an exhalative deposit related to the subjacent volcanism and plutonism.

Member X

Intermediate to mafic lavas that overlie the thin-bedded tuffs of member Y are grouped together into member X. The lower contact was observed in float southwest of the dam on Lake Michie. It shows fine tuff similar to unbedded parts of member Y injected into fractures in member X. These fractures may have resulted from rapid chilling of the lava when it came in contact with water. As fracturing took place, the water-saturated fine tuff moved upward into the fractures in response to the weight of the overlying lava. The upper contact and the presumed intrusive contact of the Moriah pluton are not exposed because of a covering of Triassic rocks to the south. The contact with the Butner stock is discordant to the northeasterly trend of member X, but was not observed in outcrop. The minimum thickness could not be ascertained because of the Triassic covering.

The rocks of member X are light green to nearly black, depending on the magnetite and chlorite content. They are commonly porphyritic with an intergranular to intersertal groundmass. Amygdules are common to all parts of member X and in some cases make up as much as 35% of the rock. Vesicles are now filled commonly by quartz, calcite, epidote, and chlorite. Amygdule diameters generally range up to 1 cm, although amygdules larger than 1 cm have been observed. The groundmass mafic minerals are chlorite, epidote, and minor actinolite. Chlorite commonly occurs in large amounts (60% to 70%) in the matrix between amygdules and phenocrysts. Epidote and calcite are common in the groundmass and make up 15% to 20% of the rock. Plagioclase occurs as highly saussuritized phenocrysts, as well as small lathlike crystals that show little alteration. Plagioclase composition is An_8, determined by the Michel-Levy method, and makes up approximately 15% of the rocks. Quartz occurs only as vesicle fillings. Accessory minerals are apatite and magnetite. The local abundance and large size of amygdules in member X implies that the depth of water present in the basin at the time of extrusion was probably less than 500 m (Moore, 1965).

Post-metamorphic Sedimentary Rocks

Arkosic sandstone and conglomerate of the Deep River Triassic basin are present in the southern part of the map area (Fig. 2). Sandstone is commonly cross-bedded and locally contains fossilized wood.

INTRUSIVE ROCKS

General Statement

A large area of the northeastern quadrant of the Durham North quadrangle is occupied by intrusive rocks (Fig. 2). The plutons named herein are the Butner stock in the southeastern part of the area and the Moriah pluton in the western part of the area. The proximity of the two intrusive bodies suggests that they may be related. Intrusive apophyses piercing the volcanic strip separating the two bodies and interpretation of magnetic anomalies in the area from an aeromagnetic map (see U.S. Geological Survey, 1973) support the relation by suggesting that the intrusives may be connected at depth. Therefore, the Moriah pluton and the Butner stock have been combined with a related hypabyssal, dacite porphyry into the herein-named Flat River complex. Other intrusive rocks to the east and northeast in the Oxford area (Hadley, 1973) were believed by us to have been comagmatic with those in the Flat River complex until isotopic dating, presented in this report, showed that at least part of the complex described by Hadley is much younger.

Pre-metamorphic mafic dikes and post-metamorphic diabase dikes of probable Triassic/Jurassic age are common in all parts of the map areas.

Flat River Complex

The rocks of the Flat River complex were described briefly by Glover and Sinha (1973) in their study of the Virgilina deformation in the Roxboro quadrangle. Subsequent mapping to the south and southwest (Wright, 1974; and this report) has extended the known outcrop of the Flat River complex 40 km to the southwestern border of the Durham North quadrangle. Granodiorite is the dominant rock type in the complex, but the range of composition includes granite, quartz diorite, and minor gabbro.

Poor exposure and little vertical relief hinders detailed examination of contact zones of the Flat River complex. However, three types of contact zones seem to be present and to bear on interpretation of emplacement conditions. The most common type of contact is found along the eastern border of the Moriah pluton in the northeast quadrant of the Durham North quadrangle and results from the chilling of the margin of the granodioritic core. The contact usually contains porphyritic dacite that, in some cases, shows evidence of protoclastic shearing. Porphyritic dacite in the margins of the Moriah pluton to the southwest of the study area is generally autoclastic and associated with intrusive breccia (Wright, 1974). A second type of contact zone occurs along the northwest and southeast borders of the Moriah. Here, the contact is very sharp, the pluton margin being only slightly finer-grained than the granodioritic core and the volcanic rocks showing minor sulfide mineralization and shearing. The third type of contact zone, along the northeastern border of the lobe of volcanic rock (unit A_1), is characterized by abundant vein quartz and rock that appears to have been altered by hydrothermal fluids. The altered rock has lost its original texture and now consists of massive quartz and feldspar.

Three facies of instrusive rock are distinguished within the Flat River complex on the basis of mineralogy and texture. The predominant facies (1) is a medium-grained, light-colored granodiorite to granite (Fig. 3) that forms the core of the Moriah pluton and is present in the eastern part of the Butner stock (Fig. 2). Other facies present are (2) dacite

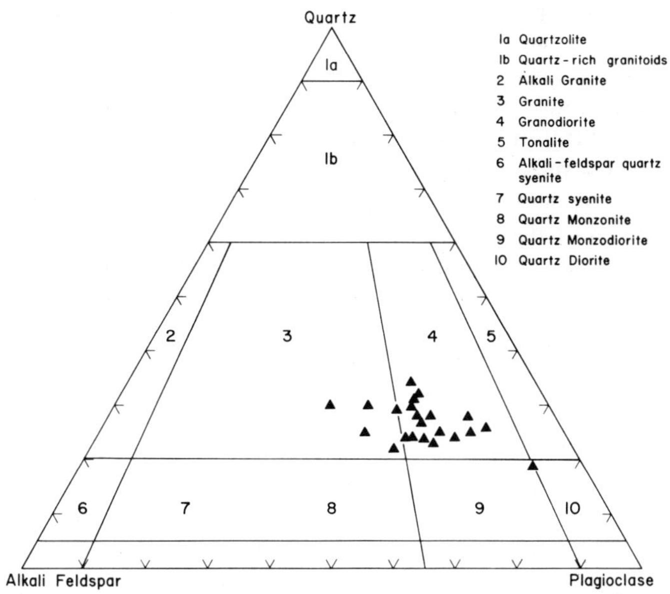

Figure 3. Graphic representation of modes from intrusive rocks of the Flat River complex. Diagram after I.U.G.S. subcommission on the systematics of igneous rocks (Streckeisen, 1973). Modes of gabbro and tonalite omitted.

porphyry, common at the borders of the Flat River complex, and (3) an intermediate to mafic facies composed of quartz diorite and gabbro.

The first facies is composed of medium-grained, predominantly porphyritic granodiorite with subordinate amounts of porphyritic granite, subequigranular granodiorite and subequigranular granite. The original mineralogy of the felsic, medium-grained facies was altered by metamorphism and hydrothermal activity. Plagioclase phenocrysts commonly make up from 30% to 40% of the granodiorite and from 18% to 30% of the granite. The phenocrysts are commonly altered to albite, epidote, and white mica. The margins of plagioclase phenocrysts are replaced by myrmekitic quartz and potassium-rich feldspar, now microperthite. Relic zoning of the plagioclase phenocrysts is defined by calcic replacement products varying in size and abundance with the original oscillatory zoning.

The groundmass consists of anhedral grains of quartz and microperthite, subhedral grains of plagioclase, and micrographic granophyric intergrowths that are believed to result from simultaneous crystallization of quartz and potassium-rich feldspar (Barker, 1970). Subhedral to anhedral plagioclase in the groundmass of the granodiorite makes up from 7% to 24% of the rock and in the quartz granite, 11% to 27% of the rock. Quartz in the granodiorite and granite occurs as individual grains and with potassium-rich feldspar in irregular to graphic intergrowths. It commonly shows undulose extinction and is rarely polygonal. Potassium-rich feldspar in the leucocratic, medium-grained, intrusive rocks occurs as individual grains and as perthitic to microperthitic graphic intergrowths with quartz. Microprobe x-ray photographs of the distribution of Na and K in the feldspar intergrowths clearly show a random patchy distribution of nearly equal amounts of albite and microcline that have exsolved under low-grade metamorphism from a single potassium-rich feldspar phase. The last emplacement and crystallization of the magma is therefore taken to be under shallow hypersolvus conditions.

Mafic minerals in the granitic rocks are biotite partly altered to chlorite, and hornblende partly altered to epidote, magnetite, and chlorite. Biotite, hornblende, and their alteration products commonly make up 10% of both the granodiorite and granite. Accessory minerals present are zircon, pyrite, sphene, apatite, magnetite, and ilmenite.

The granite and granodiorite locally incorporate slightly more mafic autoliths. The autoliths are finer grained than the host and generally consist of altered plagioclase laths in a matrix of quartz and minor potassium feldspar and mafic material. It is suggested that these autoliths represent portions of a chilled margin or roof that has been reintruded and partially assimilated by the coarser phase. A large roof pendant similar to the smaller autoliths in mineralogy and texture occurs east of the Flat River near the North Carolina State Forestry field camp. The contact zone between the two phases is sharp, but granophyric intergrowths in the groundmass of the coarser phase commonly embay the finer phases. Granophyric intergrowths also seem to be more common near the contact.

A spherulite-like texture was found in one sample of the felsic, medium-grained, intrusive rocks near the contact with the surrounding volcanic rocks. Radiating crystallites of intergrown quartz and feldspar are developed in an otherwise medium-grained porphyritic rock. The spherulitic texture may indicate that locally glass was present in the medium-grained intrusives near the contacts.

The second facies of the Flat River complex, dacite porphyry, is distinguished from the leucocratic medium-grained facies by its finer-grained groundmass. The dacite represents a chilled margin of the Flat River complex. It weathers to a pale bluff to red color and contains phenocrysts of quartz and plagioclase as much as 1 cm long. The porphyry in the study area shows less protoclastic deformation than the intensely deformed dacite several miles to the southwest along the boundary of the Flat River complex (Wright, 1974).

Plagioclase and quartz occur as euhedral to subhedral grains that make up approximately 20% of the dacite. Plagioclase phenocrysts are, in some cases, broken and rounded. Quartz phenocrysts commonly show resorption. Mid-Paleozoic metamorphism has saussuritized and albitized the original plagioclase. The groundmass in the dacite is composed of a fine-grained anhedral aggregate of quartz, feldspar, chlorite, magnetite(?), epidote, and white mica.

The third facies of the Flat River is an intermediate to mafic facies. Intermediate to mafic rocks occur along the eastern margins of both the eastern and western plutons in the study area (Fig. 2). All have been metamorphosed to greenschist facies but retain their igneous textures. Quartz diorite is the most common rock type and is associated with minor amounts of gabbro. No major occurrences of gabbro were found in the mapped area, but bodies of gabbro have been noted to the north in the Flat River complex by Glover and Sinha (1973).

The quartz diorite is commonly a medium-grained porphyritic to equigranular rock with subhedral crystals of plagioclase and amphibole comprising as much as 90% of the rock. Alteration by late-stage fluids and metamorphism was almost complete in the quartz diorite. Albitized plagioclase crystals, some with clear untwinned rims, are largely replaced by a coarse aggregate of clinozoisite, calcite, and white mica. Locally, minor amounts of potassium feldspar surround plagioclase. The primary hornblende is commonly altered to blades or tablets of actinolite or to an aggregate of chlorite, epidote, and magnetite. Anhedral quartz in the quartz diorite makes up from 7% to 16% of the rock. Other minerals present are mainly secondary. Chlorite, calcite, white mica, and epidote group minerals occur as alteration products of plagioclase and amphibole. Sphene is present as alteration rims around the opaque oxide. Apatite is also present in minor amounts.

One small body of gabbro occurs near the border between the felsic medium-grained facies and the intermediate facies of the Moriah pluton and may be a more mafic differentiate of the diorite. Metamorphic effects are similar to the above, pyroxene was uralitized, and olivine was replaced by chlorite and magnetite.

A chilled margin of porphyritic dacite is present between the intermediate to mafic facies and the felsic medium-grained facies in the southwest (Fig. 2). This dacite suggests that the intermediate to mafic facies was emplaced prior to the emplacement of the felsic facies.

Early Cambrian Intrusive Rocks

Early Cambrian intrusive rocks occur just east of the study area. Compositionally, these rocks appear to be similar to those in the Flat River complex, but mineral age data presented herein indicate that they are much younger. Hadley (1973) mapped intrusives to the north of the Early Cambrian rocks that may be correlative with them.

Pre-metamorphic Intrusive Rocks of Uncertain Age

Greenstone dikes, possible meta-lamprophyres or metabasalts, are common in the study area. The dikes have been observed in all members of the volcanic sequence in the area and in the Flat River complex. The greenstone dikes are usually deeply weathered and commonly exhibit a metamorphic foliation that is consistent with the north-northeast regional trend (Glover and Sinha, 1973; Tobisch and Glover, 1971). The dikes are composed of actinolite, chlorite, plagioclase, epidote, and minor white mica and carbonate material. The age of the dikes is probably Cambrian, or possibly Lower Ordovician. They are known by intrusive relations to be younger than 575 m.y., the age of the Roxboro pluton (Glover and Sinha, 1973), which is cut by the greenstone dikes, and older than the mid-Paleozoic, possibly Taconic, metamorphic episode.

Post-metamorphic Igneous Rocks

Post-metamorphic igneous rocks of probable Triassic/Jurassic age are common throughout the study area. They occur as diabase dikes and sills(?) and are generally less than 10 m thick, but have been observed to be as much as 70 m thick (Prouty, 1931). Compositionally, the diabasic dikes and sills(?) are composed of calcic plagioclase An-50, determined by extinction angles, pigeonite (very low 2V), minor olivine, and magnetite. Textures range from subophitic to ophitic.

ISOTOPIC DATING

General Statement

To determine the age of the Flat River complex, zircons were separated from three samples of the intrusive rocks of the study area. Two samples (K-39 and K-40) came from the Moriah pluton (Fig. 2), and the third (LGO-43) came from intrusive rocks just east of the northeastern border of the study area. Sample localities are given in Table 1 and plotted in Figure 2.

Methods

Approximately 75 lb of each sample was crushed and processed through the standard Wilfley table, magnetic separator, and heavy liquid separation processes. All samples were acid-washed twice, before and after handpicking, for 30 min in 7N nitric acid. This rendered the zircon concentrate 99% pure. Sample K-40 was split into non-magnetic and magnetic concentrates, and both the splits and the combination were analyzed. Zircons in samples K-39 and K-40 were uniformly small (1 mm), well faceted, and commonly had overgrowths and opaque inclusions. Sample LGO-43 has a somewhat more diverse zircon population with some zircons >1 mm. Zircons from each sample were then dissolved and processed through ion exchange columns in accordance with procedures described by Krogh (1972). Lead concentrate was loaded onto a rhenium filament by the silica gel–phosphoric acid method of Cameron and others (1969). Uranium concentrate was loaded onto a rhenium filament by the titanium oxide–phosphoric acid method of Krogh (1972). Analysis for lead and uranium was performed on a 34-cm solid source Avco mass spectrometer with an in-line PDP-11 minicomputer in the Isotope Geology Laboratory of Virginia Polytechnic Institute and State University.

Data

Two of the samples analyzed came from the Moriah pluton (Fig. 2) and one from a petrographically similar pluton to the east. The similarity of zircon morphology in the two samples from the Moriah pluton prevented splitting on the basis of size. However, K-40 was split magnetically. Both magnetic splits of the K-40 zircons and zircons from the other sample from the Moriah pluton (K-39) define a linear regression best-fit chord with an upper intercept (emplacement age) of 650 ± 30 m.y. (Fig. 4). The third sample analyzed (LGO-43) was originally obtained to provide a third point on the chord of the Moriah pluton; however, the analysis of the zircons from this sample gave a nearly concordant point at 520 m.y. This near concordancy probably represents a good estimate of the age of this intrusive and indicates that it is too young(?) to be comagmatic with the Flat River complex.

TABLE 1. ISOTOPIC ANALYSIS OF ZIRCONS AND NBS 983 Pb STANDARD

Sample No.	ppm U	ppm Pb	Observed Ratios Pb^{206}/Pb^{204}	Observed Ratios Pb^{207}/Pb^{206}	Observed Ratios Pb^{208}/Pb^{206}	Atomic Ratios Pb^{206*}/U^{238}	Atomic Ratios Pb^{207*}/U^{235}	Atomic Ratios Pb^{207*}/Pb^{206*}	Apparent Age (m.y.) Pb^{206*}/U^{238}	Apparent Age (m.y.) Pb^{207*}/U^{235}	Apparent Age (m.y.) Pb^{207*}/Pb^{206*}
K-40 Combined	492 ±1	33.7 ±.46	1158 ±23	.07461 ±.00036	.19373 ±.00131	.06186	.52672	.06183	387	430	668
K-40 Magnetic	888 ±11	36.1 ±.06	1335 ±27	.07180 ±.00017	.17615 ±.00028	.03738	.31115	.06044	237	275	620
K-40 Non-magnetic	386 ±7	29.3 ±.01	2474 ±49	.06659 ±.00012	.16905 ±.00024	.07079	.59087	.06061	441	471	626
K-39	361 ±3	37.0 ±.10	457 ±9	.09359 ±.00036	.25709 ±.00070	.08483	.71459	.06120	525	548	646
LGO-43	533 ±1	49.0 ±.12	1756 ±35	.06633 ±.00012	.23412 ±.00059	.08128	.64768	.05786	504	507	524
NBS9 983			2688 ±53	.07128 ±.00006	.01366 ±.00002						
NBS 983			2731 ±55	.07116 ±.00014	.01364 ±.00011						
NBS 983			2695	.07120 ±.00004	.01362 ±.00002	Absolute values (Catanzaro and others, 1968)					

Common Pb Corrections: 600 m.y. - Pb^{206}/Pb^{204} = 17.5639, Pb^{207}/Pb^{204} = 15.8776, Pb^{208}/Pb^{204} = 36.467

U^{238}/U^{235} = 137.88 500 m.y. - Pb^{206}/Pb^{204} = 17.7142, Pb^{207}/Pb^{204} = 15.8886, Pb^{208}/Pb^{204} = 36.627

Decay Constants: U^{235} = .98485 x 10^{-9} yr^{-1}; U^{238} = 1.55130 x 10^{-9} yr^{-1} (Jaffey and others, 1971)

Sample Locations:

K-40 magnetic 4½ km southeast of Bahama, N.C., Durham North Quadrangle, Figure 2.

K-40 non-magnetic Lat. 36°07'42", Long. 78°51'00"

K-39 1½ km east of Bahama, N.C., Durham North Quadrangle, Figure 2. Lat. 36°10'18", Long. 78°51'54"

LGO-43 Tar River Quarry on north side of Tar River in northwestern Creedmore, N.C., Quadrangle (located east of map area. Lat. 36°13'35", Long 78°38'08".

Emplacement

We used two sources of data for obtaining insight into the level of emplacement of the Flat River complex. They are (1) the field relations of the Flat River complex referenced to other studies where the level of emplacement has been ascertained and (2) petrographic relations referenced to studies of the mineralogy of granites and granodiorites as a function of depth of emplacement.

A first approximation to the depth of emplacement can be obtained in a general way from the zonal characteristics given by Buddington (1959) for epizonal, mesozonal, and catazonal intrusions. These are strongly overlapping depth zones because temperature and viscosity of rock vary as a function of time, composition, depth, and other factors. Characteristics of the Flat River complex, such as sharp contacts, coeval volcanic rocks, discordancy to the regional structural trends, the presence of granophyre, protoclastic margins, and autobrecciation indicate that the Flat River complex was emplaced in the epizone, 1- to 10-km depth

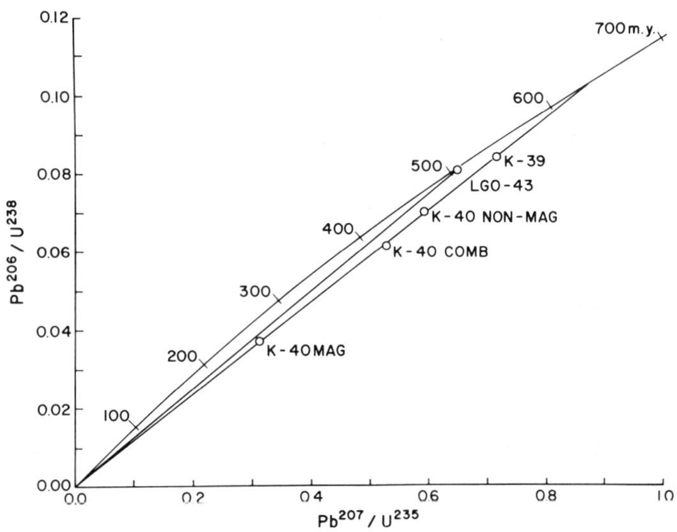

Figure 4. Concordia diagram for zircons from the Moriah pluton of the Flat River complex, North Carolina. LGO-43 is from a younger pluton just east of the study area (see text).

according to Buddington. Other field data indicate that the Flat River may have been surface breaking. Vent breccias characteristic of surface-breaking plutons (Fiske and others, 1963; Cater, 1969) are present in the Flat River complex. Wright (1974) has linked these vent breccias, which cut the complex, with the surrounding volcanic stratigraphy (in part with unit A, Fig. 2). Many of these vents are filled with comminuted volcanic and intrusive material of diverse lithologies. This explosive brecciation of both volcanic rocks and pluton implies a very shallow level of emplacement, shallow enough to allow rapid escape of volatiles to the surface as phreatic eruptions. Protoclastic marginal dacites are also characteristic of surface-breaking plutons (Cater, 1969). Protoclastic hypabyssal rocks (porphyritic dacite) are common to the Flat River complex in and southwest of the study area (Wright, 1974).

Experimental work most applicable to determining the level of emplacement is that of Tuttle and Bowen (1958) on the qz-ab-or system. Two basic requirements are necessary to apply petrography from intrusive bodies of this study: (1) the magma was granitic in composition and (2) the magma was saturated with respect to water during crystallization. In the case of the granodioritic Flat River complex, the first requirement is met by the granophyric groundmass which represents the last liquid crystallized. Modal analyses of this granitic fraction were used to compare with the results of Tuttle and Bowen and others (Fig. 5).

The second requirement is not obviously met because sharp contacts, lack of pegmatitic or aplitic material, and abundance of primary amphibole suggest that the Flat River crystallized, at least in part, as an undersaturated magma.

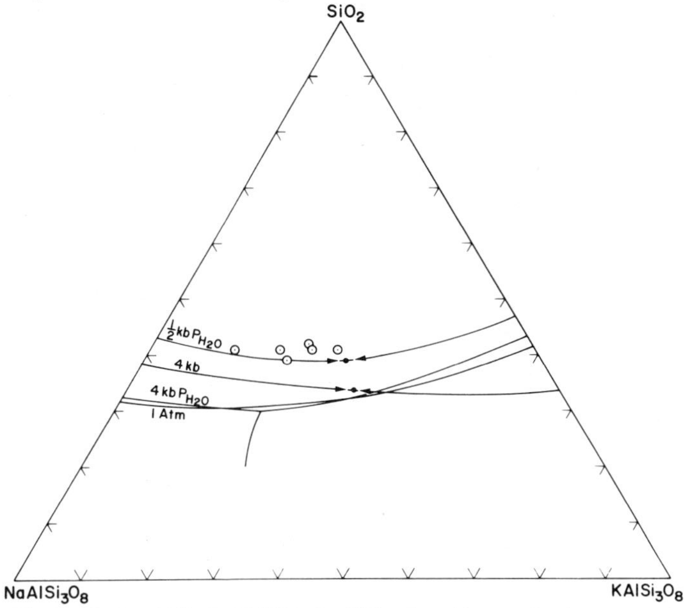

Figure 5. $NaAlSi_3O_8$-$KAlSi_3O_8$-SiO_2 plot: Granophyric compositions determined by modal analysis and microprobe compositional data on the minerals. 1/2 kb P_{H_2O} curve (Tuttle and Bowen, 1958), 1-atm curve (Schairer, 1950), 4 kb and 4 kb P_{H_2O} curves (Steiner, and others, 1975).

The presence of abundant hypersolvus granophyre, local spherulitic texture in the Flat River complex, and coeval felsic extrusives associated with the Flat River implies a hypersolvus granite-granophyre-rhyolite association as described by Tuttle and others (1964). However, a vapor phase was probably developed during the last stages of crystallization as is indicated by the position of the plots of modal analyses of the granophyric groundmass on a qz-ab-or diagram (Fig. 5). These modal analyses on stained thin sections were recalculated to weight percent using electron microprobe (compositional) data on the minerals (McConnell, 1974). Significant error is inherent in the diagram as a result of difficulty in distinguishing albite and potassium feldspar in the fine-grained microperthitic granophyric matrix. However, this error should not affect the interpretation because the quartz content is relatively certain. It can be seen from this diagram that the modes are consistent with the low-pressure saturated curves rather than the unsaturated curves. Gibbon and Wyllie (1969) stated that the West Farrington granodiorite, which is somewhat similar to the Flat River, probably did not develop a vapor phase until the last stages of crystallization during the formation of granophyric groundmass. The plot is consistent with the idea that the groundmass was formed at low pressures and correspondingly shallow depths (<1 km). Therefore, both field and petrologic data from the Flat River complex indicate that it was intruded at very shallow and probably surface-breaking levels.

The mechanism of intrusion of the Flat River rocks is somewhat complex. Only minor amounts of xenolithic and autolithic material were found in the Flat River. One roof pendant or autolith was observed in the northwest, and smaller fragments of volcanic rock occur in the dacite to the southwest (Wright, 1974). This material shows no sign of assimilation or recrystallization by the pluton. For these reasons, stoping and assimilation are not believed to have been a major factor in final emplacement of the Flat River complex. The lack of intense deformation in the contact zones or large-scale disruption of the regional trend of the country rock seems to indicate that forceful intrusion was not dominant during emplacement. Hopson and others (1970) stated that emplacement of shallow level batholiths is not explained by either shouldering aside or stoping, and that shallowly emplaced batholiths, such as the Cloudy Pass, Snoqualmie, and Tatoosh, have spread out under a thin covering of volcanic material. The broadly strata-parallel shape (Fig. 1), the lack of features suggesting other methods of emplacement, and the presence of a chilled margin in the Flat River similar to that present in the Cloudy Pass suggest that the Flat River complex was emplaced in a similar manner.

CONCLUSIONS

Volcanic textures and structures indicate that the pyroclastic rocks of unit II (Glover and Sinha, 1973), consisting of members X, Y, Z, accumulated in both marine and nonmarine environments. During part of this time, volcanic activity waned and an exhalative unit of layered iron oxide and chert accumulated in a quiet, restricted, subaqueous environment. The age of the intrusive Flat River complex implies that the pyroclastic rocks of unit II began accumulating prior to 650 ± 20 m.y. All of these rocks are of greenschist metamorphic grade.

The direct link between the volcanic rocks of unit II and the Flat River complex was established by Wright (1974); the closeness of the age of the complex (650 ± 30 m.y.) and the age of the top of unit II (620 ± 20 m.y., according to Glover and Sinha, 1973), petrographic data, and comparisons with experimental results in the qz-ab-or system confirm that the Flat River complex was emplaced at a depth of less than 1 km and was locally surface breaking. The Flat River complex can be viewed as a fossil volcanic magma chamber that erupted to furnish pyroclastic debris for part of unit II.

Perhaps many of the other similar complexes shown on geologic maps of the Carolina slate belt are also shallow subvolcanic chambers. Additional studies of this general type might lead to a more accurate understanding of the geographic trends of slate belt volcanic arcs through time and the physical conditions (i.e., marine versus nonmarine pyroclastics, shallow versus deep plutons) or eruption and emplacement of the magma. Such factors share importance from both academic and economic perspectives.

ACKNOWLEDGMENTS

This study evolved from McConnell's M.S. thesis at Virginia Polytechnic Institute and State University. The writers gratefully acknowledge partial support from the North Carolina Department of National and Economic Resources, Mineral Resources Section, and from the Department of Geological Sciences at Virginia Polytechnic Institute and State University. The manuscript has benefited from review by Clayton Newton, and especially from technical comments and assistance by J. Alexander Speer and A. K. Sinha.

REFERENCES CITED

Barker, D. S., 1970, Compositions of granophyre, myrmekite, and graphic granite: Geological Society America Bulletin, v. 81, p. 3339–3350.

Bell, H., III, Marvin, R. F., and Mehnert, H. H., 1972, Potassium-Argon ages from areas of hydrothermal alteration in South Carolina: U.S. Geological Survey Professional Paper 800-D, p. D117–D121.

Black, W. W., 1977, The geochronology and geochemistry of the Carolina slate belt of north-central North Carolina [Ph.D. dissert.]: University of North Carolina at Chapel Hill, 118 p.

Buddington, A. F., 1959, Granite emplacement with special reference to North America: Geological Society America Bulletin, v. 70, p. 671–747.

Cameron, A. E., Smith, D. H., and Walker, R. L., 1969, Mass spectrometry of nanogram-size samples of lead: Analytical Chemistry, v. 41, p. 525–526.

Catanzaro, E. J., Murphy, T. J., Shields, W. R., and Garner, E. L., 1968, Absolute isotopic abundance ratios of common, equal atom, and radiogenic lead isotopic standards: Journal of Research, National Bureau of Standards, v. 72A, p. 261–267.

Cater, F. W., 1969, The Cloudy Pass epizonal batholith and associated subvolcanic rocks: Geological Society of America Special Paper 116, 54 p.

Fisher, R. V., 1966, Rocks composed of volcanic fragments and their classification: Earth-Science Review, v. 1, p. 387–398.

Fiske, R. S., Hopson, C. A., and Waters, A. C., 1963, Geology of Mount Rainier National Park, Washington: U.S. Geological Survey Professional Paper 444, 93 p.

Gibbon, D. L., and Wyllie, P. J., 1969, Experimental studies of igneous rock series: The Farrington Complex, North Carolina, and the State Mountain Rhyolite, Texas: Journal of Geology, v. 77, p. 221–239.

Glover, L., III, and Sinha, A. K., 1973, The Virgilina deformation, a late Precambrian to Early Cambrian(?) orogenic event in the Central Piedmont of Virginia and North Carolina: American Journal of Science, v. 273-A (Cooper Volume), p. 234–251.

Hadley, J. B., 1973, Igneous rocks of the Oxford area, Granville County, North Carolina: American Journal of Science, v. 273-A (Cooper Volume), p. 217–233.

Hopson, C. A., Cater, F. W., and Crowder, D. F., 1970, Emplacement of plutons, Cascade Mountains, Washington: Geological Society of America Abstracts with Programs (Cordilleran Section), v. 2, p. 104.

Jaffey, A. H., Flynn, K. F., Glendinin, L. E., Bentley, W. E., and Essling, A. M., 1971, Precision measurement of the half-lives and specific activities of ^{235}U and ^{238}U: Physical Review, C, v. 4, p. 1887–1906.

Kish, S. A., Butler, J. R., and Fullagar, P. D., 1979, The timing of metamorphism and deformation in the central and eastern Piedmont of North Carolina: Geological Society of America Abstracts with Programs, v. 11, p. 184–185.

Krogh, T. E., 1972, A low-contamination method for hydrothermal decomposition of zircon and extraction of U and Pb for isotopic age determinations: Geochimica et Cosmochimica Acta, v. 37, p. 485–494.

McConnell, K. I., 1974, Geology of the late Precambrian Flat River Complex and associated volcanic rocks near Durham, North Carolina [M.S. thesis]: Blacksburg, Virginia Polytechnic Institute and State University.

Moore, J. G., 1965, Petrology of deep-sea basalts near Hawaii: American Journal Science, v. 263, p. 40–52.

Prouty, W. F., 1931, Triassic deposits of the Durham basin and their relation to other Triassic areas of eastern United States; American Journal of Science, v. 21, p. 473–490.

Ross, C. S., and Smith, R. L., 1961, Ash-flow tuffs; their origin, geologic relations and identification: U.S. Geological Survey Professional Paper 366, 81 p.

Schairer, J. F., 1950, The alkali-feldspar join in the system SiO_4-$KAlSi_3O_8$-SiO_2-H_2O at 4 bar: Journal of Geology, v. 58, p. 512–517.

Steiner, J. C., Jahns, R. H., and Luth, W. C., 1975, Crystallization of alkali feldspar and quartz in the haplogranite system $NaAlSi_3O_8$-$KAlSi_3O_8$-SiO_2-H_2O at 4 kb: Geological Society America Bulletin, v. 86, p. 83–98.

Streckeisen, A., 1973, To each plutonic rock its proper name: Earth-Science Reviews, v. 12, p. 1–33.

Tabor, R. W., and Crowder, D. F., 1969, On batholiths and volcanoes—intrusion and eruption of late Cenozoic magmas in the Glacier Peak area North Cascades, Washington: U.S. Geological Survey Professional Paper 604, 67 p.

Tobisch, O. T., and Glover, L., III, 1971, Nappe formation in part of the southern Appalachian Piedmont: Geological Society America Bulletin, v. 82, p. 2209–2230.

Tuttle, O. F., and Bowen, N. L., 1958, Origin of granite in the light of experimental studies in the system $NaAlSi_3O_8$-$KAlSi_3O_8$-SiO_2-H_2O: Geological Society of America Memoir 74, 153 p.

Tuttle, O. F., Luth, W. C., and Jahns, R. H., 1964, The hypersolvus granite-granophyre-rhyolite association [abs.]: American Geophysical Union Transactions, v. 45, p. 124.

U.S. Geological Survey, 1973, Aeromagnetic map of the northern parts of the Durham North and Creedmore, N.C. quadrangles: U.S. Geological Survey Geophysical Investigations Map GP-883.

Wright, J. E., 1974, Geology of the Carolina "slate" belt in the vicinity of Durham, N.C. [M.S. thesis]: Blacksburg, Virginia Polytechnic Institute and State University, 77 p.

MANUSCRIPT ACCEPTED BY THE SOCIETY JULY 8, 1982

Printed in U.S.A.

Geological Society of America
Special Paper 191
1982

Geochronological investigation of the Lincolnton metadacite, Georgia and South Carolina

Robert H. Carpenter
Carpenter Mineral Exploration Company
P.O. Box 632
Elberton, Georgia 30635

A. Leroy Odom
Department of Geology
Florida State University
Tallahassee, Florida 32306

M. E. Hartley III
C-E Minerals, Exploration Office
4020 Lexington Road
Athens, Georgia 30601

ABSTRACT

A geochronological study of the Lincolnton metadacite, a prominent unit in the slate belt along the Georgia–South Carolina border, provides indicated primary ages of 554 ± 20 m.y. (Rb-Sr) and 568 m.y. (U-Pb, zircon). On the basis of both lithological and geochronological data, the Lincolnton metadacite and associated felsic pyroclastic sequence are interpreted to be correlative with the Uwharrie Formation in the Albemarle-Denton-Asheboro area of North Carolina. On the basis of lithologic comparison alone, the Lincolnton metadacite and associated felsic pyroclastic sequence are also interpreted to be equivalent to the Persimmon Fork Formation near Columbia, South Carolina. Prominent argillite units that overlie felsic metavolcanics in all three areas are likewise considered correlative. These units include the Tillery Formation (Albemarle-Denton-Asheboro area), the Richtex Formation (Columbia, South Carolina, area), and the upper sedimentary sequence (Lincolnton, Georgia–McCormick, South Carolina, area).

Available data suggest a Cambrian age for most of these units, although some of the lowermost formations could actually be of very late Precambrian age.

INTRODUCTION

In recent years, a number of geologists have interpreted the slate belt as an ancient island-arc system. Geochronological investigations of principal volcanic units in various areas report ages that range from late Precambrian through the Ordovician, a time span of perhaps 250 m.y. to 300 m.y. Published age determinations have not been sufficiently consistent to enable interpretation of the geological history of the entire slate belt in a precise geochronological framework. This investigation, considered in conjunction with certain previous studies in North Carolina, appears to define rather precisely the age of volcanism in the southern part of the slate belt.

The present study focuses on a prominent felsic volcanic unit, termed the Lincolnton metadacite, which crops out over an area of about 35 to 40 square miles in Lincoln County, Georgia, and McCormick County, South Carolina. The results of Rb-Sr and U-Pb analyses are described. In addition, correlations of major units recognized in the slate belt of the Carolinas and Georgia are made.

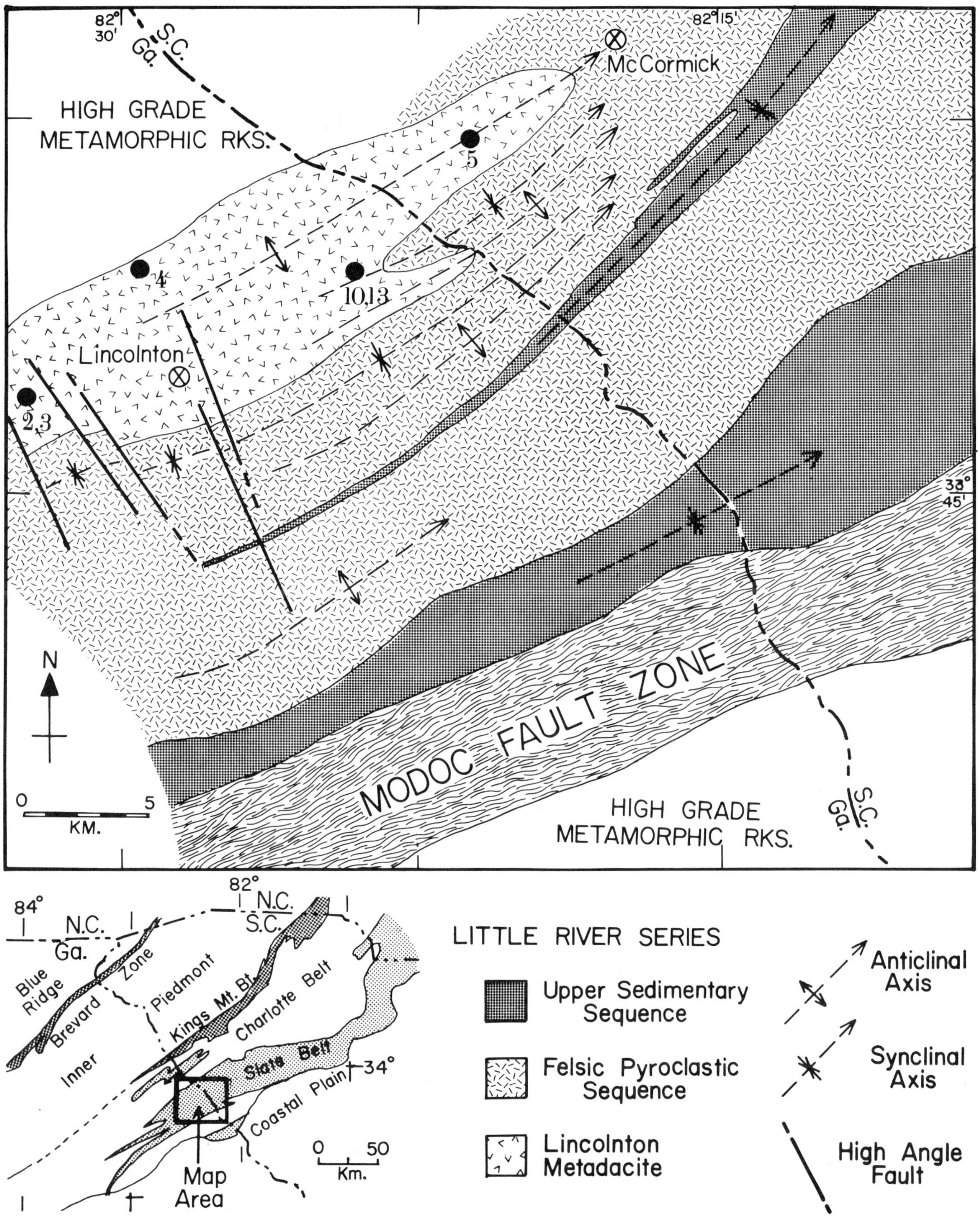

Figure 1. Geologic map showing sample locations. Inset map shows location of map area in southern Appalachians.

GENERAL GEOLOGY OF THE LINCOLNTON-MCCORMICK AREA

The location of the study area, sample localities, and general geological relationships are shown in Figure 1. The area includes the principal segment of the slate belt along the Georgia–South Carolina border. The slate belt is bounded on the southeast by the Modoc fault zone and high-grade metamorphic rocks (amphibolite facies) of the Kiokee series. It is bounded on the northwest by high-grade metamorphic rocks (amphibolite facies) of uncertain affiliation. A sequence of isoclinally folded volcanic and volcaniclastic rocks that have been only weakly metamorphosed (greenschist facies) occurs within the slate belt. The stratigraphy and structure of the slate belt has been only recently determined in this area (Carpenter, 1976; Paris, 1976a).

From oldest to youngest, the stratigraphic units are as follows: (1) the Lincolnton metadacite, (2) the felsic pyroclastic sequence, and (3) the upper sedimentary sequence.

Lincolnton Metadacite

This unit is typically a quartz porphyry consisting of large phenocrysts of blue, opalescent quartz in a cream-colored matrix of finely crystalline quartz and plagioclase. Precise petrographic nomenclature for this unit is somewhat problematical. Similar rocks elsewhere have been termed soda rhyolite, albite rhyolite, and quartz keratophyre, as well as dacite. Whitney and others (1978) have considered the various petrographic terms appropriate to this lithology, and have suggested retention of the term "dacite." Compared to more typical felsic volcanic rocks, the Lincolnton metadacite is enriched in sodium and depleted in potassium. Paris (1976a, 1976b) has recognized five petrographic subtypes of metadacite. Principal variables are size and number of quartz phenocrysts and texture and crystallinity of groundmass minerals. Along the northwestern contact, the unit shows affects of increasing metamorphic grade, which has resulted in a recrystallized, gneissose texture. Interbedded within the metadacite are minor mafic flows, mafic and felsic tuffs, and metasediments.

On the basis of the homogenous nature of the metadacite within a given stratigraphic unit and the euhedral, unbroken nature of feldspar and quartz phenocrysts, the metadacites are interpreted to represent a sequence of flows. Because of the apparent restriction of the unit to the Lincolnton-McCormick area, the outcrop is considered to expose a part of a felsic volcanic center. The aggregate thickness of flows and interbedded pyroclastics is probably in excess of 5,000 ft (Carpenter, 1976).

Felsic Pyroclastic Sequence

This unit is composed mainly of felsic vitric and vitric-crystal tuffs, but other common lithologies include lithic tuffs, agglomerates, and mafic tuffs. Also present are some mafic and felsic flows and intercalated clastic and chemical (exhalative) sedimentary lithologies, which include tuffaceous graywackes, argillites, cherts, iron formations, bedded barite, and bedded sulfide bodies. The average thickness is approximately 3,500 ft.

This unit is interpreted to be a time-transgressive sequence of pyroclastic debris derived mainly from the volcanic center represented by the Lincolnton metadacite (Carpenter, 1976).

Upper Sedimentary Sequence

This sequence consists mainly of banded argillite and thin, interbedded, mafic volcanics. Some graywacke is commonly present. It occurs most extensively in the large synclinorium that lies immediately west of the Modoc fault, but is also exposed along other synclinal axes to the west (Fig. 1). The upper part of the unit has been removed by erosion so that the total thickness is not known. An estimated minimum thickness is about 5,000 ft.

This unit is interpreted to have formed during a period of relative quiescence with respect to explosive, felsic volcanism. Persistent delicate laminae in the argillite suggests deep-water sedimentation in an unagitated environment.

Intrusive rocks in the study area are mainly sills and dikes. The rocks that exhibit a distinct metamorphic overprint include metagabbro (sills), metabasalt, meta-andesite, and metarhyodacite. These intrusives tend to be more concentrated in the Lincolnton metadacite, but become less abundant in younger sequences. This suggests that they were actually "feeder dikes" to higher level volcanics. The dikes show a pronounced tendency to occur as "swarms" with a dominant trend between N20°W and N60°W. Postmetamorphic dikes include the Triassic-Jurassic diabases that are common throughout the Piedmont province and several lamprophyre dikes.

The low-grade rocks of the slate belt are interpreted by a number of investigators (for example, Hatcher, 1972) to represent a complex synclinorium flanked by adjacent anticlinoria (the Charlotte belt and Kiokee belt). Within the slate belt, a single generation of vertical isoclinal folds with a gentle plunge to the northeast represents the predominant fold style. The folds become more open in the southeastern part of the belt. Along the boundaries of the slate belt in the more intensely metamorphosed rocks, some superimposed folds are evident, but these lose their prominence in the central portion.

METHOD OF STUDY

Sampling

The locations of sample sites in the Lincolnton metadacite are indicated in Figure 1. Samples LMD-2, 3, 5, 10, and 13 were collected from outcrops in stream beds, and sample LMD-4 was collected from a roadcut exposure. At most sites, several samples in close proximity (weighing approximately 1 kg) were collected from the freshest exposures that were amenable to sampling. Care was taken to collect each sample from an area no larger than 30 cm × 30 cm. Individual outcrops containing dikes or quartz veins were not sampled.

Analytical Methods

Rb-Sr Whole-Rock Analyses

Six whole-rock samples were analyzed for their Rb and Sr concentrations by a stable isotope dilution method. $^{87}Sr/^{86}Sr$ ratios were calculated from the spiked analyses, but with the highly enriched ^{84}spike used (NBS-SRM-988), the possible uncertainty in the $^{87}Sr/^{86}Sr$ due to the spike contribution of ^{87}Sr and ^{86}Sr is a negligible part of the 2 σ analytical uncertainty reported in these ratios of approximately 0.05%. The analytical uncertainty on the Rb and Sr concentrations are thought to be ±1.0%. The laboratory value measured on the MIT E & A $SrCo_3$ standard for $^{87}Sr/^{86}Sr$ is 0.70797. Rb and Sr blanks that were run concurrently with the samples were insignificant (Sr = 2.2×10^{-8} g; Rb = 0.9×10^{-8} g).

U-Pb Zircon Analyses

Two size fractions of zircons from one rock and two magnetic fractions from another were analyzed for their U and Pb concentrations and the isotopic composition of Pb. The procedures used were those of Krogh (1973) with minor modifications. The isotopic analyses of Pb were done on the Florida State University Isotope Geochemistry Laboratory's 12-in. NBS mass spectrometer. The temperature of the single filament was monitored with an optical pyrometer, and the extent of mass fractionation was judged by values measured for NBS-SRM983 under the same conditions. These values are $^{208}/206 = 2.1676$, $^{207}/206 = 0.91443$, and $^{204}/206 = 0.05906$. No blanks were determined with these samples, but the laboratory U blank is insignificant, and the maximum Pb blank in these analyses can be judged from the reported $^{204}Pb/^{206}Pb$ ratios.

RESULTS

The results of the analyses of the whole rocks and zircons, as well as constants used in the calculations, are given in Tables 1 and 2 and are shown graphically in the isochron (Fig. 2) and concordia (Fig. 3) diagrams.

A least-squares cubic regression of the six whole rocks define a 554 ±20 m.y. isochron with an initial $^{87}Sr/^{86}Sr$ of 0.70318 ±0.00039 (1 σ uncertainty). Regression of the four zircon data points produce a cord that intercepts concordia at an age of 566 ±15 m.y. for an episodic model.

TABLE 1. Rb-Sr CONCENTRATIONS AND ISOTOPIC RATIOS FOR WHOLE-ROCK SAMPLES OF THE LINCOLNTON METADACITE

Sample	Rb(ppm)	Sr(ppm	($^{87}Rb/^{86}Sr$)	($^{87}Sr/^{86}Sr$) (1σ)
LMD-2	15.94	181.0	0.255	0.70516 ± 0.00032
LMD-3	15.24	349.0	0.126	0.70436 ± 0.00016
LMD-4	34.29	150.7	0.683	0.70896 ± 0.00030
LMD-5	23.71	190.6	0.360	0.70596 ± 0.00018
LMD-10	30.34	167.9	0.523	0.70692 ± 0.00017
LMD-13	28.39	148.4	0.553	0.70729 ± 0.00033

Note: Constants used in calculations: $^{85}Rb/^{87}Rb = 2.5930$; $^{86}Sr/^{88}Sr = 0.11940$; $\lambda^{87}Rb = 1.42 \times 10^{-11}$ yr^{-1}.

DISCUSSION

The good agreement between Rb-Sr and U-Pb ages, the zircon morphology, and the quite low initial $^{87}Sr/^{86}Sr$ ratio leads us to believe that the eruption and crystallization of the Lincolnton magma occurred close to 560 m.y. ago. The low initial $^{87}Sr/^{86}Sr$ ratio is typical of island-arc volcanics and seems to be characteristic of Cambrian meta-igneous rocks of the southern Piedmont. The age and initial ratio are indistinguishable from Cambrian rocks found in the high-grade Kiokee belt to the southeast (G. S. Russell, personal commun.). This is consistent with the idea that the Kiokee rocks are high-grade equivalents of the slate belt.

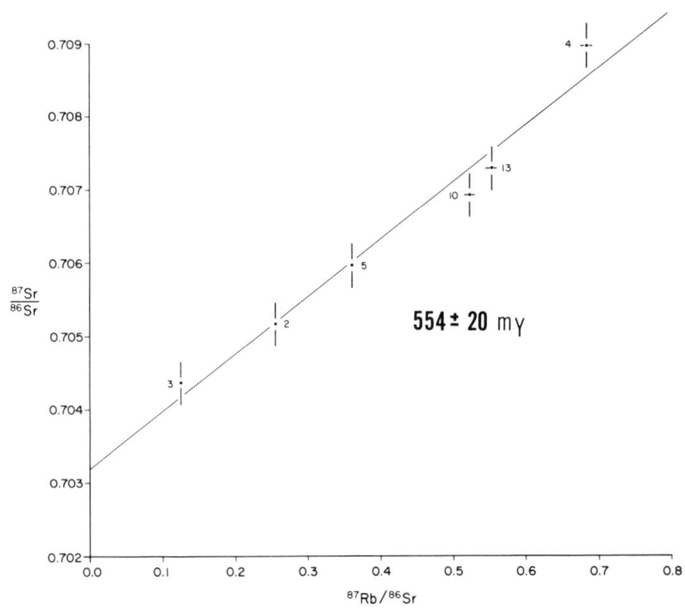

Figure 2. Isochron summarizing Rb-Sr isotopic data for samples 2, 3, 4, 5, 10, and 13 collected from the Lincolnton metadacite. The indicated age is 554 ±20 m.y., with an $^{87}Sr/^{86}Sr$ initial ratio of 0.70318 ±0.00039.

TABLE 2. U-Pb ANALYTICAL DATA FOR ZIRCON SAMPLES OF THE LINCOLNTON METADACITE

Sample	Concentrations		Observed ratios			Ages (10^6 years)		
	U (ppm)	Pb (ppm)	$^{208}/206$	$^{207}/206$	$^{204}/206$	$^{206}Pb/^{238}U$	$^{207}Pb/^{235}U$	$^{207}Pb/^{206}Pb$
LMD-3 200m	509	32.70	0.1737	0.06193	0.000325	383	399	498
LMD-3 200m	540	35.88	0.20146	0.06170	0.000360	383	395	469
LMD-4 most mag	350	26.61	0.21367	0.06435	0.000440	438	452	526
LMB-4 least mag	228	19.82	0.2201	0.06417	0.000410	496	503	536

Note: Concordia Intercept = 566 m.y.

Constants used in calculations:
$\lambda^{235}U = 9.85 \times 10^{-10}\ yr^{-1}$; $\lambda^{238}U = 1.55 \times 10^{-10}\ yr^{-1}$; $^{238}U/^{235}U = 137.88$. In common lead: $^{208}/204 = 38.42$, $^{207}/204 = 15.72$, $^{206}/204 = 18.51$.

Comparison with Investigations in North Carolina

From a stratigraphic point of view, the Albemarle-Denton-Asheboro area in central North Carolina has provided the most complete documentation of the stratigraphic succession of any other area in the slate belt (Conley and Bain, 1965; Stromquist and Sundelius, 1969). The succession determined here, together with available geochronological information is summarized in Figure 4A. Three major cycles of felsic volcanism are indicated. These include the Uwharrie Formation, the Flat Swamp Member of the McManus Formation, and the Morrow Mountain rhyolite. Rb-Sr whole-rock ages (recalculated with an ^{87}Rb decay constant of $1.42 \times 10^{-11}\ yr^{-1}$) reported by Hills and Butler (1969) and Fullagar (1971) for the Uwharrie Formation and Morrow Mountain rhyolite are 553 ± 49 m.y. and 514 ± 15 m.y., respectively. Wright and Seiders (1980) reported a U-Pb zircon age of 586 ± 10 m.y. for the upper part of the Uwharrie Formation in the vicinity of Asheboro, North Carolina. Black (1978) obtained a 540 ± 7 m.y. Rb-Sr whole-rock age for metadacite from the Badin greenstone. Stromquist and Sundelius (1969) suggested that the Morrow Mountain rhyolite and Badin greenstone which comprise the Tater Top Group of Conley and Bain (1965) may actually be local volcanic feeder systems and flows within the Albemarle Group. If this is the case, the Albemarle Group would represent the youngest volcanic rocks in central North Carolina.

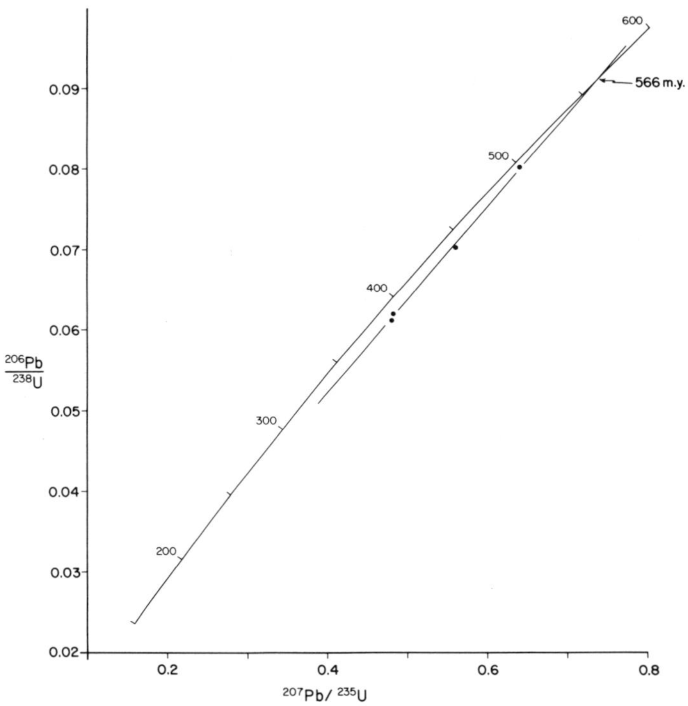

Figure 3. Concordia diagram summarizing U-Pb isotopic data for samples collected from the Lincolnton metadacite. For an episodic model, discordant ages define a cord that intercepts concordia at 566 ± 15 m.y.

The validity of these ages is supported by the occurrence of a trilobite (Paradoxides) in the upper part of the McManus Formation. According to St. Jean (1973), the most probable age range is earliest Middle Cambrian to middle Middle Cambrian, although an Early Cambrian age is also possible. Based on the geologic time scale compiled by Van Eysinga (1975), the Cambrian extends from about 570 m.y. to approximately 500 m.y. Accordingly, radiometric ages noted previously indicate that the Uwharrie Formation is latest Precambrian–Early Cambrian, and the Morrow Mountain rhyolite is Late Cambrian. The location of Paradoxides at an intermediate position in the section is consistent with a Middle Cambrian age.

The most probable correlative units in the Lincolnton-

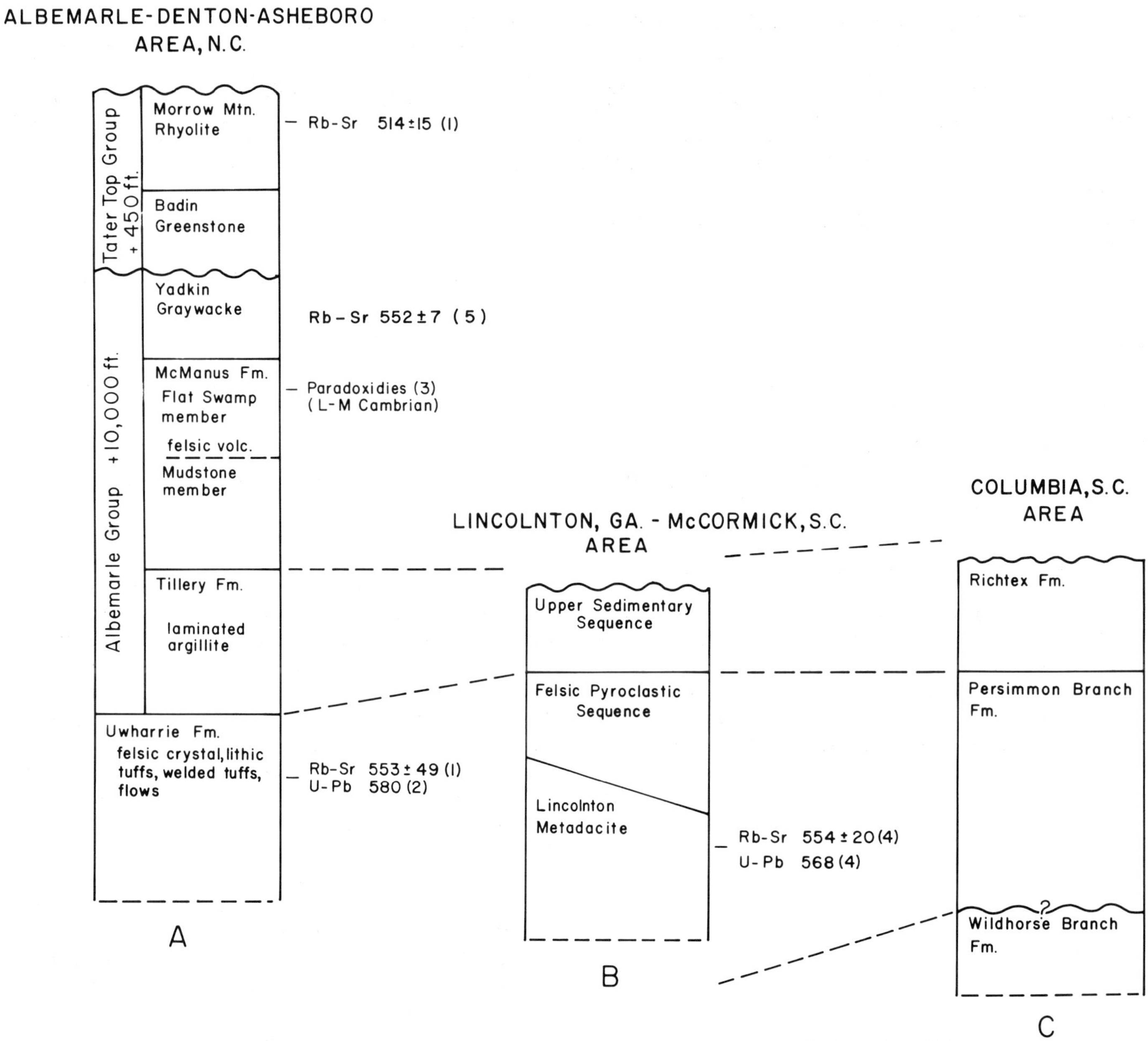

Figure 4. Representative stratigraphic sections of the slate belt and compilation of published geochronological data. References in ( ) are as follows: (1) Hills and Butler, 1969; Fullagar, 1971; (2) Wright and Seiders, 1980; (3) St. Jean, 1973; (4) present study.

McCormick area are indicated in Figure 4. The Uwharrie Formation is probably equivalent to the Lincolnton metadacite and the felsic pyroclastic sequence. Both are composed mainly of felsic volcanics and provide comparable ages. The banded argillites of the Tillery Formation are remarkably similar in appearance to those in the upper sedimentary sequence in the Lincolnton-McCormick area. The resemblance has been previously noted by Bell (1973). Because the banded argillites of the upper sedimentary sequence occur in the cores of synclines, younger units are not exposed, either because of erosion or nondeposition. Consequently, units identified in the Lincolnton-McCormick area are considered equivalent to the lowest units in the North Carolina section and would appear to be mainly of Early Cambrian age.

Northeast of Asheboro, North Carolina, in the Roxboro–Chapel Hill area, several investigations have indicated an older sequence of metavolcanic rocks in the slate belt. Glover and Sinha (1973) have determined U-Pb ages in the Roxboro, North Carolina, area for a folded and faulted sequence of metavolcanic rocks and a granodiorite pluton that postdates an early episode of deformation. The primary age for the volcanic rocks (lithic lapilli tuffs) is reported to be 620 ± 20 m.y., and the age for the granodiorite pluton is

575 ±20 m.y. Accordingly, the volcanic rocks were folded and faulted prior to intrusion of the pluton.

Wright and Seiders (1980) have reviewed isotopic and paleontologic ages of units in the Albemarle-Denton-Asheboro and the Roxboro-Durham areas of the Carolina slate belt. Considering both radiometric ages and lithostratigraphic relations, they suggested three plausible correlations of units between the two areas. Wright and Seiders' arguments have shown that when one attempts to accommodate not only the lithostratigraphic sequences and radiometric ages but also the Virgilina deformation (Glover and Sinha, 1973) as well, no reasonable correlation between the two areas is entirely satisfactory. The difficulty in part might be in trying to constrain the correlations to accommodate differences in isotopic ages determined by different methods.

Comparison with Investigations in the South Carolina Area

The geologic section determined for the slate belt in the vicinity of Columbia, South Carolina, by Secor and Wagener (1968) is shown in Figure 4C. The stratigraphic and structural relationships in this area have been reviewed recently by Costello and others (1981).

The Richtex Formation is a banded argillite which, according to Secor and Wagener (1968), is probably equivalent to the Tillery Formation in North Carolina. We suggest that it is also equivalent to the upper sedimentary sequence in the Lincolnton-McCormick area. The Persimmon Fork Formation is composed mainly of felsic pyroclastics and is probably equivalent to the Lincolnton metadacite and felsic pyroclastic sequence in the Lincolnton-McCormick area and the Uwharrie Formation of the North Carolina slate belt. The Wildhorse Brance Formation is somewhat problematical. It consists of a mixture of felsic and mafic volcanics with abundant interbedded graphitic sedimentary rocks. Comparable rocks are not recognized in either the Lincolnton-McCormick area or the Albemarle-Denton-Asheboro area. This is possibly the oldest unit (late Precambrian?) in the southernmost slate belt.

The only recent isotopic age determinations for the slate belt of South Carolina are for metarhyolites near Pageland, South Carolina, just south of the North Carolina boundary. Here, Butler and Fullagar (1975) reported Rb-Sr ages of 511 ±23 m.y. The Cambrian age is consistent with fossil and isotopic ages described previously for the Albemarle-Denton-Asheboro area of North Carolina.

CONCLUSIONS

A geochronological investigation of the Lincolnton metadacite, a prominent unit in the slate belt along the Georgia–South Carolina border, provides indicated primary ages of 554 ±20 m.y. (Rb-Sr) and 566 m.y. (U-Pb, zircon). On the basis of both lithological and geochronological data, the Lincolnton metadacite and associated felsic pyroclastic sequence are interpreted to be correlative with the Uwharrie Formation in the Albemarle-Denton-Asheboro area of North Carolina. On the basis of lithologic comparison alone, the Lincolnton metadacite and associated felsic pyroclastic sequence are also interpreted to be equivalent to the Persimmon Fork Formation near Columbia, South Carolina. Prominent argillite units that overlie felsic metavolcanics in all three areas are likewise considered correlative. These units include the Tillery Formation (Albemarle-Asheboro-Denton area), the Richtex Formation (Columbia, South Carolina, area), and the upper sedimentary sequence (Lincolnton-McCormick area).

The slate belt of Georgia and the Carolinas appears to be composed of at least two different chronological sequences of metavolcanic rocks. From the Albemarle-Denton-Asheboro area in central North Carolina to Georgia, volcanic rocks are interpreted to be mainly of Cambrian age. In the Roxboro–Chapel Hill area in north-central North Carolina, metavolcanic rocks were folded and faulted prior to 575 m.y.

If our stratigraphic correlation between the Lincolnton-McCormick and the Albemarle-Denton-Asheboro areas is correct, then of the three alternative correlations between the central and northern North Carolina slate belt offered by Wright and Seiders (1980), their third alternative (see their Fig. 4C) now would seem to be preferred. This implies that the Roxboro–Chapel Hill area of the slate belt is older than that of central North Carolina and Lincolnton, Georgia, and the possibility of a major angular unconformity between the central northern areas of the North Carolina slate belt.

On the other hand, we readily can accept Wright and Seider's suggestion that the Virgilina deformation (Glover and Sinha, 1973) was a local disturbance. The Virgilina disturbance was an "orogenic event" only in the sense that the slate belt was an active orogene. Local buckling and verticle and horizontal displacements are part of the tectonics of modern orogenes such as those of the western Pacific. It is perhaps too much to expect that the Virgilina deformation was so widespread as to be an important element in consideration in regional correlations.

From a comparison of the Lincolnton-McCormick, Albemarle-Denton-Asheboro, and Roxboro–Chapel Hill areas of the slate belt, it appears that the best lithostratigraphic correlations can not be accommodated by the isotopic dates. On the other hand, perhaps it is unrealistic to attempt such lithostratigraphic correlations over large distances within volcanic arcs where similar stratigraphies associated with different volcanic centers need not be time equivalents.

REFERENCES CITED

Bell, H., III, 1973, Some results of geochemical sampling in McCormick County, South Carolina: U.S. Geological Survey Bulletin 1376, 22 p.

Black, W. W., 1978, Chemical characteristics and Rb/Sr ages of metavolcanics from the Carolina slate belt of North Carolina: Geological Society of America Abstracts with Programs, v. 10, p. 162–163.

Butler, J. R., and Fullagar, P. D., 1975, Lilesville and Pageland plutons and the associated meta-rhyolites, eastern Carolina slate belt: Geological Society of America Abstracts with Programs, v. 7, p. 475.

Carpenter, R. H., 1976, General geology of the Carolina slate belt along the Georgia–South Carolina border, *in* Chowns, T. M., ed., Stratigraphy, structure, and seismicity in the slate belt rocks along the Savannah River: Georgia Geological Society Guidebook 16, p. 16–20.

Conley, J. F., and Bain, G. L., 1965, Geology of the Carolina slate belt west of the Deep River–Wadesboro Triassic basin, North Carolina: Southeastern Geology, v. 6, p. 117–138.

Costello, O. P., Jr., Secor, D. T., Jr., and Snoke, A. W., 1981, Structural relationships as a key to stratigraphic sequence in the Carolina slate belt, Lake Murray, South Carolina: Southeastern Geology, v. 22, p. 139–147.

Fullagar, P. D., 1971, Age and origin of plutonic intrusions in the Piedmont of the southeastern Appalachians: Geological Society of America Bulletin, v. 82, p. 2845–2862.

Glover, Lynn, III., and Sinha, A. K., 1973, The Virgilina deformation, a late Precambrian to Early Cambrian(?) orogenic event in the central Piedmont of Virginia and North Carolina: American Journal of Science, v. 273-A, p. 234–251.

Hatcher, R. D., Jr., 1972, Developmental model for the southern Appalachians: Geological Society of America Bulletin, v. 83, p. 2735–2760.

Higgins, M. W., 1971, Correlation of metavolcanic rocks in the Maryland, Delaware, and Virginia Piedmont: Geological Society of America Abstracts with Programs, v. 3, p. 320.

Higgins, M. W., Sinha, A. K., Zartman, R. E., and Kird, W. S., 1977, U-Pb zircon dates from the central Appalachian Piedmont: A possible case of inherited radiogenic lead: Geological Society of America Bulletin, v. 88, p. 125–132.

Hills, F. A., and Butler, J. R., 1969, Rubidium-strontium dates for some rhyolites from the Carolina slate belt of the North Carolina Piedmont: Geological Society of America, Abstracts for 1968, Special Paper 21, p. 445.

Krogh, T. E., 1973, A low-contamination method for hydrothermal decomposition of zircon and extraction of U and Pb for isotopic age determinations: Geochimica et Cosmochimica Acta, v. 37, p. 485–494.

Paris, T. A., 1976a, The geology of the Lincolnton 7.5′ quadrangle, Georgia–South Carolina [M.S. thesis]: Athens, University of Georgia, 191 p.

———1976b, The Lincolnton metadacite, *in* Chowns, T. T., ed., Stratigraphy, structure, and seismicity in slate belt rocks along the Savannah River: Georgia Geological Society Guidebook 16, p. 13–15.

Secor, D. T., and Wagener, H. D., 1968, Stratigraphy, structure, and petrology of the Piedmont in central South Carolina: Division of Geology, South Carolina State Development Board, Geologic Notes, v. 12, p. 67–84.

St. Jean, J., 1973, A new Cambrian trilobite from the Piedmont of North Carolina: American Journal of Science, 273-A, 196–216.

Stromquist, A. A., and Sundelius, H. W., 1969, Stratigraphy and structure of the Albemarle Group of the central Carolina slate belt in central North Carolina: U.S. Geological Survey Bulletin 1274-B, 22 p.

Van Eysinga, F.W.B., 1975, Geologic time table: Amsterdam, Elsevier Scientific Publishing Company, 3rd edition.

Whitney, J. A., Paris, T. A., Carpenter, R. H., and Hartley, M. E., III, 1978, Volcanic evolution of the southern slate belt of Georgia and South Carolina: A primitive oceanic island arc: Journal of Geology, v. 86, p. 173–193.

Wright, J. E., and Seiders, V. M., 1980, Age of zircon from volcanic rocks of the central North Carolina Piedmont and tectonic implications for the Carolina volcanic slate belt: Geological Society of America Bulletin, v. 91, Part 1, p. 287–294.

MANUSCRIPT ACCEPTED BY THE SOCIETY JULY 8, 1982

Printed in U.S.A.

Geological Society of America
Special Paper 191
1982

Ore deposits of the northern parts of the Carolina slate belt, North Carolina

P. Geoffrey Feiss
Department of Geology,
University of North Carolina at Chapel Hill
Chapel Hill, North Carolina 27514

ABSTRACT

Modelers of the Carolina state belt (CSB) have come to view the province as a remnant of a late Precambrian–early Paleozoic island arc. Thus the metallogeny and tectonic setting of the stratabound massive sulfides, iron/manganese formations, and barite deposits, of the Au, W, and Cu veins, of disseminated Cu and Mo porphyry-like deposits, and of the kyanite-andalusite-pyrophyllite deposits are of considerable significance. On the basis of a characterization of these deposits and their regional setting, it can be concluded that (1) the eastern parts of the CSB in North Carolina were dominated by subaerial to shallow marine conditions with volcaniclastic debris shed westward into a deepening back-arc basin, (2) the major folds of the area were syndepositional, (3) the nonfoliated, postmetamorphic granites are typical of those related to post-tectonic plutonism, (4) the Kings Mountain belt appears to have developed in a similar tectonic setting to the CSB, (5) the aluminosilicate deposits mark significant linear zones of subaerial to shallow submarine volcanism which terminate basinward in volcanic centers, and (6) the western parts of the CSB were dominated by relatively deep water, quiet conditions, distal to volcanic centers.

INTRODUCTION

In regional reviews, it is unusual to discuss metallogeny. Mineral deposits are most often viewed in isolation from the regional setting after the major problems of an area have been solved. This derives from the long-held prejudice that most mineral deposits, being epigenetic, are a footnote to the geologic history of a region. With increasing acceptance of syngenetic models in ore deposition and with the recognition that ore deposits should be viewed in their tectonic setting (Mitchell and Bell, 1973), it is a reasonable proposition that in a collection of articles, such as contained in this volume, we should include a look at those mineral deposits that are clues to the geologic history of the area—especially an area as sketchily known as the Carolina slate belt.

The Carolina slate belt is not a major metallogenic province. Few metallic deposits have been exploited, and there are no active metal mines in the slate belt today. This is not to ignore the historical significance of the slate belt mineral deposits. From 1802 to 1848, essentially all the gold mined in the United States was produced from gold mines in the Appalachian Piedmont. The first American gold rush occurred in Cabarrus County, North Carolina, with the discovery of placer and "soft outcrop" gold deposits in the slate belt. The original discoveries, near the Reed mine, were already "legendary" by the time Dennison Olmstead visited the area in the early 1820s (Olmstead, 1825). The Charlotte mint, founded in 1838, produced gold coins until the competition from California led to its decline in the 1850s. During that time an estimated $15 million worth of gold (at $20 an ounce) was mined from the slate belt and the adjacent Charlotte belt.

Not all the historically active mines in the slate belt are gold mines. Silver, copper, lead, and zinc, in addition to gold, were mined from massive sulfide deposits. In addition, the only economic tungsten mineralization east of the Rockies is in a vein-type deposit along the eastern margin of the slate belt, the Hamme mine in North Carolina. Several copper and molybdenum porphyry prospects have been investigated over the past few decades. Nonmetallic minerals such as barite, pyrophyllite, andalusite, and kyanite have also been and are currently being mined. Over one-half of the U.S. production of kyanite and related aluminosilicate refractories comes from the slate belt or correlative sequences.

This paper is not intended for those who wish detailed descriptions of slate belt mineral deposits. It is the goal of this contribution to set those deposits that are of tectonic significance in a regional framework and to treat them as lithostratigraphic units and to relate them to models of slate belt evolution. The author feels that the occurrence of bedded, polymetallic massive sulfide deposits in the slate belt is as significant in recreating the regional geologic history as is the presence of finely laminated argillites and volcaniclastic rocks. I will view the ore deposits as paleogeographic indicators in a region where extensive rock exposures are rare and the mineral deposits are often better known than the surrounding country rocks.

The deposits to be discussed are limited to those that have an intrinsic tectonic value rather than necessarily an economic one. The area of interest is essentially restricted to North Carolina where the slate belt is at its largest areal extent and the stratigraphic relations are somewhat clearer from a regional point of view.

GEOLOGIC SETTING

The papers in this volume review the state of knowledge of the geology of a number of related areas in the Carolina slate belt. Thus, the slate belt lithologies of North Carolina will only be quickly summarized.

Figure 1 is a general geologic map of the Carolina slate belt in North Carolina and immediately adjacent states. The slate belt consists of an estimated 9,000 m (Conley and Bain, 1965) to 26,000 m (Seiders and Wright, 1977) of metavolcanic and metasedimentary rocks, ranging in age from 520 to 650 m.y. A number of areas in North Carolina have been described in some detail (e.g., Conley and Bain, 1965; Stromquist and Sundelius, 1969; Glover and Sinha, 1973; Seiders and Wright, 1977). On the whole, the metamorphic rank is low, ranging from essentially unmetamorphosed to chlorite and more rarely biotite grade. To the east, the slate belt is either in fault contact with the Triassic rift basins or is overlain by onlapping coastal-plain sediments. East of the Triassic basins, a broad, apparently fault-bounded anticlinorium of high-grade schists and gneisses, the Raleigh and Kiokii belts in North Carolina and Georgia, respectively, is often present. In North Carolina, this high-grade zone is, in turn, flanked to the east by additional slate belt rocks, the eastern slate belt. Slate belt lithologies are recognized in deep wells from many regions on the coastal plain in North Carolina and Georgia (Sundelius, 1970), suggesting that slate belt rocks may well provide the basement on which the Mesozoic-Cenozoic coastal-plain onlap lies.

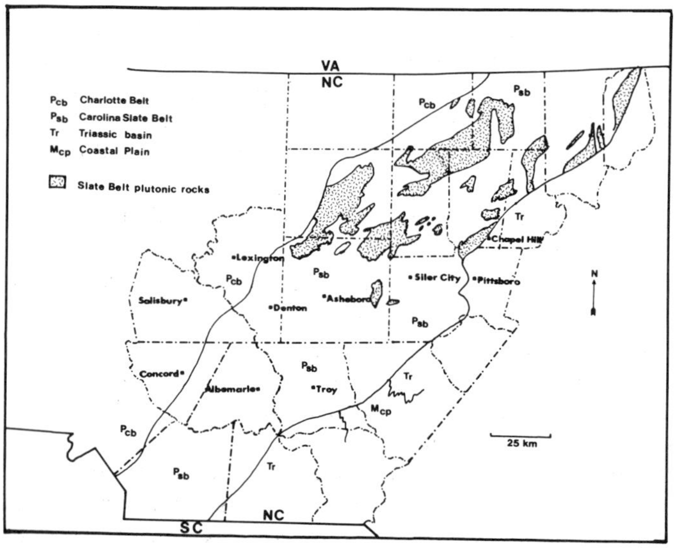

Figure 1. Generalized map of the North Carolina slate belt showing boundaries with adjacent geological provinces.

To the west, the nature of the slate belt–Charlotte belt boundary is equally as equivocal. The Charlotte belt is a plutonic-metamorphic belt, with significantly higher metamorphic grades than the adjacent slate belt rocks. The plutonic rocks are similar in age, composition, and petrology to many plutons in the slate belt. The two belts appear to be separated by a shear zone, bounded by the Gold Hill and Silver Hill faults in North Carolina. This leaves the nature of this western contact somewhat uncertain as to whether it is in fact a fault contact, a suture, a metamorphic gradient, or all of the above. Best and others (1973) presented gravity evidence across this western border of the slate belt which suggests that there is a major density contrast in the underlying crust. Recent COCORP studies have simply argued the presumed allochthonous nature of the whole Piedmont without characterizing the Charlotte belt–slate belt boundary (Cook and others, 1979).

Along strike, the complexity of facies changes within the volcanic basin that became the slate belt, combined with the incomplete state of mapping, makes regional correlation impossible. Major problems in the relative age and sequence of units remain to be worked out. (Seiders and Wright, 1977).

Structurally, the slate belt is a series of northeast-southwest–striking anticlines, synclines, and monoclines. Major structures, as shown in Figure 2, were active during sedimentation (Conley and Bain, 1965) and, thus, have significant control over the distribution of the various lithostratigraphic units, including the syngenetic mineral deposits. In terms of deep structure, the COCORP data suggest that the whole of the Appalachian Piedmont may well be allochthonous (Cook and others, 1979) and, thus, any discussion of the underlying crust at the time of formation of the volcanics is moot.

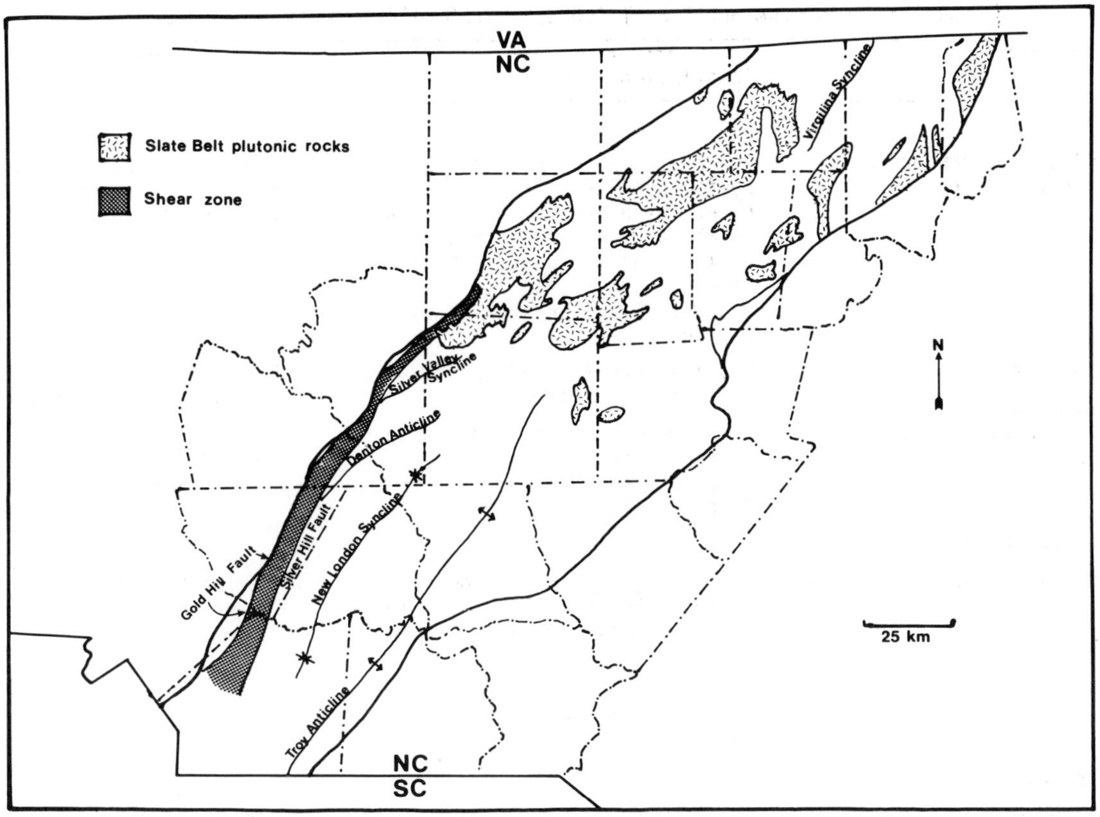

Figure 2. Major structures within the Carolina slate belt of North Carolina.

The rocks of the Carolina slate belt in North Carolina vary from coarse fragmental volcanics to finely laminated argillites. Generally, to the south, the volcanics are bimodal with a felsic to mafic ratio of approximately 4:1. In the north, more intrusive rocks are exposed with a general absence of mafic units (Butler, 1979). With the exception of volcanic debris, there is essentially no clastic sedimentation. Put dramatically, "In central North Carolina not one sand grain or pebble of a non-volcanic rock type (excepting mudstone pebbles of intraformational conglomerate) has been identified" (Seiders and Wright, 1977, p. 8). Rhyolite flows and areas that have been mapped as volcanic centers (Stromquist and Conley, 1959; Seiders and Wright, 1977) are also known.

There is agreement that the Carolina slate belt was generated in a convergent tectonic environment. The chemical bimodality of the volcanics, the lithostratigraphic character of the slate belts rocks, the clear linear character of the Carolina slate belt, and the presumed tectonic setting all indicate that the Carolina slate belt is an island-arc system. The most widely accepted model correlates this arc development with the late Precambrian closing the Iapetus (Proto-Atlantic) Ocean. Other details, such as the polarity of the subduction zone(s) and the nature of the crust on which this arc was developed, lack any current consensus. It is likely that many of the contributions to this collection of articles will serve to answer these last questions.

It is in this somewhat imperfectly known province that the mineral deposits that will be discussed in this paper are found.

TYPES OF MINERAL DEPOSITS

Table 1 briefly summarizes the types of deposits that will be discussed in this paper. Some of the classifications would not be accepted without argument by other workers. Nonetheless, as shall become apparent, this classification, in combination with what regional geology is known, is of considerable use in defining the tectonic setting of the Carolina slate belt. The individual deposit types are discussed in greater detail below, and the location of typical examples is shown in Figure 3.

Polymetallic massive sulfide deposits

As a class, these deposits are bedded, conformable sulfide bodies with thicknesses ranging up to 4.5 m and with strike lengths of 100 to 500 m. Typically, the massive sulfide lenses pinch and swell and form a complex, anastomosing cluster of deposits. Sulfides are directly hosted by argillaceous to

TABLE 1. CAROLINA SLATE BELT DEPOSIT TYPES DISCUSSED IN THIS PAPER

Type of deposit	Description	Slate belt example
SYNGENETIC		
Polymetallic, volcanogenic, massive sulfide deposits	Bedded, massive, stratabound pyrite-sphalerite-chalcopyrite-galena with abundant chert and exhalites. Au and Ag values. Primary depositional features common, unless obscured by recrystallization and remobilization	Silver Hill-Silver Valley, NC Gold Hill district, NC Haile mine, SC Lincolnton-McCormick area, SC-GA
Barite deposits	Massive, stratabound barite lenses in quartz-sericite schists	Barite Hill, SC Carolina "Barite Belt," NC
Banded iron and manganese formations	Stratabound magnetite-hematite, ankerite, or iron silicates (e.g., talc). Psilomelane or Mn-oxides in cherty quartzites	Arvonia area, VA Kings Mountain area, NC Mt. Athos Fm., VA
EPIGENETIC, POSTMAGMATIC		
Molybdenum (-copper?) porphyries	Very low grade, disseminated molybdenite or chalcopyrite in granitic rocks, usually post-tectonic	Halifax County, NC Brewer mine(?), SC
Tungsten veins	Sheeted quartz-fluorite-sericite veins with heubnerite, scheelite, and minor sulfides	Hamme district, NC
METAMORPHIC-REMOBILIZED		
Vein gold deposits	Cm- to m-thick quartz veins with free Au and minor sulfides. Concordant and discordant to local bedding and cleavage	Reed mine, NC
Copper veins in epidosite	Quartz-calcite-albite-epidote-chlorite-sericite veins with chalcocite and bornite in mafic volcanics	Virgilina district, VA
Pyrophyllite-kyanite-andalusite deposits	Massive concordant to discordant aluminosilicates with, variously, sericite, quartz, pyrite, specular hematite, fluorite, and lazulite	Willis Mtn., VA Graves Mtn., GA NC pyrophyllite district

cherty sediments contained within units that are predominantly submarine, intermediate to felsic, pyroclastics and epiclastics. Along strike with the sulfides, cherty units and iron formations, both sulfide and oxide facies, are common. The iron formations are less common in North Carolina, but are seen in the Arvonia syncline (Gair, 1978) and in isolated occurrences elsewhere. Barite is described in the McDuffie district (Howell and Weisenfluh, 1978), the McCormick district (Bell and others, 1980) and in the so-called Carolina barite belt. Figure 4 is a schematic representation of a typical slate belt massive sulfide deposit as it probably looked immediately following deposition.

The mineralogy of the typical slate belt massive sulfide consists of pyrite and sphalerite with varying amounts of galena and chalcopyrite. The deposits have been variously mined for silver, gold, zinc, lead, and copper. A variety of associated minerals have also been observed, including pyrrhotite, magnetite, pentlandite, tetrahedrite, pyrargyrite, arsenopyrite, bismuthinite, native bismuth, chalcocite, covellite, and electrum (Indorf and Feiss, 1978). The sulfide beds commonly contain sedimentary structures, including cross-bedding, graded bedding, slump structures, scour and fill, and other soft-sediment features. They are always conformable.

Their volcano-sedimentary origin is clear. Details of their origin, such as their relationship to the various phases of volcanism, their age, and their stratigraphic position, remain to be worked out. At present, none larger than several million tons has been described.

Barite Deposits

Barite is not abundant in the Carolina slate belt. As indicated previously, there are barite beds associated with

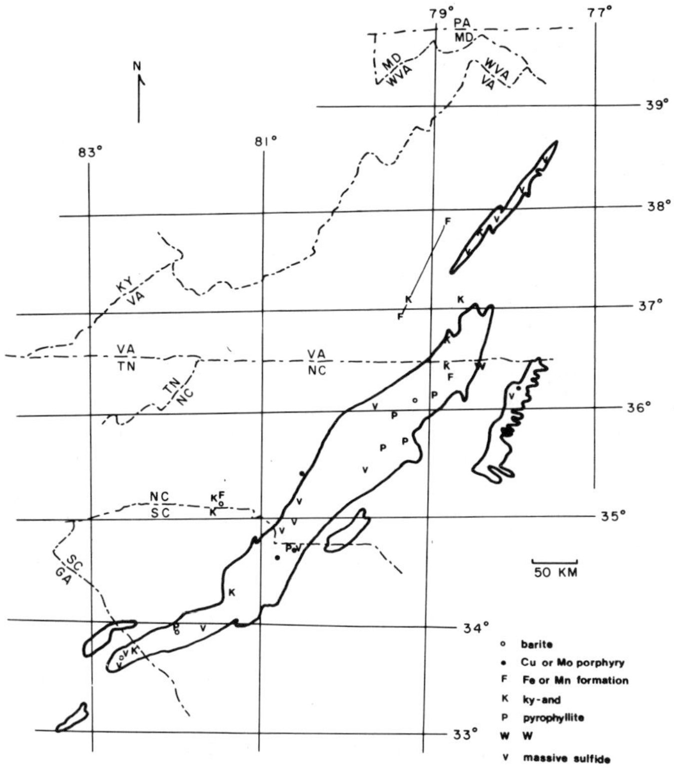

Figure 3. Location of the major economic mineral deposits of the Carolina slate belt.

the polymetallic massive sulfides. Howell and Weisenfluh (1978) and Luethe and Windisch (1978) considered the barite-bearing massive sulfide deposits of South Carolina and Georgia to be analogs of Kuroko types of massive sulfides. These deposits are in the McDuffy gold belt and the Lincolnton "metadacite" of Whitney and others (1978).

To the west of the slate belt, along the eastern margin of the Kings Mountain belt, are a series of barite deposits with associated sphalerite, pyrite, and galena, which are conformable with the schistosity of the so-called Battleground Schist (Keith and Sterrett, 1931). These deposits consist of a series of en echelon lenses contained within quartz-sericite schists and associated with dolomitic marbles, quartzites, and kyanite quartzites. Lenses range in width from 1 to 5 m, and individual beds have strike lengths up to 60 m. The whole zone of en echelon lenses is approximately 1 km long. These deposits have been mined for barite at various times, though their limited size and the complex intercalation with the metamorphic host rocks required expensive, labor-intensive methods of mining. It appears that these deposits are syngenetic, but as they do not occur in the slate belt proper, they are difficult to fit into the slate belt regional setting without becoming involved in the equally difficult problem of the relationship of the Kings Mountain belt to adjacent Piedmont geologic provinces. Nonetheless, their apparently stratabound nature and spatial association with gold, Fe formation (both oxide and sulfide facies), and kyanite suggests a volcanogenic origin.

Iron and Manganese Deposits

In many parts of the slate belt, as well as adjacent areas containing slate belt lithologies, one finds iron formations. In the Arvonia syncline, Gair (1978) indicated a number of stratabound, quartose and cherty, banded, oxide iron formations associated with massive sulfide deposits. These units are, in hand specimen, superficially identical to Precambrian banded iron formations, though obviously their limited size makes economic exploitation unlikely. Such deposits are found on both limbs of the Arvonia syncline, in the Cambrain and Ordovician rocks of the Evington Group.

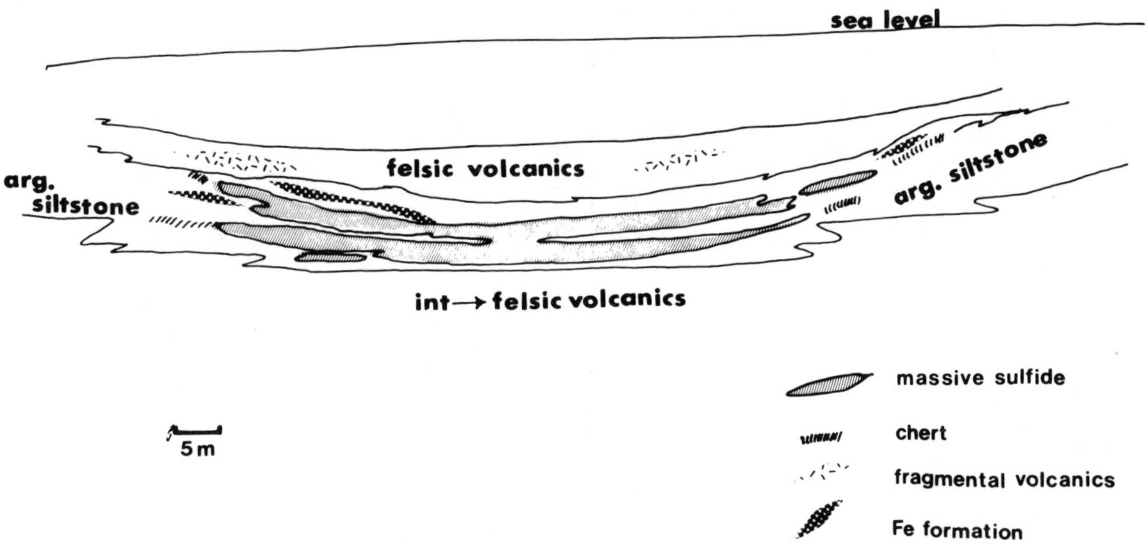

Figure 4. Schematic cross-section of a typical Carolina slate belt massive sulfide deposit following deposition. Depth of water unknown.

Their association with massive sulfide bodies, as well as their stratigraphic setting, suggests a volcanic, exhalative origin.

In the slate belt in North Carolina, fewer examples of banded iron formation are seen, although a number of hematitic iron deposits are described in the literature. Nitze and Hanna (1896) quoted J. P. Lesley as indicating that

> The beds [in Guildford County, North Carolina] were deposited, with the rest of the rocks, in water; in the same age with the rocks which hold them—are in fact rock deposits highly charged with iron, and they differ from the rest of the rocks only in this respect—that they are *more highly charged with iron.* In fact all our primary (magnetic and other) iron ores obey this law.

In the same area as the Kings Mountain belt barite deposits are a number of oxide and sulfide iron formations, which suggest a volcanic, exhalative process. Figure 5 is from Nitze and Hanna's (1896) study of this area and emphasizes (1) the very strong, northeast trend of the iron formations over a 15 to 20 mile strike length; (2) the association of the iron formations with barite bodies; and (3) the association of the banded iron formations with gold deposits and pyrite bodies. The host rocks in this area are micaceous schists with interbedded slate, quartzite, and limestone. Emmons (1856) suggested the association of the iron formations with ferruginous slates, talcose slates, and massive quartzites. The iron formations themselves are talc, chlorite, quartz, actinolite schists with granular hematite and magnetite bodies of a lenticular shape, often in en echelon clusters.

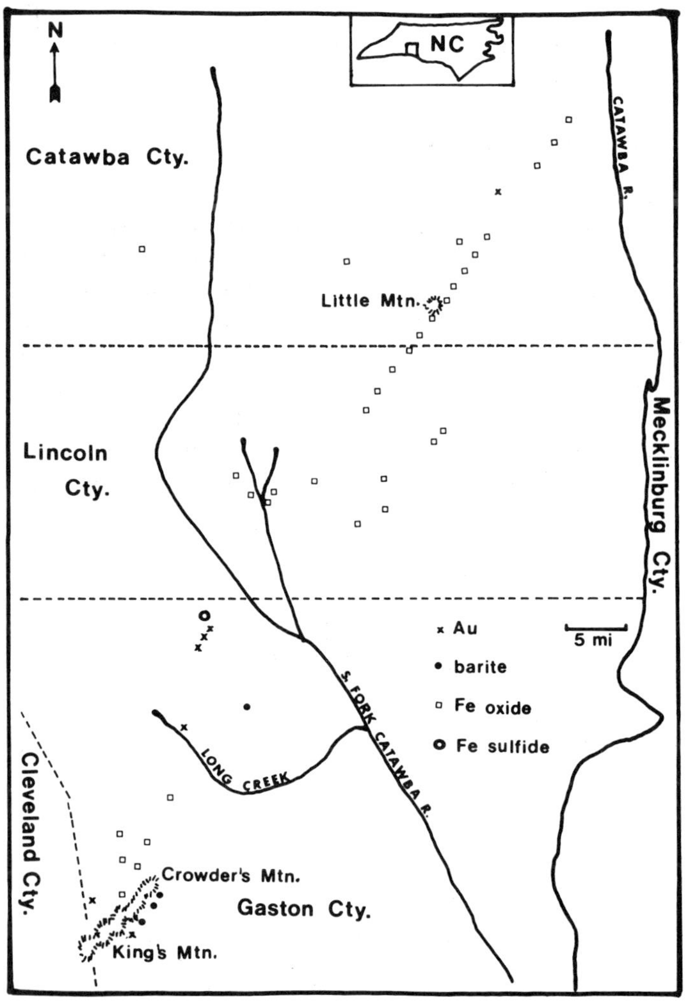

Figure 5. Location of gold, barite, banded iron formation, and pyrite deposits in the Kings Mountain–Crowder's Mountain area of south-central North Carolina (from Nitze and Hanna, 1896).

Many of the polymetallic massive sulfide deposits are overlain by pyritic quartzites which may represent the development of a late sulfide facies of iron formation.

Manganese deposits are associated with massive sulfide deposits in a number of districts. Figure 6 shows a number of manganese deposits in the Mount Athos Formation of the Evington Group (Espenshade, 1954). Cherty beds with psilomelane and various manganese oxides are conformable within quartzite members of the Mount Athos. The manganiferous rocks are associated with specularite, limonite-rich iron formations and with a few barite pods and copper shows. While Espenshade attributed these deposits to residual mechanisms, he noted their strongly conformable nature, and it appeared more likely that they are similar to the manganiferous "umbers" commonly found in association with classic volcanic-exhalative deposits, such as those of the Troodos complex. Mangenese-rich formations are much less common south of the central Virginia region. Beck (1946) described pyrolusite and psilomelane with

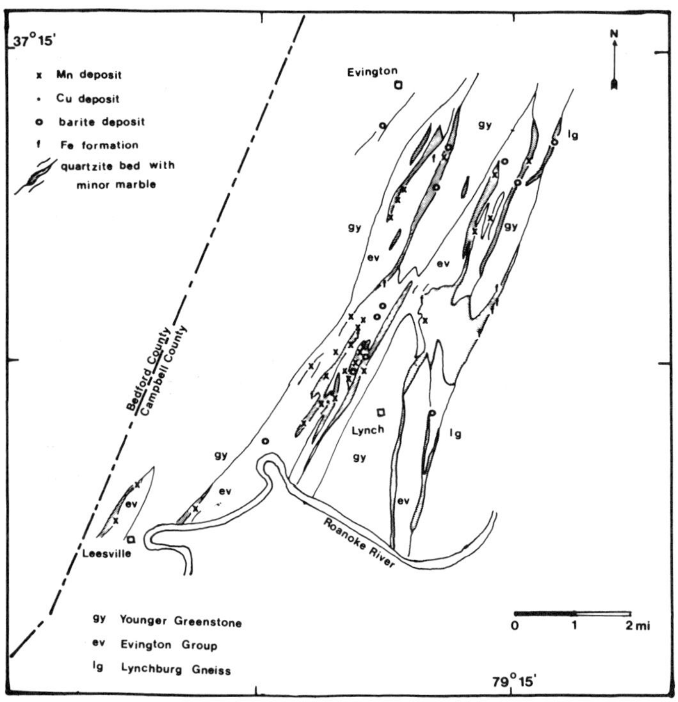

Figure 6. Manganese, barite, iron formation, and copper deposits of the Evington Group in southern Virginia (from Espenshade, 1954).

quartz as massive conformable lenses in the metavolcanics associated with the Dorn gold mine near McCormick, South Carolina.

Molybdenum and Copper Porphyry Deposits

No significant copper or molybdenum porphyry-type mineralization is present in the Carolina slate belt. Schmidt (1978) has discussed several promising prospects and reiterated the fact that many characteristics of the southeastern United States Piedmont present attractive targents for porphyry-type mineralization, particularly by analogy with Maine and the Maritime Provinces. The Haile-Brewer area in Lancaster and Chesterfield Counties, South Carolina, is an area of extensive hydrothermally altered felsic volcanics (Bell and others, 1980). Significant topaz mineralization, as well as tin, bismuth, molybdenum, and arsenic anomalies, has been described (Pardee and others, 1937). In addition, molybdenum enrichment in a number of post-tectonic, 300-m.y.-old granites of Halifax and Nash Counties, North Carolina, and in Lancaster and Kershaw Counties in South Carolina, have porphyry affinities, though they are not as highly altered. Speer (1978) suggested that these are more deeply emplaced than the typical western U.S. porphyry molybdenum deposit. It is also possible that these granites are analagous to "molybdenitic porphyries" (Schmidt, 1978), that is, those granites that are molybdenum enriched relative to crustal averages, but clearly, are not ore-grade.

Tungsten Deposits

About 3 miles north of Townsville, North Carolina, is the only economic tungsten deposit in the eastern United States. The deposit is a 1,600-m-long, up to 10-m-wide, series of quartz-huebnerite veins which cut diagonally across the contact between a large, 520- to 620-m.y.-old granodiorite pluton and the phyllites and greenstones of the Carolina slate belt. Since the beginning of production in 1941, approximately 1 million short tons of WO_3 have been produced (Gair and others, 1975). The veins are often sheared and contain, in addition to quartz and huebnerite, minor sulfides, scheelite, molybdenite, tetrahedrite, and tellurides, as well as a gangue of fluorite, sericite, and rare carbonates (Slack and others, 1978). Slack and others suggested a premetamorphic origin by analogy with Pasto Bueno (Landis and Rye, 1974). The analogy is enhanced by the similarity to the famous sheeted wolframite-quartz veins of Kiangsi Province, China (Horsnail and Parker, 1979).

Gold Deposits

The gold deposits of the slate belt have been studied about once a generation (Olmstead, 1825; Emmons, 1856; Nitze and Hanna, 1896; Pardee and Park, 1948; McCauley and Butler, 1966; Worthington and Kiff, 1970). Aside from gold mined from clearly volcanogenic, massive sulfide deposits, slate belt gold is found in quartz veins ranging in width from a few centimeters to 3 m and in length up to hundreds of meters. Many of these quartz veins are conformable with bedding in the host phyllites, slates, and schists. However, an equal number appear to be disconformable and suggest that the gold-quartz veins were emplaced after consolidation. Worthington and Kiff (1970) argued on the basis of (1) the absence of granitic sources and hydrothermal alteration, (2) the conformable nature of many of the "veins," and (3) the preferential localization of the gold deposits in the stratigraphically lowest felsic volcanic units that the deposits are volcanogenic. The discordant nature of many gold veins, cross-cutting both bedding and foliation, suggests that the mobilization of some vein-filling material was late, possibly associated with waning metamorphism. The evidence for this metamorphic remobilization is moot. Discordant quartz veins are abundant in many parts of the Carolina slate belt and presumed to be metamorphic in origin. The cogent observation of Worthington and Kiff that mineralized quartz veins are stratigraphically controlled suggests that while the present configuration of these veins is metamorphically controlled, it is more than likely that the gold was originally disseminated in the volcanic host rocks via a volcanogenic, exhalative mechanism. Thus, the metallogenesis is, in fact, a two-stage process, possibly separated by several hundred million years. Recent lead isotopic studies (Kish and others, 1980) confirm this suggestion.

Copper Veins in Epidosite

The Virgilina district of southern Virginia and northern North Carolina lies west and north of the Hamme district in lithologies similar to slate belt lithologies. Weed (1911) and Laney (1917) described the general geology and the twenty-odd mines located in the district. The vein mineralization is confined to quartz veins in the Virgilina greenstone. Mineralization consists of quartz, calcite, albite, epidote, chlorite, and sericite with chalcocite and bornite. Minor covellite, malachite, azurite, cuprite, and native copper are also present. Stein and Kish (1978a) indicated that adjacent to the veins the greenstone is converted to epidosite masses depleted in copper relative to the rest of the greenstones. They suggested a metamorphic mobilization of the copper into the veins, which they tentatively dated by Rb/Sr whole-rock methods at 350 m.y. (Stein and Kish, 1978b).

Porphyllite-Andalusite-Kyanite Deposits

The slate belt produces a significant proportion of the U.S. supply of aluminosilicates. This production is from two types of deposits. The first are the large kyanite deposits such as Graves Mountain, Georgia, and Willis Mountain, Virginia. The second are pyrophyllite-andalusite deposits, predominantly from the eastern portion of the slate belt in North Carolina.

The Graves Mountain and Willis Mountain occurrences are large monadnocks of interbedded lenses of kyanite quartzites and sericitic schists. The kyanitic rocks are 30% to 40% kyanite. Radcliffe (1978) suggested, by analogy with kyanite-bearing sands in Florida, that this occurrence represents the metamorphism of a low-energy, coastal or lagoonal sedimentary facies whose provenance was in high-grade metamorphic rocks of the Piedmont or Blue Ridge. This is in contrast to various hydrothermal and replacement theories as summarized by Espenshade and Potter (1960).

Similar origins have been proposed for the pyrophyllite-andalusite deposits in North Carolina. These deposits are, in general, concordant, lens-shaped masses of pyrophyllite-quartz-sericite rock with varying amounts of diaspore, topaz, andalusite, pyrite, and kyanite. They are contained within tuffaceous units of pyroclastic or epiclastic origin. Sykes and Moody (1978) described the Hillsborough deposit, Orange County, North Carolina, in considerable detail and suggested that these highly aluminous rocks originated syngenetically by the action of geothermal brines which leached the volcanics near vent areas during subaerial hot spring activity.

The similarity of the analyses given by Radcliffe (1978) and Sykes and Moody (1978) suggests a similar origin for the kyanite massifs of Graves Mountain and the pyrophyllite deposits of the slate belt. Figure 7 is a plot of Al_2O_3 versus SiO_2 for Sykes and Moody's aluminosilicate-bearing schists, quartzites, and breccias and for Radcliffe's typical Graves Mountain kyanite and for a Florida kyanite-quartz

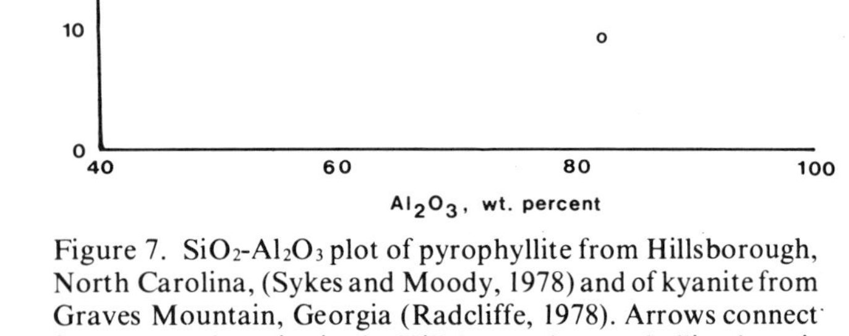

Figure 7. SiO_2-Al_2O_3 plot of pyrophyllite from Hillsborough, North Carolina, (Sykes and Moody, 1978) and of kyanite from Graves Mountain, Georgia (Radcliffe, 1978). Arrows connect ferrugenous breccias in the Hillsborough pyrophyllite deposit (the iron is a late, secondary breccia [bx] cement) with iron corrected analyses calculated by subtracting secondary iron and recalculating analyses to 100%. Ky = kyanite.

sand. The strong negative correlation (r = −0.9458) suggests a silica-alumina mixing line with the major variable being the amount of detrital quartz and kyanite in the case of a sedimentary origin or of hydrothermal silica in the volcanogenic case. The possiblity exists that the large, kyanite massifs of Willis Mountain and Graves Mountain are either high-grade equivalents of the slate belt pyrophyllite deposits or are, themselves, derived locally from the erosion of subaerial, sinter-type deposits similar to the Hillsborough and other North Carolina occurrences.

It may be that the kyanite quartzites associated with the barite, massive sulfides, and iron formation deposits of the Kings Mountain belt are of a similar origin. The close spatial relationship of the deposits implies a proximal origin, perhaps similar to the Lincolnton-McCormick area (Bell and others, 1980).

TECTONIC SETTING OF MINERAL DEPOSITS

The types of deposits described in Table 1 and the preceding section are significant to the tectonic models of the slate belt. Mitchell and Bell (1973) summarized the characteristics of the ore deposits of convergent plant boundaries. Numerous other studies, such as Sawkins (1972, 1976), Hutchinson (1973), and Sillitoe (1976), have expanded upon these concepts and made economic geologists familiar with the metallogenic implications of plate tectonics.

The syngenetic ore deposits of the Carolina slate belt, that is, the polymetallic massive sulfides, the barite deposits, and the iron-manganese formations, are characteristic of convergent boundaries, commonly associated with tholeiitic to calc-alkaline volcanism. As shown in Table 2, from Large (1977), many of the variations in metal ratio, mineralogy, host-rock composition and texture, and even in morphology may be attributed more to proximity to volcanic centers than to broader features of regional tectonic setting.

The presence of both molybdenitic, granodiorite intrusives and the tungsten mineralization of the northern North Carolina slate belt are characteristic of late orogenic, LIL-rich, granitic magmas of peraluminous affinities. The presence at the Haile mine of significant topaz, fluorite, and apatite and the occurrences at other slate belt deposits, such as the Phoenix and Furniss mines in Cabarrus County, North Carolina, suggest that mineralized plutons of the slate belt bear affinities to late, postorogenic, anatectic granites such as one might expect from a continental margin–island arc sequence, or at least from an island arc that was building on a continentlike substrate.

The progenitors of the metamorphic deposits, especially the gold deposits and the pyrophyllite-kyanite-andalusite deposits, indicate a shallow marine to subaerial volcanic environment. Except for the similarity of the gold deposits to many epithermal gold veins in island-arc settings and the presence of fluorine in many of the aluminosilicate-bearing

TABLE 2. DIFFERENCES BETWEEN PROXIMAL AND DISTAL VOLCANOGENIC MASSIVE SULFIDE DEPOSITS

	Proximal	Distal
Metal content	Cu-rich; possible Zn and Au	Pb-rich; Cu-poor
Alteration	Distinct alteration pipe in footwall	No distinct alteration pipe
Iron sulfides and oxides	Pyrite and pyrrhotite are dominant sulfides; magnetite in footwall	Pyrite dominant; no phrrohotite; magnetite in hanging wall
Form	Pipelike or mushroom shaped; massive and cross-cutting	Well-banded, strataform, and blanket-shaped
Zoning	Zoned with CU at base, Zn/Cu increasing upward; Pb generally enriched in hanging wall	Unzoned
Volcanic setting	Toward top of volcanic sequence of massive pyroclastics, fragmental volcanics, and lavas	Within mixed sedimentary/volcanic pile

Note: Information from Large (1977).

deposits, there is little *a priori* reason to associate these deposits with any unique tectonic setting.

DISCUSSION

It still remains to attempt to use the deposits to view slate belt tectonism on a regional scale. Figure 8 shows a number of different ore deposits and, in addition, a number of unique geologic features of these same deposits and the adjacent slate belt rocks.

The distribution of syngenetic massive sulfide deposits and the remobilized gold deposits in the Carolina slate belt of North Carolina is not random. With the exception of a few deposits along the axis of the Virgilina synclinorium, no significant deposits are known in the northern half of the belt. This correlates with several other observations. First,

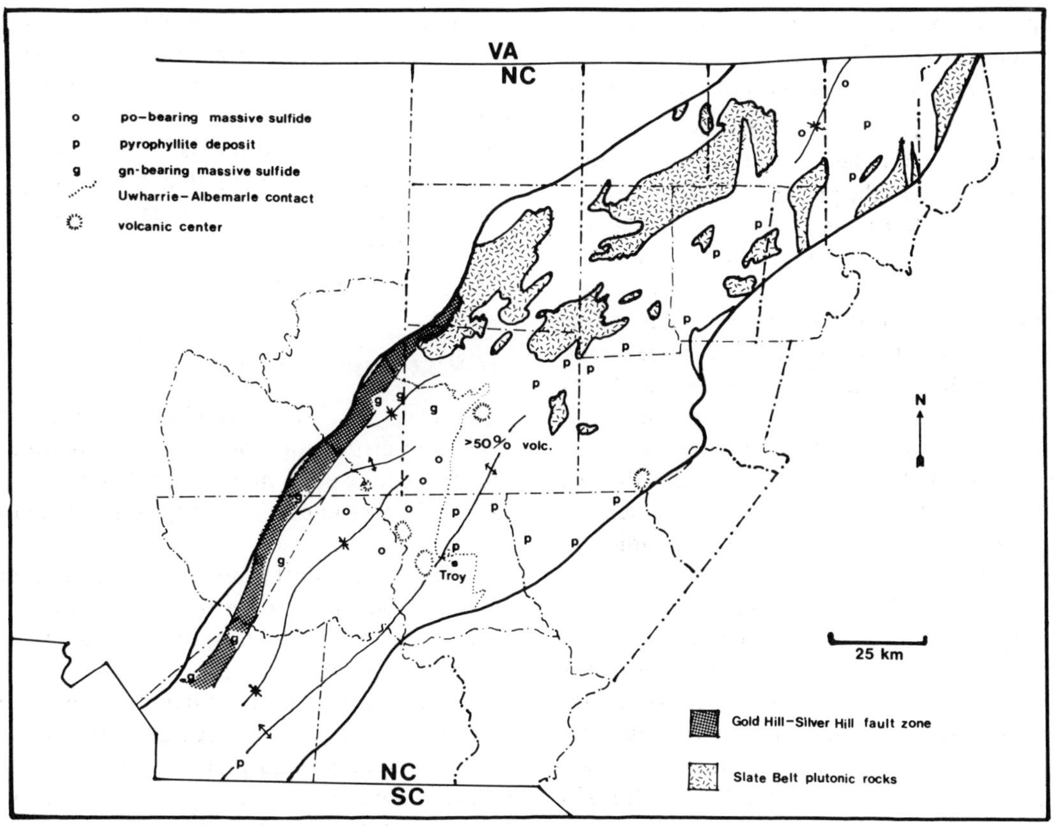

Figure 8. Summary map of the ore deposits of the Carolina slate belt in North Carolina in the regional context of their volcanic setting as described in the text. Po = pyrrhotite; gn = galena.

the northern half of the belt is dominated by plutonic rocks—unlike the southern half. This strongly suggests either shallower levels of intrusion or deeper erosion in the north. Furthermore, syngenetic deposits are by definition associated with predominantly submarine volcanism. In several areas, especially near Chapel Hill, Siler City, and Pittsboro, North Carolina, the volcanics are dominantly subaerial (Hauck, 1977; Green, 1977; and Wilkinson, 1978). In addition, the massive sulfide deposits are associated almost exclusively with fine-grained epiclastics and volcaniclastics and are not found in areas of abundant fragmental volcanics. Two explanations for this are possible. One is that this is a function of the discontinuous nature of the brine-exhalation process, such that during episodes of active volcanism or in areas characterized by relatively constant volcanism, such brines are not produced. This seems less likely than the argument that the brines require relatively quiescent conditions to accumulate in sufficient quantity to produce an ore deposit. The boundary line that runs north and west of Troy, North Carolina, (Fig. 8) is the approximate dividing line between slate belt formations that are greater than 50% volcanic debris (the Uwharrie Formation) and those that are dominantly volcanic mudstones and graywackes (the Albemarle Group). It is not clear whether the Uwharrie is subaqueous or submarine, though evidence is presented by Seiders and Wright (1977) that suggests a subaqueous origin for parts of the Uwharrie Formation. However, even if the Uwharrie is subaqueous in origin, note that no significant deposits are found east of the Uwharrie-Albemarle contact.

The pyrophyllite-andalusite deposits of North Carolina are also shown in Figure 8. In the previous section these are proposed to be siliceous sinter or geothermal vent areas and to be exclusively of subaerial to very shallow subaqueous origin. This is corroborated by their location east of the Uwharrie-Albemarle line and in areas known to be dominated by subaerial volcanic debris. In addition, these deposits define two linear belts (one east-northeast and one roughly east-west) that suggest major zones (lineaments?) of subaerial, geothermal activity. The hatchured areas on Figure 8 represent known volcanic centers, based either on detailed mapping or on the abundance of rhyolitic flow rocks. It is interesting to note that both trends of aluminosilicate deposits terminate, at their western end, in known volcanic centers within the Albemarle Group. No significant massive sulfide deposits have been found adjacent to these volcanic centers, though a number of small deposits and shows are reported. This is in apparent contrast to the association of massive sulfide mineralization and aluminosilicates in the Lincolnton-McCormick area (Bell and others, 1980). It appears that the pyrophyllite-andalusite deposits in North Carolina define linear zones of volcanic activity and geothermal vents (or both) that terminate to the west in a developing submarine, possibly back-arc basin.

It appears that the anticlinal axes, such as the Troy and Denton anticlines, are negative loci for massive sulfide deposits, whereas the synclinal axes appear to localize deposits. This is strongly suggestive of syndepositional growth of the regional structural elements. In addition, the only pyrrhotite-bearing massive sulfides in the North Carolina slate belt are located along the axes of the New London synclinorium. Pyrrhotite is often viewed as metamorphic in origin; however, the absence of any clear evidence for higher metamorphic grades in these areas and the fact that these deposits also contain significant pyrite with no clear textural evidence for replacement suggest that the pyrrhotite is not the result of thermal desulfidation. Plimer (1978) proposed that in unmetamorphosed massive sulfide deposits of this type, the presence of pyrrhotite can be attributed to higher temperatures and lower P_{S_2} of the brines that accumulate in deep basins as, for example, at Sullivan or Outukumpu. If we assume that the pyrrhotite in the Carolina Slate belt is primary, then, adopting Plimer's model, the syndepositional character of the major fold structures is reconfirmed.

Jambor (1979), in reviewing the characteristics of proximal versus distal massive sulfide deposits in New Brunswick, corroborated the proposal of many investigators that the most distal deposits are lead-rich relative to the copper- and zinc-rich deposits closer to the brine source. In Figure 8, the galena-bearing massive sulfide deposits are shown, and it is apparent that they lie exclusively in the western slate belt. This region is well west of the volcanic centers and the Uwharrie-Albemarle contact. Thus, the massive sulfide deposits are consistent with an eastward volcanic island arc with volcaniclastic sediments being shed westward into a subsiding back-arc basin. It is noteworthy that even the deposits lying west of the Silver Hill fault fit this model, implying that this shear zone represents predominantly transverse movement.

CONCLUSIONS

It is clear that by a careful investigation of the mode of origin and distribution of the mineral deposits in the Carolina slate belt, we can make a number of assumptions about slate belt tectonics. The important conclusions are as follows:

1. The eastern part of the slate belt was predominated by subaerial volcanic activity and by volcanic centers that shed their erosional products into a relatively quiet back-arc basin to the west.
2. The major folds in the slate belt were active during the deposition of the volcano-sedimentary units.
3. The postmetamorphic granites are characteristic of late, tensional tectonics and suggest an igneous source that, if not continental, at least allowed the concentrations of LIL elements and late, hydrothermal activity characteristic of post-tectonic plutonism.

4. The character of the ore deposits along the eastern Kings Mountain belt strongly suggests a similar environment of formation of the Kings Mountain rocks to the slate belt formations to the east, across the Charlotte belt.

5. The pyrophyllite-andalusite-kyanite deposits appear to follow significant linear zones in the slate belt, marking areas of predominantly subaerial volcanism or locally derived, volcanic sedimentation (or both) and terminating basinward in volcanic centers.

6. The syngenetic deposits themselves appear to be indicators of relatively quiescent, shallow to moderately deep water sedimentation in basins supplied predominantly by local volcanic source areas. The presence and absence of galena and pyrrhotite appear to correlate with water depth (bearing in mind the effect of metamorphic grade on pyrite stability) and to relative proximity to brine sources.

ACKNOWLEDGMENTS

The ideas presented here have been nurtured by many discussions with many different people. I wish particularly to thank Robert Butler, Steven Kish, Steven Hauck, Richard Kazda, Terry Bottrill, Robert Carpenter, Jack Gair, John Slack, Jim Craig, Henry Bell, and Mason Banks, though clearly they cannot be held in any way responsible for any errors herein. This work has been partially supported by the University of North Carolina Research Council.

REFERENCES CITED

Beck, W. A., 1946, Exploration of the Piedmont manganese belt, McCormick County, South Carolina, and Wilkes County, Georgia: United States Bureau of Mines Report of Investigation 3858, 5 p.

Bell, H., Carpenter, R. H., and Feiss, P. G., 1980, Volcanogenic ore deposits of the Carolina slate belt, *in* Frey, R. W., ed., Excursion in southeastern geology, Volume 1: American Geological Institute, p. 149–178.

Best, D. M., Geddes, W. H., and Watkins, J. S., 1973, Gravity investigations of the depth of source of the Piedmont gradient in Davidson County, North Carolina: Geological Society of America Bulletin, v. 84, p. 1213–1216.

Butler, J. R., 1979, The Carolina slate belt in North Carolina and northeastern South Carolina: A review: Geological Society of America Abstracts with Programs, v. 11, p. 172–173.

Conley, J. F., and Bain, G. L., 1965, Geology of the Carolina slate belt west of the Deep River–Wadesboro Triassic basins, North Carolina: Southeastern Geology, v. 6, p. 117–138.

Cook, F. A., Albaugh, D. S., Brown, L. D., Kaufman, S., Oliver, J. E., and Hatcher, R. D., 1979, Thin-skinned tectonics in the crystalline southern Appalachians; COCORP seismic-reflection profiling of the Blue Ridge and Piedmont: Geology, v. 7, p. 563–567.

Emmons, E., 1856, Geological report of the midland Counties of North Carolina: New York, George P. Putnam and Sons, 351 p.

Espenshade, G. H., 1954, Geology and mineral deposits of the James River–Roanoke River manganese districts, Virginia: U.S. Geological Survey Bulletin 1008, 155 p.

Espenshade, G. H., and Potter, D. B., 1960, Kyanite, sillimanite, and andalusite deposits of the southeastern states: U.S. Geological Survey Professional Paper 336, 121 p.

Gair, J. E., 1978, Sulfide deposits and related geology, southeast limb of Arvonia syncline and adjacent folds, *in* Gair, J. E., ed., Massive sulfides of Virginia field trip guidebook: International Geological Correlation Program, Project 60, p. 53–56.

Gair, J. E., Windolph, J. F., Jr., and Wright, N. A., 1975, Preliminary results of geochemical soil survey, Hamme Tungsten district, North Carolina: U.S. Geological Survey district, North Carolina: U.S. Geological Survey Circular 711, 19 p.

Glover, L., III, and Sinha, A. K., 1973, The Virgilina deformation, a late Precambrian to Early Cambrian(?) orogenic event in the central Piedmont of Virginia and North Carolina: American Journal of Science, v. 273-A, p. 234–251.

Green, G. B., Jr., 1977, The geology of the slate belt rocks of the Goldsboro and Bear Creek quadrangles, North Carolina [M.S. thesis]: North Carolina State University, 68 p.

Hauck, S. A., 1977, Geology and petrology of the northeast one quarter of the Bynum quadrangle, Carolina slate belt, North Carolina [M.S. thesis]: University of North Carolina at Chapel Hill, 146 p.

Horsnail, D. F., and Parker, P. D., 1979, Geology of tungsten deposits: Institute of Mining and Metallurgy Transactions, Ser. A, v. 88, p. A116–A120.

Howell, D. E., and Weisenfluh, J. A., 1978, The McDuffie gold belt: Regional geology of volcanogenic sulfide deposits associated with the Carolina slate belt, South Carolina–Georgia: Geological Society of America Abstracts with Programs, v. 10, p. 174.

Hutchinson, R. W., 1973, Volcanogenic sulfide deposits and their metallogenic significance: Economic Geology, v. 68, p. 1223–1246.

Indorf, C. P., and Feiss, P. G., 1978, Mineralogy of the Silver Hill Zn-Pb-Ag deposit, Carolina slate belt, Davidson County, North Carolina: Geological Society of America Abstracts with Programs, v. 10, p. 172.

Jambor, J. L., 1979, Mineralogical evaluation of proximal-distal features in New Brunswick massive-sulfide deposits: Canadian Mineralogist, v. 17, p. 649–664.

Keith, A., and Sterrett, D. B., 1931, Description of the Gaffney and Kings Mountain quadrangles. U.S. Geological Survey Atlas, Folio 222, 13 p.

Kish, S. A., Feiss, P. G., and Fullagar, P. D., 1980, Age and origin of metallic ore deposits in the southern Appalachian Piedmont—lead and strontium isotopic studies: Geological Study of America Abstracts with Programs, v. 12, p. 463.

Landis, G. P., and Rye, R. O., 1974, Geologic, fluid inclusion, and stable isotope studies of the Pasta Bueno tungsten–base metal ore deposit, northern Peru: Economic Geology, v. 69, p. 1025–1059.

Laney, F. B., 1917, The geology and ore deposits of the Virgilina district of Virginia and North Carolina: Virginia Geological Survey Bulletin 14, 176 p.

Large, R. R., 1977, Chemical evolution and zonation of massive sulfide deposits in volcanic terrains: Economic Geology, v. 72, p. 549–572.

Luethe, R. D., and Windisch, T. C., 1978, Geology and sulfide deposits of the McCormick, South Carolina–Lincolnton, Georgia, mineral district: Geological Society of America Abstracts with Programs, v. 10, p. 174.

McCauley, C. K., and Butler, J. R., 1966, Gold resources of South Carolina: South Carolina Division of Geology Bulletin 32, 78 p.

Mitchell, A. H., and Bell, J. D., 1973, Island-arc evolution and related mineral deposits: Journal of Geology, v. 81, p. 381–405.

Nitze, H.B.C., and Hanna, G. B., 1896, Gold deposits of North Carolina: North Carolina Geological and Ecnomic Survey Bulletin 3, 200 p.

Olmstead, D., 1825, On the gold mines of North Carolina: American Journal of Science, v. 9, p. 5–15.

Pardee, J. T., and Park, C. F., Jr., 1948, Gold deposits of the southern Piedmont: U.S. Geological Survey Professional Paper 213, 156 p.

Pardee, J. T., Glass, J. J., and Stevens, R. E., 1937, Massive, Low-fluorine topaz from the Brewer mine, South Carolina: American Mineralogist, v. 22, p. 1058–1064.

Plimer, I. R., 1978, Proximal and distal stratabound ore deposits: Minera-

lium Deposita, v. 13, p. 345–353.
Radcliffe, D., 1978, Geology of kyanite: Twelfth Forum on the Geology of Industrial Minerals: Georgia Geological Survey Information Circular 49, p. 69–78.
Sawkins, F. J., 1972, Sulfide ore deposits in relation to plate tectonics: Journal of Geology, v. 80, p. 377–397.
——1976, Massive sulfide deposits in relation to geotectonics: Journal of Geology, v. 80, p. 377–397.
Schmidt, R. G., 1978, The potential for porphyry copper-molybdenum deposits in the eastern United States: U.S. Geological Survey Professional Paper 907-E, 31 p.
Seiders, V. M., and Wright, J. F., 1977, Geology of the Carolina volcanic slate belt in the Asheboro, North Carolina, area, *in* Burt, E. R., ed., Field guides for Geological Society of America, Southeastern Meeting, Wiston-Salem, North Carolina, p. 4–25.
Sillitoe, R. H., 1976. Andean mineralization: A model for the metallogeny of convergent plate margins: Geological Society of Canada special Paper 14, p. 59–100.
Slack, J. F., Foose, M. P., and Casadevall, T., 1978, Huebnerite textures and related fold structures and their bearing on ore genesis at the Hamme (Tungsten Queen) mine, North Carolina: Geological Society of America Abstracts with Programs, v. 10, p. 198.
Speer, J. A., 1978, Molybdenum mineralization in the Liberty Hill and Winnsboro plutons, South Carolina: Economic Geology, v. 73, p. 558–561.
Stein, H. S., and Kish, S. A., 1978a, The Virgilina mining district of North Carolina and Virginia—preliminary evidence for a copper deposit of metamorphic origin: Geological Society of America Abstracts with Programs, v. 10, p. 198.
1978b, The age and origin of ore deposits in Carolina slate belt: Geological Society of America Abstracts with Programs, v. 10, p. 497.
Stromquist, A. A., and Conley, J. F., 1959, Geology of the Albemarle and Denton quadrangles, North Carolina: Carolina Geological Society Field Trip Guidebook, 36 p.
Stromquist, A. A., and Sundelius, H. W., 1969, Stratigraphy of the Albemarle Group of the Carolina slate belt in central North Carolina: U.S. Geological Survey Bulletin 1274-B, 22 p.
Sundelius, H. W., 1970, The Carolina slate belt, *in* Fisher, G. W., Pettijohn, F. J., Reed, J. C., Jr., and Weaver, K. N., eds., Studies in Appalachian geology: Central and southern: New York, Interscience Publishers, Inc., 351–367.
Sykes, M. L., and Moody, J. B., 1978, Pyrophyllite and metamorphism in the Carolina slate belt: American Mineralogist, v. 63, p. 96–108.
Weed, W. H., 1911, Copper deposits of the Appalachian states: U.S. Geological Survey Bulletin 455, 166 p.
Whitney, J. A., Paris, T. A., Carpenter, R. H., and Hartley, M. E., III, 1978, Volcanic evolution of the southern slate belt of Georgia and South Carolina: A primitive oceanic island arc: Journal of Geology, v. 86, p. 173–192.
Wilkinson, S. E., 1978, The geology of the northeast quarter of the Silk Hope quadrangle, Carolina slate belt, North Carolina [M.S. thesis]: University of North Carolina at Chapel Hill, 56 p.
Worthington, J. E., and Kiff, I. T., 1970, A suggested volcanogenic origin for certain gold deposits in the slate belt of the North Carolina Piedmont: Economic Geology, v. 65, p. 529–537.

Manuscript Accepted by the Society July 8, 1982

Typeset by CGE Services, Denver, Colorado
Printed in U.S.A. by Malloy Lithographing, Inc., Ann Arbor, Michigan